Biotechnology of Biopolymers

Biotechnology of Biopolymers

Editor

Federico Noguerra

scitus
academics

Biotechnology of Biopolymers
Edited by **Federico Noguerra**

ISBN: 978-1-68117-534-8
Library of Congress Control Number: 2016942934

www.scitusacademics.com
Box No. 4766, 616 Corporate Way,
Suite 2, Valley Cottage,
NY 10989

Printed in United States of America on Acid Free Paper

Preface

Biopolymers are polymers that occur in nature. For example, proteins and carbohydrates are biopolymers. Many biopolymers are currently commercially produced on large scales, although they are usually not used in plastics production. Even if only a small proportion of the already produced biopolymers were used in the production of plastics, our dependence on manufactured, non-renewable resources would be significantly reduced. For billions of years these polymers have been on Earth. It is older than synthetic polymers, such as plastics. The most important thing for humans is probably the DNA biopolymer. The entire structure of the body and the genetic behavior from parents to children are based on it. Both RNA and DNA are composed of nucleic acids, which alternate with large amounts of genetic data in certain patterns. Cellulose is the most popular biopolymer. It is also the richest organic compound on this earth—it accounts for 33% of the entire plant component on Earth. These polymers play an important role in nature. They are very useful in carrying out functions such as energy storage, cellular transmission and preservation and genetic information.

This book covers occurrence, synthesis, production and isolation, applications and properties, modification and biodegradation, relevant methods of analysis to reveal biopolymer structures and properties, mathematical and experimental models. This guide is hopefully compassionate for many scientists, doctors, pharmaceuticals, other professionals and engineers in a wide range of disciplines, academia and industry.

Table of Contents

Chapter 1

The role of polysaccharides in the molecular evolution of biopolymers

Aleksander N. Zimnitskii[1*], Sergey A. Bashkatov[2], Vadym N. Urazbaev[3], Aleksey V. Chemeris[4], Renat S. Yamidanov[1]

[1]Research and Production Association "Plazan", Moscow, Russia; [*]Corresponding Author: zanol@yandex.ru

[2]Bashkir State University, Ufa, Russia

[3]Institute of Organic Chemistry, Russian Academy of Sciences, Ufa, Russia

[4]Biochemistry and Genetics, Institute of Russian Academy of Sciences, Ufa, Russia

ABSTRACT

The empirical (biochemical, the PCR method) and computational (quantum-chemical, PM3) methods have shown the ability of polysaccharides to catalyze the polymerization of amino acids and nucleotides within the physiologically acceptable temperature range. The possibility of nucleotide aminoacylation in the presence of polysaccharides has been established. The suggestion has been made that abiogenic aminoacylation of nucleotides by polysaccharides served as the prototype of the original aminoacylpre-tRNA-synthetase activity and subsequently determined the formation of the modern mechanism of genetic information transfer via three biopolymer types—nucleic acids, proteins and polysaccharides.

KEYWORDS

Molecular Evolution; Aminoacylation; Polysaccharides; Amino Acids; Nucleic Acids; Guantum Chemistry; PCR

1. INTRODUCTION

Chemical and molecular evolution of biopolymer monomers—proteins and nucleic acids (NA) is the fundamental issue in the understanding of life origin on our planet. The renowned works by S. Miller [1] demonstrated the possibility in principal of nucleic acid (NA) formation from the simplest chemical compounds—methane, ammonia, hydrogen and water vapor. Furthermore, the research conducted by the academician V. N. Parmon

[2] made it possible to prove the possibility of practically selective ribose and simplest sugar synthesis from formaldehyde in the presence of phosphate anions (natural apatite). Abiogenic formation of the monomers of modern biopolymers was verified not only by the model chemical experiments, but also by the study of interstellar and alien matter composition [3]. Thus, today it is beyond any doubt that amino acids, sugars and nucleotides were formed abiogenically at the initial stages of the evolution both inside and outside the primitive atmosphere of the Earth. Besides, S. W. Fox [4] showed experimentally that proteinaceous multi-molecule systems can be formed under achievable in the natural conditions thermal influence on amino acid mixture. The so-called thermal proteinoids of S. W. Fox and oligonucleotides of L. Orgel were effectively formed at temperatures above 100°C. These processes were, thus, localized only in the volcanic activity zone in the conditions of the Earth's primitive atmosphere, which led to the destruction of polymers under the influence of the excessive thermal energy. At the same time, a great number of scientists [6] believe that life origin on our planet is connected with the period of its cooling to temperatures below 100°C and the formation of primary hydrosphere elements. Therefore, high temperatures apparently could not play the leading role in the system of biomolecule polymerization. A reaction catalyst is needed to promote this process in the physiological temperature range in the absence of excessive thermal energy. This catalyst should not only promote polymer assembly, but also have an information intensive structure, which is in its turn able to change and evolve. Clays which, according to J. Bernal [7], could have played this part, are not able to function as catalysts of monomer polymerization,

despite their ability to adsorb biopolymers. Moreover, the ability of these minerals to evolve is out of the question, even taking into account the great variety of their crystal forms. Therefore, the search of the catalyst of the polymerization process of biologically significant monomers in the physiological temperature range is a vital issue in the understanding of the initial stages of biopolymer molecular evolution.

As discussed earlier, it is generally agreed that biomolecules were represented by sugars, nucleotides as well as amino acids at the initial stages of evolution [1-7]. According to the academician V. N. Parmon [2], sugar synthesis was more advantageous, since only sugars (as opposed to NA and amino acid (AA) possess the autocatalytic synthesis system working successfully in highly diluted solutions (the “formose” reaction of Butler, sugar synthesis from formaldehyde in the presence of calcium or magnesium ions). Moreover, it is well-known that calcium and magnesium ions form a complex with a sugar molecule to which a formaldehyde molecule is then joined. The formation of complexes from several sugar molecules was observed [2], suggesting the possibility of abiogenic oligoand polysaccharide (PS) formation in aqueous media. Thus, oligosaccharide appearance in the conditions of Earth’s primary atmosphere was more likely than oligopeptide or oligonucleotide appearance. Taking into account that a number of researchers earlier established the information capacity (microheterogeneity) of a polysaccharide chain [8-10], while we established the complementarity of NA to PS [11-14], the aim of the present work was to reveal the ability of PS to accelerate oligopeptide and oligonucleotide polymerization processes abiogenically.

2. MATERIALS AND METHODS

2.1. Quantum-Chemical Calculations

Theoretical calculations for the formation of nucleotide-polysaccharides (PS) complexes in this work were carried out by the PM3 semiempirical method [15] with the complete optimization of geometrical parameters employing the GAMESS program [16]. A global minimum of the total electron energy was sought by the Newton-Raphson method with an energy gradient of 0.1 kcal/mol ΔE, starting from various initial approximations of the complex structure.

2.2. Methods of Work with Amino Acids, Peptides and Polysaccharides

PS-AA complex was formed by dissolving 100 mg of PS (amylose—Am or polyuronides—PU) in 400 ml of water and adding 20 biogenic amino acids, 0.05 mm each. The mixture was made up to 500 ml with water. The solution was incubated at 40°C for 1 hour, and then it was evaporated to 50 ml in vacuum. The complexes were precipitated by adding triple volume of ethanol at +4°C for 12 hours. The precipitate was collected by centrifugation and dried out. The quantity of AA was determined in the precipitate.

The amino acid composition was determined by dissolving the AA-PS precipitates in 0.1 N HCl. The obtained solution was filtered, and AA were determined in the filtrate by a standard method on the Automatic Amino Acid Analyzer T-339 (Czech Republic). For this purpose, 100 µ of the sample was introduced by a sample loop onto a 31.5 cm high ion-exchange column containing ion-exchange resin LG ANB Ostion LB. Then a stepwise elution was conducted with pH 3.15, pH 5.25 and pH 9.25 buffered solutions. Acidic AA were eluted with the first buffer, neutral AA—with the second one, and basic AA—with the third one. AA were quantified photometrically after the ninhydrin reaction in a capillary reactor. The peaks were identified and quantified with reference to a standard AA mixture containing 2.5 mM/ ml of each amino acid. Assay sensitivity was 0.05 nM AA. PS mixtures precipitated by ethanol without AA served as the control.

The abiogenic formation of peptide bonds under thermal influence was carried out in the following way. 20 amino acids, 0.05 mM each, were dissolved in 50 ml of distilled water containing 100 mg of PS (Am or PU). The samples were evaporated to dryness in vacuum and the obtained crystals were exposed to different temperatures (20°C, 40°C, 60°C, 120°C, 130°C) for 6 hours. The mixture of AA-PS crystals was then dissolved in water. The presence of peptide bonds was determined by the biuret method [17]. The control was amino acid mixture dissolved in 50 ml of distilled water without PS treated in the same way as the test sample. Amino acids were obtained from "Fluka" company.

2.3. Methods of Work with Nucleotides, Polynucleotides and Polysaccharides

As a rule, acidic polysaccharides and polyuronides in plants are represented by the monomers of two hexuronic acids—D-glucuronic acid and D-galacturonic acid. Dglucuronic acid is common in hemicellulose polysaccharides, whereas D-galacturonic acid in the form of galacturonans is the main component of pectic substances. It is worth noting that these polysaccharides can be easily obtained from oat seeds and apple pulp. Thus, acidic oligosaccharides were extracted from apples according to a standard method described earlier for oat seeds [11-14]. The extracted PU was additionally treated with deoxyribonuclease and proteases to prevent contamination with endogenous DNA. The enzymes from PU samples were then eliminated by three-fold treatment with phenol-chloroform mixture followed by three-fold precipitation by triple volume cooled ethanol [18].

Human placenta glycosaminoglycans were obtained by a standard method [11-14]. The purity of PS used in the experiment was controlled by electrophoresis on cellulose acetate bands in 0.1 N HCl.

5' nucleotide monophosphates (sodium salt) in ribo-, deoxyribo-form and oligo (dT)-cellulose were obtained from "Sigma-Aldrich", amylose (neutral polysaccharide, NS) was obtained from "Serva Feinbiochemica GMBH".

The procedure of adenosine monophosphate (AMP) polymerization (sodium salt) was as follows. 10 mg of AMP and 10 mg of PU were dissolved in 50 ml of H_2O. Then the solution was evaporated to dryness in vacuum at a temperature of 40°C and the obtained nucleotide PS crystals were exposed to different temperatures (20°C, 40°C, 60°C, 120°C, 130°C) for 6 hours. Poly-A fragments of NA were extracted on oligo (dT)-cellulose by a standard method [19].

Deoxynucleotide polymerization in the presence of polyuronides (PU) was carried out by dissolving 10 mg of PU in 40 ml of water and adding 2.5 mg of each of the four nucleotides. The mixture was made up to 50 ml with H_2O. The solution was incubated at 40°C or 60°C for 1 hour, then it was evaporated to dryness in vacuum and dissolved in 5 ml of water.

The complexes were precipitated by adding triple volume of ethanol at +4°C for 12 hours. The precipitate was collected by centrifugation and dried out. The quantity of nucleotides was estimated spectrophotometrically by measuring the absorbance at 260 nm. The correspondence of the absorbance to the quantity of nucleotides was determined by a calibration curve individually for each nucleotide. PS alcohol precipitation in the absence of nucleotides served as the control.

Polymerase chain reaction (PCR) was conducted with short oligonucleotides in non-specific conditions of hybridization (the DNA-RAPD method). The amplification process was preceded by a preliminary denaturation at 94˚C for 2 min and consisted of 40 cycles, each one including the following steps: denaturation at 94˚C for 20 sec, annealing of the primers at 30˚C for 20 sec, and elongation at 72˚C for 20 sec. The following oligonucleotides were used for DNA-amplification: (AFK-3— acggtggacg, AFK-1—gcgtccaattc, ADE-1—tcgaatagc).

2.4. Gel Filtration

Gel-filtration on Sephadex G-25 was carried out on a column 30 cm long, 1.5 cm in diameter. The volume of the applied sample was 5 ml, fraction size was 3 ml. PU was detected spectrophtometrically by measuring the absorbance at 220 nm, whereas NA was detected by measuring the absorbance at 260 nm.

3. RESULTS

3.1. AA Polymerization on PS

The purity of PS used in the experiment was determined by biochemical analysis and was confirmed by electrophoresis. The results of the research are shown in **Figure 1** and **Table 1**.

The formulas of PS used in the experiments can be represented as a recurrent dimer with the sequence AA, AB, BA, BB, where A is neutral sugar (hydroxymethyl group) and B is acidic sugar (carboxyl group). Bonding ability between amino acid radicals and monosaccharides was analyzed using the quantum-chemical method. The results are shown in **Table 2**.

As **Table 2** demonstrates, end groups of amino acid radicals are able to form quite strong hydrogen bonds both with carboxyl and hydroxymethyl groups of monosaccharides. The energies of bonds between AA radical end groups and monosaccharides are not equal (**Table 2**). Five amino acids (Leu, Ile, Pro, Val, Ala, Gly) have practically zero bond energy between the radical and functional sugar groups. Three amino acids (Cys, Phe, Trp) are characterizes by a weak bond energy, 0 to 2 kcal/mol. Four amino acids (Tyr, Met, Thr, Ser) have a medium bond energy, 2 to 5 kcal/mol, whereas seven amino acids (Arg, Lys, Gln, Glu, Asp, Asn, His) have a high bond energy with the interaction strength at the carboxyl group being 5 to 10 kcal/mol.

It is worth noting that the bond energy between water and the carboxyl and hydroxymethyl groups of saccharides is −5.72 and −3.57 kcal/mol, respectively. In view of this fact the estimates of AA and PS radical bonds are more accurate during the dehydration of PS-AA complexes. Furthermore, aminocarboxymethyl groups of amino acids, themselves, are capable of forming bonds with the carboxyl and hydroxymethyl saccharide groups, ΔE of which are −18.33 and −5.72 kcal/mol, respectively. These values are significantly higher than those of the bond energies formed in aqueous medium. Thus, PS are able to form hydrogen bonds with AA in aqueous medium and, at the same time, position the carboxyl group of the AA part in an optimal way in relation to the amino group of another AA to form random linear peptide bonds in the process of dehydration. These calculations testify that in solution the bulk of amino acids are bound to PS by their aminocarboxymethyl group. For some AA (two last classes), however, the bond energy of radical end-group is quite high. In this case the aminocarboxymethyl group of the given AA remains free, at a certain distance from the PS chain.

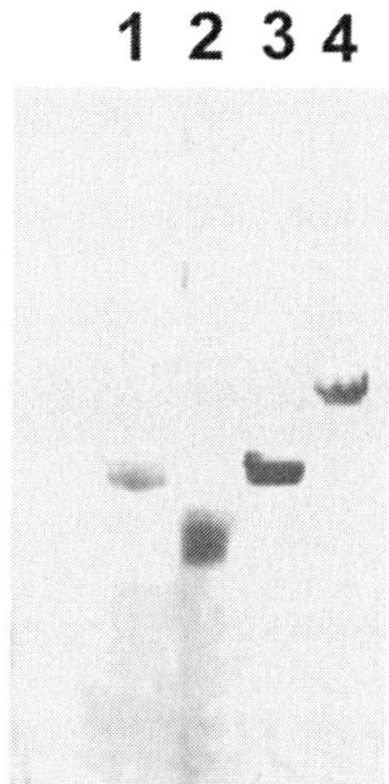

Figure 1. PS electrophoregram on cellulose acetate bands in 0.1 N HCl. 1—polyuronides, 2—heparin sulfate, 3—chondroitin sulfate, 4— hyaluronic acid.

Table 1. Characteristics of PS used in the experiments.

Monosaccharides Polysaccharides	Ratio, %				Mcg MS per 1000 mg PS			
	UA	NS	AAH	OMS	UA	NS	AAH	OMS
Polyuronide (PU)	75	3	0	22	747	33	0	220
Hyaluronic acid (HA)	54	0	46	0	535	0	465	0
Chondroitin sulfate (CS)	41	0	59	0	413	0	587	0
Amylose (Am)	0	100	0	0	0	1021	0	0

Note: UA—uronic acid identified with the Dische test, NS—neutral sugars, identified by the anthrone method, acetylamino hexoses (AAH) for GAG—non-sulphated for HA and sulphated for CS, using the formula AAH = 1000 − UA, other monosaccharides (OMS)—using the formula OMS = 1000 − UA − NS.

Table 2. Bond energy of the end-group of L-amino acid radical with the carboxyl or hydroxymethyl group of saccharides (ΔE, kcal/mol). Quantum-chemical calculation, the PM3 method.

Amino acid	COOH	CH_2OH
Arg	−5.63	−4.54
Lys	−5.90	−2.27
Tyr	−2.15	−0.61
Trp	−0.41	−0.45
Gln	−6.16	−2.85
Met	−2.23	−1.10
Glu	−10.23	−3.35
Phe	−0.55	−0.83
Asp	−9.47	−1.63
Asn	−8.09	−2.64
His	−7.73	−4.69
Leu	-	-
Ile	-	-
Thr	−4.80	−3.15
Ser	−4.93	−3.46
Pro	-	-
Cys	−1.75	−0.21
Val	-	-
Ala	-	-
Gly	-	-

Note: 1) Blank (-) means that the bond of AA radical end-group with the functional group of a polysaccharide is insignificant. As a result, during optimization, AA rotates in such a way that AA amino carboxymethyl group interacts with the functional group of the PS chain, but not with the radical; 2) Pentoses are used as monosaccharides in the quantum-chemical calculations.

In order to verify the bonding ability between polysaccharides and amino acids, an experiment was conducted on polysaccharides in the solutions of amino acids present in equal proportions. It will be recalled that amylose is a homoglycan of monosachharides with the hydroxymethyl group (neutral polysaccharide, NS) and polyuronide is mainly a homoglycan of polysaccharides with the carboxyl group (acidic polysaccharide, AS). In fairly concentrated PS and amino acid solutions, PS monomers must bind with AA, if such a process exists. The subsequent ethanol precipitation of PS insolubilizes only those amino acids that have formed quite strong complexes with PS in the process of ethanol dehydration and leaves water-soluble and alcohol-soluble AA in the solution. It is worth noting that amino acids were not detected in the original PS samples. This fact is indicative of a high degree of PS purification from possible protein contaminants. The results of the experiment for the coprecipitation of PS with 20 amino acids are shown in **Table 3**.

The presented data demonstrate that all AA used in the experiment interacted with PS at different levels of efficiency. The fact that the process of bond formation depends largely on the AA radical is beyond question. It has been established that homoglycans of acidic sugars (PU) bind AA more efficiently than homoglycans of neutral sugars (amylose) (124.46 and 32.20 mcg/mg PS, respectively). The experimental results are in good agreement with the quantum-chemical calculations (**Table 2**), which showed that the bond energy of the complexes AA radical end group—acidic saccharide carboxyl group is significantly higher (on average −5.00 kcal/mol) than the bond energy of the complexes AA-neutral saccharide hydroxymethyl group (on average −2.27 kcal/ mol). This, apparently, accounted for a more efficient formation of the AA-PU complexes. The adsorption rate of AA from the first class was up to 2.6 mcg/mg PU, whereas 7 AA from the fourth class showed the maximum PU affinity (on average 9.28 mcg/mg), which also confirms the results of the quantum-chemical calculations.

Acidic amino acids (Asp, Thr, Ser, Glu, Pro, Ala) bind more effectively to amylose (about 33% to AM vs. 23% to PU). In contrast, basic amino acids (His, Lys, Arg) were detected more often in the complexes with PU (about 30% with PU vs. 17% with Am). Neutral AA form complexes with PU in equal proportion (about 50%).

The results of the experiment for the abiogenic formation of peptide bonds between AA in the presence of PS are shown in Figures 2 and 3.

The presented data (Figures 2) testify that in the control (amino acids without PS), temperatures within the range 40˚C to 130˚C promote peptide bond formation in the process of AA dehydration. Maximum peptide bond formation was detected at temperatures over 100˚C (the Fox thermal proteinoids). The formation of peptide bonds was not detected in the control sample at 20˚C. However, temperature increase over 120˚C leads to the decrease in peptide bond formation. The presence of PS, especially acidic sugar homopolymers (PU), dramatically improved the effectiveness of this process (approximately by an order of magnitude) in the physiologically acceptable temperature range (20°C - 60°C), which indicates a significant catalytic activity of PS in terms of peptide bond formation. Gel-filtration (**Figure 3**) confirmed the fact of amino acid polymerization with the formation of peptide bonds. The peak of AA elution was detected in fractions 25 - 28, whereas the maximum of peptide bond was detected as early as in fraction 15. This attests to the formation of polypeptides, which are dozens of AA monomers in length. Peptide bonds were practically not formed in the absence of PS at temperatures below 40°C.

Table 3. Composition of amino acids detectable in the ethanol precipitate of PS (mcg/mg).

	L-amino acids	PU	Am
1	Asp	4.86 ± 0.5	2.68 ± 0.3
2	Thr	3.16 ± 0.3	2.46 ± 0.3
3	Ser	8.76 ± 1.0	1.96 ± 0.2
4	Glu	4.70 ± 0.5	1.72 ± 0.2
5	Pro	3.28 ± 0.4	0.82 ± 0.15
6	Ala	2.86 ± 0.35	0.84 ± 0.15
7	Val	2.34 ± 0.3	1.74 ± 0.2
8	Ile	2.52 ± 0.3	1.58 ± 0.2
9	Leu	2.26 ± 0.3	1.62 ± 0.2
10	Tyr	2.02 ± 0.2	0.84 ± 0.2
11	Phe	2.2 ± 0.3	1.32 ± 0.3
12	His	11.62 ± 1.3	2.06 ± 0.25
13	Lys	10.44 ± 0.2	1.3 ± 0.15
14	Arg	14.84 ± 1.5	2.26 ± 0.3
15	Met, Cys, Gly, Asp, Gln, Trp	32.0 ± 3.5	9.2 ± 1.3
	Total	124.46	32.20

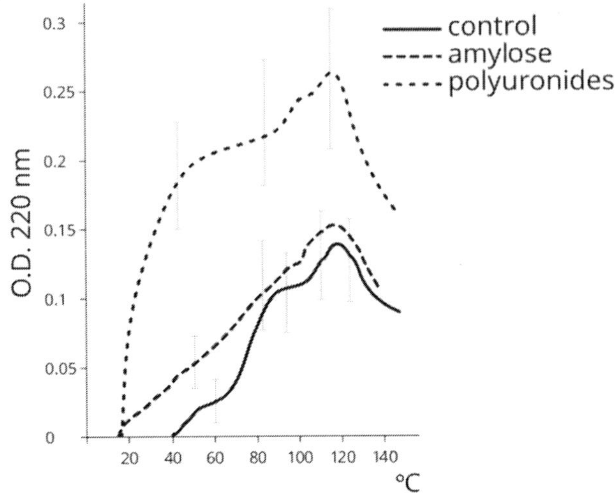

Figure 2. Dynamics of peptide bond formation by means of the biuret reaction in the mixture of 20 amino acids under various temperature regimes, in the absence of PS (control) and in the presence of PS (PU, Am).

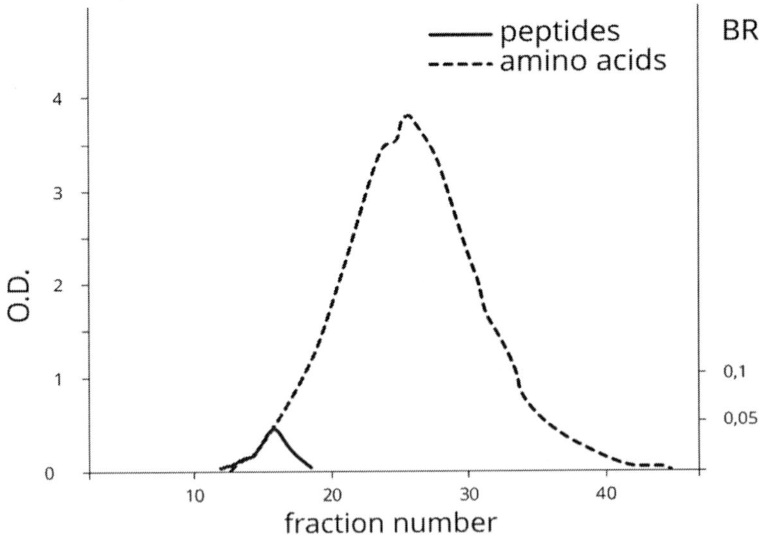

Figure 3. Gel filtration of the products obtained by temperature treatment (60°C) on AA mixture in the presence of PS (PU) on Sephadex G-25. (Peptides detection is based on OD of the biuret reaction (BR) and AA detection is based on OD at 280 nm.)

3.2. NA Polymerization on PS

It was found that both acidic and neutral polysaccharides are capable of forming hydrogen bonds with nucleic acid bases. **Table 4** demonstrates energy estimates of hydrogen bonds between the carboxyl group of polyuronides and the hydroxymethyl group of neutral saccharides with nucleotides obtained by the PM3 method. The computations show that the energy of hydrogen bonds between nucleotides and acidic polysaccharides is more than 3 times higher than the corresponding estimates for neutral polysaccharides, thus, theoretically predetermining their higher capacity for nucleotide adsorption. The calculations were confirmed experimentally (**Table 5**).

Indeed, homopolymers of acidic saccharides (PU) adsorb 3.5 times more nucleotides as compared to homoglycans NS (amylose).

As can be seen from the above, the specific adsorption of nucleotides on PS is beyond any doubt, according to the presented results and our earlier publications describing the physicochemical principles of these processes [11-14]. However, the question of nucleotide polymerization on a linear PS molecule has been undetermined. Our in silico study of mononucleotide interaction with a polysaccharide molecule showed that nucleotide monomers on PS are ideally oriented to form covalent bonds between them [12]. This should significantly stimulate the process of polynucleotide formation from monomers bound with PS. This assumption was verified in the series of experiments for the dehydration of PU complexes with AMP as it was described in the experimental section. These experiments were aimed at obtaining poly(A) oligomers with their further isolation using the specific oligo(dT) cellulose sorbent.

The control of our experiment was the approach used by L. Orgel for the synthesis of polynucleotides under the influence of thermal energy [5]. Polyuronides were used in the empirical sample in parallel with AMP. The results of the experiment are shown in **Figure 4**.

The results shown in **Figure 4** demonstrate that AMP polymerization is practically undetected in the control sample dehydrated at a temperature below 100°C, which corresponds to the data provided by L. Orgel. Indeed, the free AMP polymerizes at a temperature above 100°C, while the nucleotide in the presence of PU shows the maximum NA polymerization rate in the temperature range 30°C to 80°C. Further temperature increase in the empirical sample resulted in the destruction of the polynucleotide.

Table 4. Energy of hydrogen bonds between the carboxyl (PU) and hydroxymethyl groups (NS) of saccharides in a complex with nucleotides (PM3 method, kcal/mol).

Nucleotide polysaccharides	A	G	T	C	Average value
Acidic polysaccharides (PU)	7.9	8.5	6.4	7.8	7.65
Neutral polysaccharides (NS)	1.6	1.8	2.9	2.0	2.08

Table 5. Quantity of deoxynucleotides detectable in the ethanol precipitation of oligosaccharides (mcg NA/mg PS).

Nucleotide polysaccharides	A	G	T	C	Average value
Acidic polysaccharides (PU)	80 ± 11	85 ± 10	77 ± 8	95 ± 15	84.3
Neutral polysaccharides (NS)	24 ± 3	23 ± 3	26 ± 2	25 ± 3	24.5

As can be seen from the above, PU demonstrated a significant catalytic activity in the process of nucleotide polymerization within the physiological temperature range (20˚C - 60˚C), and apparently acted as the template for the arrangement of NA polymer molecules. It seems fair to assume that, that PU do not only adsorb nucleotides, but also facilitate the process of their polymerization when dehydrated. Abiogenic polymerization of nucleotides is about 10 times more intensive in the presence of PU. NA concentration is increased in the PU-NA complexes in the subsequent processes of solution and repeated dehydration of the complexes with the mixture of nucleotides, as it is demonstrated in**Table 6**.

Our next objective was to isolate NA fragments formed on the PU by chromatography for their further analysis using the PCR method. As it is shown in **Figure 5**(A), the results of PU-NA complex chromatography on Sephadex G-25 demonstrate that high-polymer PU components were eluted predominantly before fraction 15. Low-polymer components were eluted between fractions 16 and 22 in smaller quantities than the high-polymer ones. The monomers eluted after the 23th fraction predictably were not represented in the samples, since ethanol precipitation of PS-NA leaves nucleotide monomers in the solution.

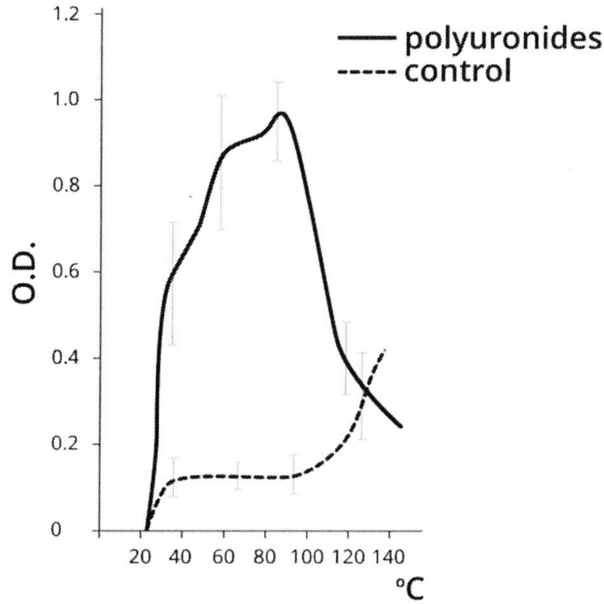

Figure 4. The dynamics of poly(A) NA formation (chromategraphy on the oligo(dT) cellulose column) at different temperatures in the absence (control) and presence of PU (OD at 260 nm). Note: PU dehydration was carried out only in the presence of AMP.

As it is shown in **Figure 5**(B), the spectrophotometric analysis of the eluted complexes demonstrates that the low-polymer fraction is rich in nucleotides, as the OD ratio A260/A220 fluctuates within the range of 0.8 - 1.0 in these samples, whereas for the high-polymer fraction it is 0.5. These results suggest the presence of abiogenically formed DNA in the fractions containing low polymer fragments.

To verify the presence of “correct” DNA chains recognized by specific enzymes in the test sample and its characteristic, we carried out an analysis using PCR. The NA polymers of fractions 16 - 22 after the chromatography oh Sephadex G-25 were used in the experiment. The results of the experiment are shown in **Figure 6**.

As **Figure 6** demonstrates, the primers themselves (Lanes 1) can serve as the template for the synthesis of high-polymer DNA. The reaction conditions facilitate the accumulation of the products at initially low template concentration due to the properties of the primers and to the presence of some DNA oligomers in the reaction mixture. Moreover, the introduction of an additional control—human DNA (Lane 2) to the experiment allows us to deny the presence of human DNA in the tested samples and to observe low ability of the primers to "self-prime" in the presence of the given DNA.

The acceleration of the polymerase activity is observed in the test sample (Lane 3) in the presence of AFK-3 primer. The new product, which is a few hundreds of nucleotide pairs in length, cannot be explained by human DNA contamination (most likely in the PCR process). Since the DNA product in the tested sample is formed in the presence of only one of the primers, it can be assumed that the amplicon, and therefore, the original DNA have low "complexity", i.e. have limited number of primary structure variants. The latter, as is well-known, is typical for DNA tandem repeats.

Table 6. Optical characteristics of PS-NA complexes subjected to dehydration once (I), twice (II), three times (III) (260/220 nm).

Dehydration frequency polysaccharides	I	II	III
PU (AS)	0.5	0.6	0.8
Am (NS)	0.9	1.0	1.1
HA (NS + AS)	0.9	1.0	1.1

The results of PCR with AFR-3 primer on PU and on the isolated NA fragments synthesized on PU are shown in **Figure 7**.

The **Figure 7** demonstrates that the primer itself (Lane 1) can serve as the template for high-polymer DNA synthesis as per the above procedure, promoting the product accumulation at initially low concentration of the template due to the properties of the primer itself and the presence of the admixture of DNA fragments of an unidentified origin.

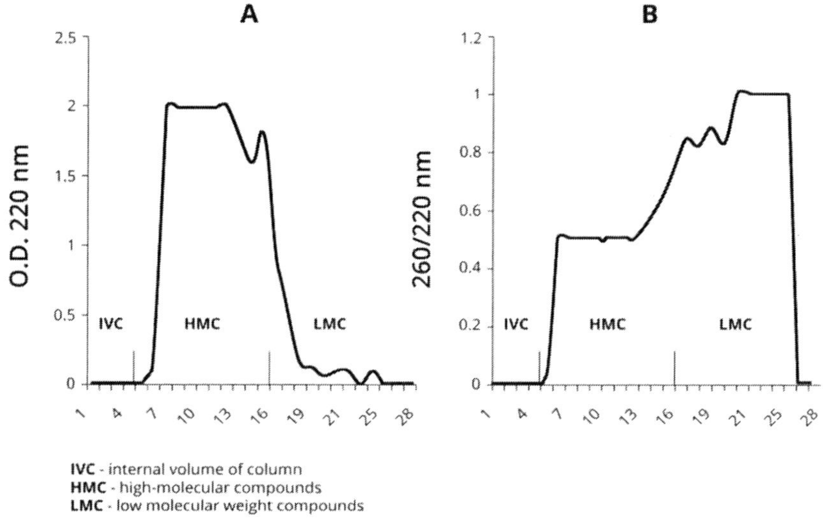

Figure 5. PU-NA complex chromatography on Sephadex G-25. A—represents detection based on OD at 220 nm, and B—represents detection based on OD 260/220 nm measurements.

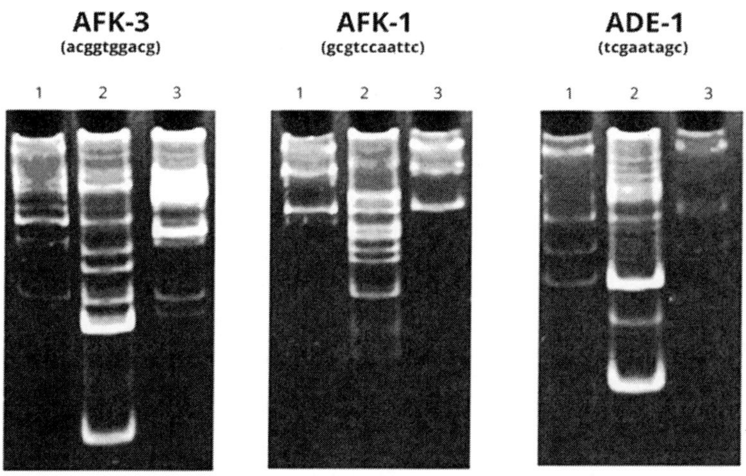

Figure 6. 1. The results of DNA-RAPD with the DNA preparation synthesized on PU. Control (primer); 2. Human DNA + primer; 3. DNA synthesized on PU, + primer.

The activity of the Taq-polymerase used in the PCR reaction was not detected in the samples containing PU and nucleotide (Lane 3) as well as in the pure PU (Lane 2). It may suggest the absence of NA fragments in the original PU samples and the ability of PU in high concentrations to suppress enzyme activity. Alternatively, the PU from fractions 16 - 22 (**Figure 5**(A)) stimulated enzyme activity. This fact attests the presence of DNA molecules in the given polysaccharide fractions (Lane 4). The synthesized product, which is a few hundred nucleotide pairs in length, does not result from human DNA contamination, as was mentioned above (**Figure 6**). No visible quantities of the product are detectable in Lane 5 demonstrating the PCR result in the conditions of selfpriming of isolated NA synthesized on PU. It can be caused by significant variations throughout the length of the analyzed NA sequences formed on PU.

As can be seen from the above, the PCR method confirmed the presence of DNA oligomers of varying sizes synthesized on PU. According to preliminary data they have a "simple", most probably tandem structure. We found that that the given NA sequences do not relate to human DNA. The nature of the given NA calls for further research [14].

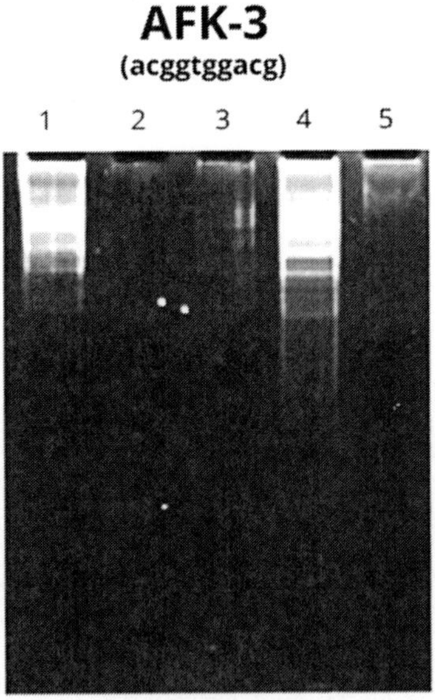

Figure 7. DNA-RAPD results with the preparations of NA synthesized on PU Lanes: 1—control (AFK-3 primer); 2— AFK-3 primer + PU; 3—AFK-3 primer + PU-nucleotide mixture; 4—AFK-3 primer + oligo nucleotides from PU-NA complex after dehydration; 5—isolated oligonucleotides from PU-NA complex in the absence of the primer.

3.3. NA Aminoacylation in the Process of AA and NA Interaction with PS

Since PS can simultaneously bind both nucleotides and amino acids, there is a need to examine possible interactions of AA with NA that are in complex with PS. When hydrogen bonds are formed between NA and PS bases, a complex is created at a distance of 5.7 - 7.5 Å from the zone of hydrogen bond formation between NA and PS to the carbon at the 2nd, 3rd and 5th position of the saccharide in nucleotide A.

Apart from the charge, another significant characteristic of AA radicals is the size. All proteinogenic amino acids can be subdivided into three groups: those with a “long” radical, 6.7 - 7.4 Å (Arg, Phe, Lys, Tyr), a “medium” one, 4.0 - 5.9 Å (Trp, Gln, Glu, Met, Asn, Asp, His, Leu, Ile, Val), and a “short” one, 2.1 - 3.8 Å, (Ser, Thr, Pro, Cys, Ala, Gly). The size, as well as the charge, can influence bond formation process. If the end group of AA radical forms hydrogen bonds with the PS monosaccharide, then the aminocarboxymethyl group of the amino acid can be located in the zone of the OH group, at the 2’, 3’ carbon of nucleotide A ribose (**Figure 8**) or phosphate, at the 5’ carbon.

The carboxyl group of AA located in this way in the PS-NA-AA complex enables the aminoacylation process at the above mentioned hydroxyls or phosphate of NA saccharide carbon ring. This assumption was verified by the quantum-chemical calculations (**Table 7**), according to which the formation of the AA-NA complex becomes possible when interacting with OH groups of the 2’ and 3’ carbons of NA ribose, in complex with PS. A fragment of RNA CCA in complex with AA and tetrasaccharide NS-NS-AS-AS was used in the calculations.

As can be seem from the above (**Table 7**), abiogenic aminoacylation of the terminal A nucleotide ribose is hypothetically possible in terms of the length of the formed hydrogen bonds (the bond length is less than or equal to 1.87 Å) for six amino acids—Arg, Lys, Tyr, Gln, Glu, Met. Furthermore, Met and Gln are able to carry out this process at the hydroxyls 2 and 3 of the ribose. Energy indices of the hydrogen bonds formed between the end group of AA radical and the carboxyl group of the saccharide also confirm the possibility of this process, since it goes with heat release (**Table 8**).

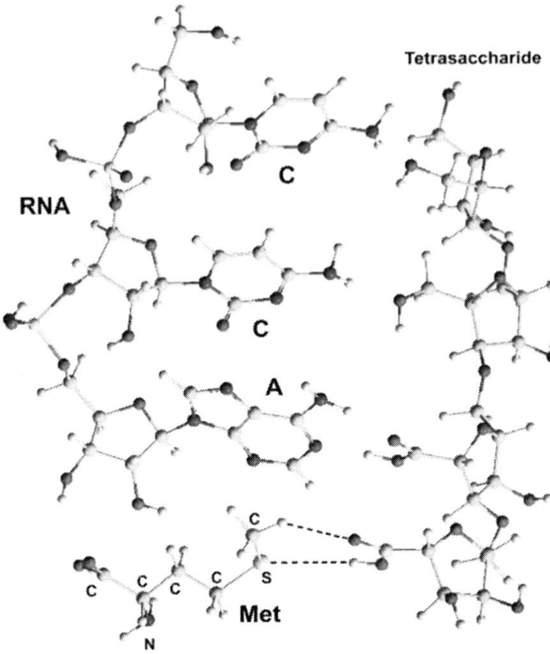

Figure 8. Quantum-chemical model of the complex RNA (CCA)- Methionine-Tetrasaccharide (NS-NS-AS-AS) (calculations by the PM3 method).

It is worth noting that there is a possibility of aminoacylation for Tyr at C_2 hydroxyl of the ribose, whereas there is no such possibility for Trp. Since these two AA can be easily detected in the complexes by the spectral analysis, we conducted additional calculations to test the ability of the given AA to aminoacylate the phosphate at the C_5-position of the ribose.

The results of the calculations are given in **Table 9** (Figures 9 and 10), which shows that theoretically Tyr is able to form bonds with the phosphate at C_5. In contrast, Trp has no such possibility in the AMP variant. However, in the ADP and ATP variants Trp amynoacylation is quite possible.

In order to prove empirically the possibility of abiogenic aminoacylation in accordance with the performed computations, dehydration of the PS—poly-A RNA-AA complex was conducted. The results of the experiment as exemplified by two AA are shown in **Table 10**. The polyA RNA fragments (Section 3.2) obtained abiogenically were re-dehydrated in the presence of PU and amino acids (Tyr or Trp) at a temperature of 60°C. Then the

Table 7. Calculation of the distances between the interacting atoms of the AA radical end group and the saccharide carboxyl group in the complex NA-AA-PS (Å)(AA is covalently bound to the hydroxyl of NA ribose at the C_2 and C_3.

AA	Carbon atom in a ribose	The distance from AA to the carboxyl group of a saccharide in PS (Å)	Possibility of hydrogen bond formation
Arg	C-2	1.86	+
	C-3	2.94	-
Trp	C-2	3.76	-
	C-3	3.67	-
Lys	C-2	1.83; 2.89	+
	C-3	2.77	-
Tyr	C-2	1.79; 1.85	+
	C-3	2.92; 4.08	-
Gln	C-2	1.78; 1.82	+
	C-3	2.53; 2.32	-
Glu	C-2	1.78; 1.78	+
	C-3	2.59; 2.32	-
Met	C-2	1.86; 2.90	+
	C-3	1.87; 3.32	+
Phe	C-2	1.93; 2.41	-
	C-3	3.99	-
Asn	C-2	3.51; 4.55	-
	C-3	4.56; 3.87	-
Asp	C-2	3.35; 4.55	-
	C-3	4.46; 3.87	-
His	C-2	2.81	-
	C-3	3.73	-
Leu	C-2	2.82	-
	C-3	2.74	-
Ile	C-2	3.02	-
	C-3	5.08	-
Val	C-2	3.94	-
	C-3	3.85	-
Ser	C-2	4.58	-
	C-3	4.85	-
Thr	C-2	4.12	-
	C-3	5.42	-
Pro	C-2	4.80	-
	C-3	4.61	-
Cys	C-2	4.06; 4.92	-
	C-3	4.97; 5.73	-
Ala	C-2	4.71	-
	C-3	6.78	-
Gly	C-2	5.51	-
	C-3	9.33	-

Note: + means that the distance between the interacting atoms is less than or equal to 1.87 Å, which suggests the possibility of stable hydrogen bonds in the complex of AA-NA with PS. In the complexes PS-NA and NA-NA the lengths of the formed hydrogen bonds do not exceed 1.86 Å [13,20].

Table 8. Total energy of hydrogen bonds, which coordinate AA as a separate molecule when the end group of the AA radical interacts with the carboxyl group of the saccharide in the process of the NA-AA-PS complex formation.

Amino acid	Hydrogen bond energy, kcal/mol
Glu	−11.43
Gln	−7.44
Lys	−6.57
Arg	−4.48
Tyr	−4.36
Met	−3.96

Table 9. Energy and length of H-bond hydrogen (AA)—oxygen (phosphate at C_5 of the ribose) in the system AA-nucleotide-disaccharide (ΔE, kcal/mol and r, Å).

Nucleotide AA	AMP		ADP		ATP	
	ΔE	r	ΔE	r	ΔE	r
Tyr	−3.76	1.80	−4.11	1.80	−4.34	1.80
Trp	−0.36	6.09	−3.03	1.84	−4.36	1.80

NA-AA complexes were isolated by an affinity chromatography on oligo (dT) cellulose columns by the poly-A fragment.

Samples of poly-A RNA of the corresponding AA dehydrated in the absence of PU served as the control. The presence of AA in the structure of RNA polymers was detected spectrophotometrically (**Table 10**). In contrast to AMP (260 nm), the above mentioned AA have the maximum absorbance at 280 nm (**Figure 11**).

As can be seen from the table above, a significant increase in absorbance at the wavelength of 280 nm is observed only in the oligo-A sample, subjected to dehydration in the presence of PU and tyrosine (OD ratio 280/ 260 equals to 1.06). This suggests the presence of AA in the RNA. Tryptophan did not form a complex with oligoA in the presence of PU. These data verify the conclusions about the abiogenic aminoacylation of nucleotide A ribose based on the quantum-chemical calculations. In the experiment outlined above, nucleotide A retained its ability to bind to nucleotide T of the sorbent (oligo (dT) cellulose), which suggests that the base does not participate in forming bonds with AA. Consequently, AA detected in the RNA complexes is connected to the ribose of the nucleotide at C_2 or C_5. Therefore, quantum-chemical calculations attest that aminoacylation of nucleoside monophosphates by tyrosine is possible both at the C_2 and C_5positions. Tryptophan is not able to aminoacylate nucleoside monophosphates either at the C_2 or C_5positions. It is most likely that the AA determined in the tyrosine-oligonucleotide complexes utilized the opportunity to aminoacylate the nucleotide at the C_2 or C_5 positions in the ribose, which is supported by the experimental results. It should be noted that PS aminoacylate phosphoric acid residue more efficiently at the C_5 position of nucleoside triphosphates.

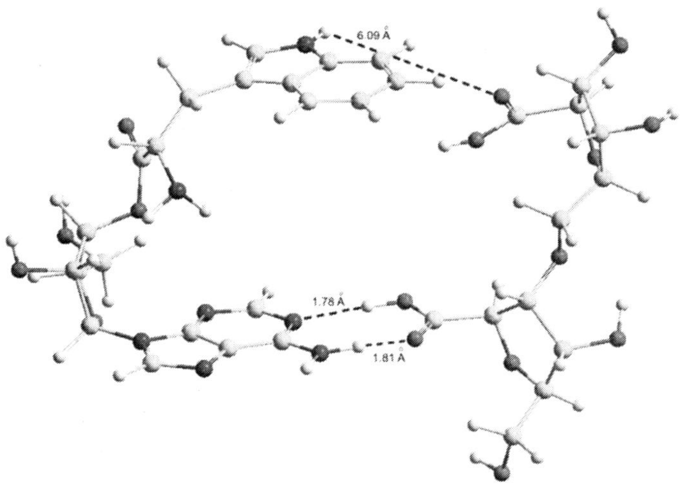

Figure 9. Quantum-chemical model of the 5′ AMP-Trp-Disaccharide complex (AA at C_5 of the ribose).

As can be seen from the above, experimental and calculation data demonstrate that there are quite strict patterns of PS interaction with AA and NA. PS are not only able to conduct adsorption and polymerization of the above noted monomers of biopolymers, but also, under certain conditions, to perform abiogenic aminoacylation of a nucleotide ribose both at the C_2-C_3 and C_5 positions.

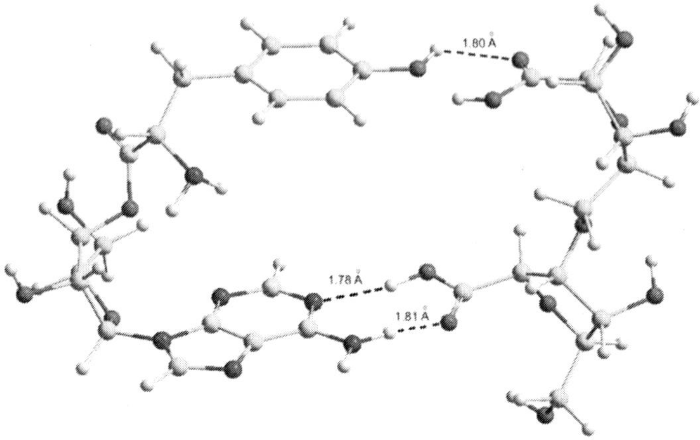

Figure 10. Quantum-chemical model of the 5' AMP-Tyr-Disaccharide complex (AA at C_5 of the ribose).

Table 10. Characteristic of abiogenically obtained poly-A RNAAA complexes under dehydration at a temperature of 60°C (OD ratio 280/260 nm).

	Tyr	Trp
control	0.80 ± 0.03	0.83 ± 0.06
PU	1.06 ± 0.06	0.80 ± 0.08

4. DISCUSSION

Regardless of the opinion diversity, it is highly plausible that sugars, amino acids, phosphorous and nitrogen compounds (NA bases) existed at the beginning of the molecular evolution. Sugar oligomers are formed in an auto-catalytic manner, abiogenically, in aqueous solutions, since they are synthesized from formaldehyde and polymerized without dehydration [2]. Amino acid and NA polymerization, in its turn, proved to be rather simple under the conditions of dehydration, under the influence of thermal energy, in the presence of PS, in the physiologically acceptable temperature range, as it was demonstrated in the present work. S. W. Fox [4] had previously established the possibility in principle of AA polymerization when their crystal mixture is exposed to temperatures above 100°C. Abiogenic formation of thermal nucleotides at temperatures above 100°C was also suggested by L. Orgel [5]. Due to such high temperatures of biopolymer formation, these abiogenic processes were restricted only to volcanic activity zones, where temperature conditions did not contribute to the stabilization of biogenic systems. Moreover, the excess of thermal energy could lead to polymer destruction. For the origin of primary living systems, similar to coacervates, a catalyst was needed to promote the polymerization process in the temperature range that would not destroy the forming polymers. It was oligosaccharides which, due to their abundance in the initial conditions at the beginning of the molecular evolution of biopolymers [2], could initiate the above-mentioned processes, which, in their turn, enabled the described reactions to prevail at the initial stages of the Earth atmosphere formation. It is hardly probable that clays and minerals could perform this function. Since minerals are characterized by a significantly lower adsorption activity as compared to PS, they are not capable of evolving in terms of their structure and do not show a catalytic activity in AA and NA polymerization. As we have demonstrated in this work, PS do not have such disadvantages.

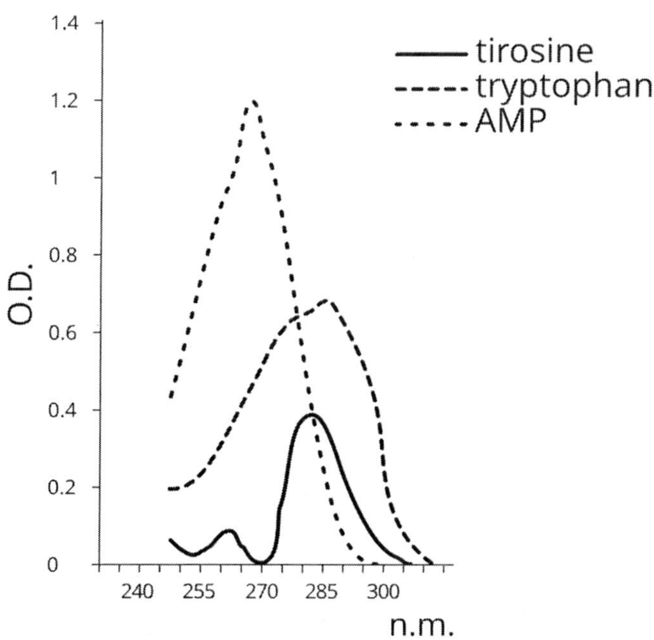

Figure 11. Absorbance spectra of AMP, tyrosine and tryptophan.

Since polysaccharides adsorbed both AA and nucleotides (the latter demonstrated selectivity in bond formation similar to NA complementarity), it was PS that could act as the primary information system, which helped to combine these two groups into one, thus, initiating the abiogenic aminoacylation of nucleotides. According to the results of the quantum-chemical calculations and empirical research, it was established that for some amino acids this process could be carried out abiogenically, not involving enzyme systems, at 2-3' end of the ribose in the nucleotides. Aminoacylation with the help of PS is possible practically for all biogenic amino acids at the 5' end of nucleoside triphosphates. Minispiral forms of RNA (the precursors of the modern tRNA [21]), whose terminal nucleotide is attached to the amino acid, were formed on PS on the principle of PS-NA complementaryty. This process could be a starting point for the template assembly of primary polypeptides abiogenically on oligosaccharide and nucleotide templates. It was PS that were able to function as the catalyst that promoted the formation of both NA and AA polymers and the modern triplet genetic code. Therefore, at the beginning of molecular evolution PS were responsible for combining nucleotides and amino acids into a single system of genetic information formation via three types of biopolymers—NA, protein and PS.

Further research in this direction may bring us closer to the understanding of the mechanisms of genetic code formation in the course of molecular evolution of biopolymers and, furthermore, to evaluate the importance of oligoand polysaccharides in modern biosystems, since in vivo synthesis of at least some of them are templatebased. The obtained data may help answer one of the principal questions about life origin on the Earth—in what way, how and when NA combined with polypeptides to form the modern system of accumulation, storage, transfer and implementation of genetic information.

ACKNOWLEDGEMENTS

This work was supported by the "Plazan" Closed Joint-Stock Company, Moscow.

REFERENCES

1. Miller, S.L. and Urey, H.C. (1955) Production of some organic compounds under possible primitive Earth conditions. Journal of the American Chemical Society, 77, 2351.

2. Parmon, V.N. (2002) Prebiotic phase of life origin. Vestnik Rossiskoi Akademii Nauk, 72, 976-983.

3. Rezanov, I.A. (1985) Great calamities in the Earth history. Mir, Moscow.

4. Fox, S.W. and Dose, K. (1972) Molecular evolution and the origin of life. S.A. Freeman Co., San Francisco.

5. Orgel, L.E. (1998) The origin of life—A review of facts and speculations. Trends in Biochemical Sciences, 23, 491-495.

6. Oparin, A.I. (1968) Life, its nature, origin and development. Nauka, Moscow.

7. Bernal, J. (1969) The origin of life. Mir, Moscow.

8. Zimina, N.P., Rykova, V.I. and Arkhipov, I.A. (1992) Animal tissue proteoglycans as irregular biopolymers: Informativeness of structure and control of biosynthesis. Uspehi Sovremennoi Biologii, 112, 571-588.

9. Silbert, J.E. (1982) Structure and metabolism of proteoglycans and glycosaminoglycans. Journal of Investigative Dermatology, 79, 31-37.

10. Sugahara, K. and Kitagawa, H. (2000) Recent advances in the study of the biosynthesis and functions of sulfated glycosaminoglycans. Current Opinion in Structural Biology, 10, 518-527.

11. Zimnitskii, A.N., Bashkatov, S.A. and Urazbaev, V.N. (2005) DNA tandem repeats and the concept of template synthesis of proteoglycans. Labirint, Moscow.

12. Zimnitskii, A.N., Bashkatov, S.A. and Urazbaev, V.N. (2007) Interaction between nucleic acids and glycans: In silico and in vitro. Biophysics, 52, 282-287. http://nanoderm.ru/img/userimg/statyi/magazine_ biofizika_ eng.pdfhttp://dx.doi.org/10.1134/S0006350907030049

13. Zimnitskii, A.N. (2012) Concept of template synthesis of proteoglycans. In: Dr. D. N. Karunaratne, Ed., The Complex World of Polysaccharides. http://www.intechopen.com/books/the-complex-world-of-polysaccharides/concept-of-template-synthesis-of-proteoglycans/ http://dx.doi.org/10.5772/48085

14. Zimnitskii, A.N., Bashkatov, S.A., Chemeris, A.V., Yamidanov, R.S. and Urazbaev V.N. (2014) Polysaccharides as initiators of nucleic acid polymerization. American Journal of Molecular Biology, 4, 20-25.

15. Sterwart, J.J.P. (1989) Optimization of parameters for semiempirical methods. II. Applications. Journal of Computational Chemistry, 10, 221-261.

16. Schmidt, M.W., Baldridge, K.K., Boatz, J.A., Elbert, S.T., Gordon, M.S., Jensen, J.H., Koseki, S., Matsunaga, N., Nguyen, K.A., Su, S.J., Windus, T.L., Dupuis, M. and Montgomery, J.A. (1993) General atomic and molecular electronic structure system. Journal of Computational Chemistry, 14, 1347-1363.

17. Orekhovich, V.N. (1968). Modern methods in biochemistry. Meditsina, Moscow.

18. Osterman, L.A. (1981) Protein and nucleic acid research methods. Electrophoresis and ultracentrifugation. Nauka, Moscow.

19. Gilyazetdinov, S.Y. and Zimnitskii, A.N. (1986) Composition of poly(A+)-RNA population in the roots of corn heterotic hybrids and their original strains. Physiology of Plants, 33, 769-777.

20. Singer, M. and Berg, P. (1998) Genes and genomes. Mir, Moscow.

21. Moras, D. (1993) Structural aspects and evolutionary implications of the recognition between tRNA and aminoacyl-tRNA-synthetases. Biochemie, 75, 651-657.

Chapter 2

Synthesis and Characterization of Biopolymeric Chitosan Derived from Land Snail Shells and Its Potential for Pb2+ Removal from Aqueous Solution

Joshua N. Edokpayi [1,*], John O. Odiyo [1], Elizabeth O. Popoola [2], Oluwagbemiga S. Alayande [3] and Titus A. M. Msagati [4]

[1] Department of Hydrology and Water Resources, University of Venda, Private Bag X5050, Thohoyandou 0950, South Africa

[2] Department of Chemical Sciences, Yaba College of Technology, P.M.B. 2011, Yaba, Lagos 101212, Nigeria

[3] Centre for Energy Research and Development, Obafemi Awolowo University, P.M.B. 13, Ile Ife 220, Nigeria

[4] Nanotechnology and Water Sustainability Research Unit, College of Science, Engineering and Technology, Florida Science Campus, University of South Africa, 1710 Roodepoort, Johannesburg 2000, South Africa

ABSTRACT

Pb^{2+} is considered to be a very toxic pollutant in the aquatic environmental media. Biopolymeric chitosan synthesized from snail shell has been studied for its potential to remove heavy metals from aqueous solution. The experiments were conducted in the range of 1–50 mg/L initial Pb^{2+} concentration at 298 K. The effects of pH, adsorbent dosage and contact time on the adsorptive property of the adsorbent were investigated and optimized. The derived chitosan was characterized using Fourier transform infrared spectrometer (FT-IR) and X-ray florescence (XRF). The experimental data obtained were analysed using the Langmuir and Freundlich adsorption isotherm models. The Langmuir model and pseudo second order kinetic model suitably described the adsorption and kinetics of the process with regression coefficients of 0.99 and 1.00, respectively. Sodium hydroxide was a better desorbing agent than hydrochloric acid and de-ionized water. From the results obtained, it is concluded that synthesized biopolymers from land snail shells has the potential for the removal of Pb^{2+} from aqueous solutions.

KEYWORDS

adsorption; chitosan; kinetics; snail shells

1. INTRODUCTION

Heavy metals are a group of pollutants occurring naturally in the earth crust. The pollution of environmental matrices has been of global concern due to their relative persistence in the environment [1]. Rapid industrialization coupled with population growth and increased urbanization has led to an increase in heavy metal concentrations [2,3]. Pb^{2+} is known to be part of the composition of effluents that are found in both domestic and industrial sources. Apart from the background concentration of Pb^{2+} in water, its presence can be attributed to the corrosion of lead pipes, the use of gasoline containing lead and the release of domestic and industrial effluents [4]. The toxic effect of Pb^{2+} is well known and several governmental and non-governmental agencies like the World Health Organization (WHO), United States Environmental Protection Agency (US EPA) and the South African Department of Water and Sanitation have established threshold values on the concentration of Pb^{2+} in drinking water and wastewaters [5]. Pregnant women and children are the most vulnerable to Pb^{2+}contamination even at low concentrations if exposed for a very long period of time [6,7].

Some conventional methods have been reported for the removal of heavy metals from water and wastewater. These include chemical precipitation [2,8,9], ion exchange [2], liquid membrane extraction [10], chemical coagulation and flocculation [11]. These methods are ineffective at low metal concentrations and generate a lot of sludge that cause disposal problems [9]. Other methods like adsorption using commercial activated carbon [12,13], microfiltration [14], reverse osmosis [9,14] and membrane techniques [9] are effective, but the cost of installation and maintenance are expensive which limit their use. These drawbacks have led to a continuous search for a cheap, renewable and more economic method for the removal of unwanted metals from aqueous solution.

Chitosan is an important natural biopolymer usually derived from the deacetylation of chitin from the shells of shrimp, crab, and other arthropods and has been widely used in pharmaceutical industries for drug delivery [15]. Sewvandi and Adikary [16] studied the removal of Cr^{6+} from wastewater using chitosan derived from shrimp shells. Similarly, Mohanasrivinidasa *et al.* [17] investigated the removal efficiencies of metals and antimicrobial activity of chitosan derived from shrimp shells. Inference from their studies suggests the potential use of crustacean shells as adsorbents for heavy metal remediation [17]. Jatto *et al.* [18] investigated the coagulative property of land snail shell in reducing water quality parameters of wastewater from food industries; their studies showed a reduction in the concentrations of nitrates, sulphates and chemical oxygen demand.

Land snails are widely consumed in different parts of the world. The shells are often disposed indiscriminately and therefore constitute a nuisance to the environment. Several degrees of injuries to children caused by discarded snail shells have been observed. This study explores the potential of preparing chitosan (a natural, biodegradable, biocompatible, bio-adhesive polymer) from the shells of land snails and using it as an adsorbent for the removal of Pb^{2+} from aqueous solution.

2. EXPERIMENTAL

2.1. Materials

Discarded land snail shells were collected from a local market at Ile-Ife, Nigeria. The shells were thoroughly washed to remove leaves, sand, dirt and other impurities. They were dried at 100 °C and pulverized into fine powder. The powder was sieved using an analytical sieve of 250 micron. Analytical grade chemicals were used in this study. Hydrochloric acid (32%) and sodium hydroxide were supplied by Sigma Aldrich (Johannesburg,

South Africa) while 1000 mg/L of Pb^{2+} was supplied from Merck (pty) Ltd. (Johannesburg, South Africa).

2.2. Preparation of Chitosan

The method reported by Mohanasrinivasan *et al.* [17] was employed with slight modification. Eighty grams of the powder was weighed into a conical flask and 100 mL of 4% NaOH was added. The mixture was boiled and stirred at 100 °C for 2 h in a water bath. After boiling and stirring, it was filtered and washed with distilled water. Red litmus was used to check if the base was completely washed away. After washing, the mixture was filtered and the residue was scraped gently into the petri dishes and then dried in the oven at 100 °C for 3 h.

After deproteinization, the weight of the sample was 62 g. Thirty milliliters of 5% 1 M HCl was added to the deproteinized sample. The mixture was boiled and stirred for 45 min at 100 °C in a water bath. Subsequent washing was done with distilled water followed by filtration. The mixture was examined with blue litmus to check the acidity of the mixture. The residue (chitin) obtained from above was scarped into the petri dish and dried in the oven at 100 °C for 2 h.

Deacetylation reaction was used to convert chitin to chitosan according to a revised procedure of Coughlin *et al.* [19]. Briefly, the isolated chitin was soaked in 510 mL of 50% NaOH (weighing 50 g of NaOH pellets and dissolved in 100 mL of distilled water), boiled at 100 °C for 2 h in water bath and cooled for 30 min at room temperature. The mixture was placed on a magnetic stirrer at 30 °C for 4 h, filtered, washed and examined with red litmus to check if the base was completely washed away. The mixture was filtered to retain the solid matter which is chitosan. The chitosan was oven dried at 90 °C for 24 h.

2.3. Characterization of the Synthesized Chitosan

Elemental studies were performed with a Rigaku ZSX Primus II X-ray Fluorescence spectrometer (USA). Fourier transform infrared (FT-IR) spectra of the synthesized chitosan from snail shells were obtained using a Perkin Elmer 100 FT-IR (Waltham, MA, USA) with accessories. The sample pellets were prepared by using a KBr press (Spectra Lab, Mumbai, India). The spectra was scanned over the wave number range of 4500 to 400 cm^{-1}. The synthesized chitosan were coated with a thin layer of carbon and the surface morphology was analyzed using a scanning electron microscope (SEM) (TESCAN, VEGA 3 SBU, Brno, Czech) and irradiated with a beam of electrons at 20 kV. Surface area and pore width were determined by N_2 gas Brunauer-Emmett-Teller method of analysis using a Micrometrics Chemisorption ASAP 2020 supplied by Norcross, GA, USA. A Perkin Elmer thermal analyzer (Waltham, MA, USA) was used for thermal degradation studies of the adsorbent.

2.4. Adsorption and Kinetics

A Stuart reciprocal shaker with a speed of 250 rpm and a temperature of 298 K was employed for all the experiments in this study. 0.1 M NaOH and 0.1 M HCl were used to adjust the pH of the solution to the desired value. The effect of adsorbent dosage was studied by varying the mass of the synthesized chitosan in the range of 0.05 to 2.0 g using a 40 mL of 10.7 mg/L Pb^{2+} solution. Optimum pH values and contact time were obtained by varying the pH value from 2 to 12 while the equilibration time was varied from 5 to 120 min. The effect of initial Pb^{2+} concentration was performed using different Pb^{2+} concentrations between 1–50 mg/L. In all cases, after equilibration, the solution was centrifuged for 3 min and filtered using a 0.45 μm filter membrane. The filtrates were subsequently analysed using Atomic Absorption Spectrometer.

2.5. Desorption Studies

Regeneration of the adsorbent was performed using de-ionized water, 0.1 M HCl and 0.1 M NaOH. One gram of the adsorbent was agitated with 10.7 mg/L of Pb^{2+} on a Stuart reciprocal mechanical shaker for 60 min at 250 rpm. After equilibration, the adsorbent was separated by centrifugation and subsequently rinsed three times using de-ionized water to remove any unadsorbed Pb^{2+} on the surface. The already used adsorbent was agitated with the three desorbing agents at the same experimental conditions. The filtrates were analyzed using flame Atomic Absorption Spectrometer.

3. RESULTS AND DISCUSSION

3.1. Characterization of the Synthesized Chitosan

The band observed at 3322 cm^{-1} (Figure 1) can be attributed to $-NH_2$ or $-OH$ groups stretching vibration [20]. The peak observed at 2915 cm^{-1} indicated alkane –C-H stretching vibration or –OH stretch of carboxylic acids, while that of 2524 cm^{-1}can be assigned to –C≡C stretching vibration of alkynes [21]. The characteristic –NH band of chitosan was observed at 1644 cm^{-1} [22]. The bands observed at 1376 cm^{-1} and 1305 cm^{-1} can be attributed to –NO stretch of nitrogen containing compounds and –CO stretch of carbonyl compounds, respectively [22]. The observed peak at 1028 cm^{-1} is assigned to –CN stretch of aliphatic amines [22]. The elemental analysis of the synthesized chitosan (Table 1) showed calcium as the major element making up to 98.2%.

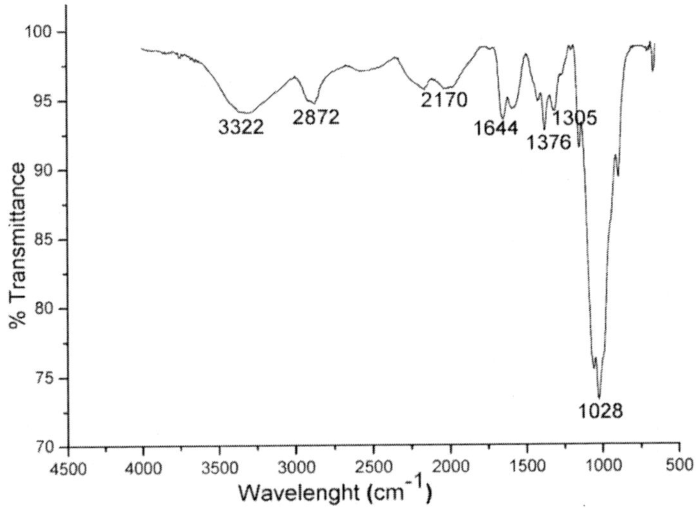

Figure 1. Infrared spectra of the adsorbent.

SEM analysis is a useful tool for investigation of the surface area and structure morphologies of various adsorbents [23]. The SEM micrograph of the synthesized chitosan (Figure 2) shows a rough surface with significant pores and remarkable irregularities suitable for the adsorption of metals. BET surface area of 1.7998 m^2/g was determined for the adsorbent. The pore volume (1.492×10^{-2} cm^3/g) and average pore diameter (21.66 nm) determined is characteristic of a mesoporous material [24,25].

Table 1. Chemical analysis of synthesized chitosan.

Component	wt % Composition
Na_2O	0.58
MgO	0.03
Al_2O_3	0.15
SiO_2	0.34
P_2O_5	0.07
SO_3	0.10
Cl	0.02
K_2O	0.04
CaO	98.20
Fe_2O_3	0.10
SrO	0.40

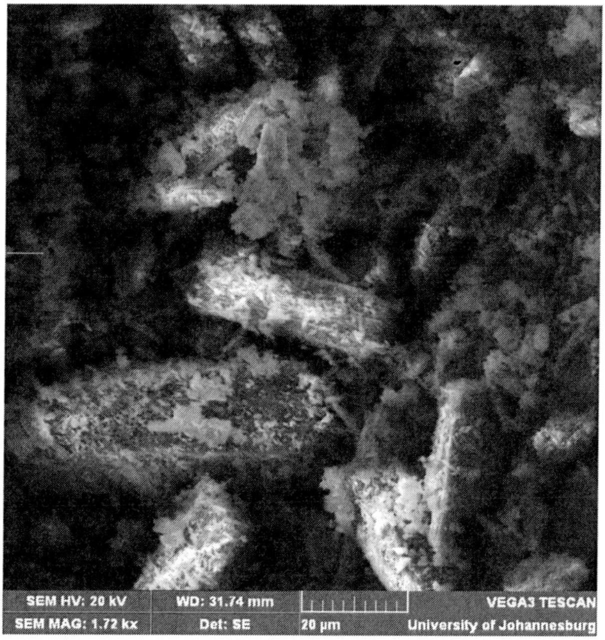

Figure 2. SEM micrograph of the adsorbent.

The synthesized chitosan (8.5 mg) was heated from 30 °C to 900 °C at 20 °C·min^{-1} using a thermogravimentary analyzer. The biosorbent was stable up to 610 °C (Figure 3) although a little loss in mass was recorded between 100 and 200 °C, which can be attributed to loss of bound water molecules in the samples. The sample degraded at temperatures between 610 and 813 °C, which accounted for about 43% of the initial biosorbent. A slight degradation was observed after this point.

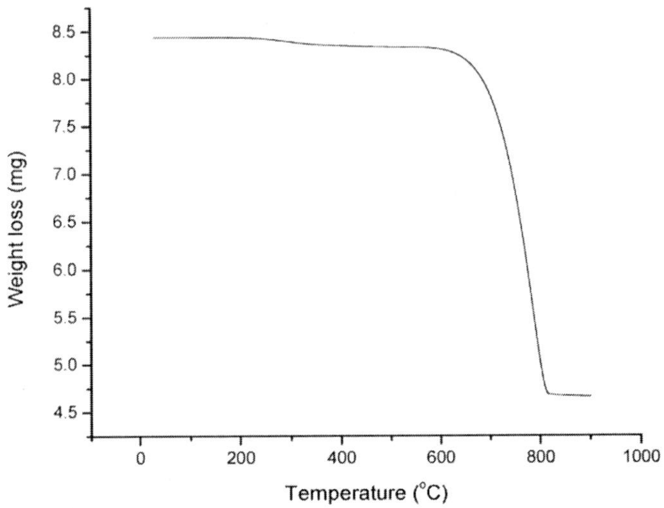

Figure 3. TGA profile of synthesized Chitosan.

3.2. Effects of Experimental Conditions on Adsorption Efficiency

The percentage uptake of Pb^{2+} by the synthesized chitosan was found to increase with increasing quantity of the adsorbent from 0.05 to 2.0 g/L (Figure 4). This finding can be attributed to the increase in binding sites on the adsorbent as the number of available binding sites is expected to increase with increased quantity of the adsorbent [25,26]. However, the amount of Pb^{2+} adsorbed per unit mass of the adsorbent decreases at increasing dosage due to overlapping of the adsorption sites by Pb^{2+}. From the experimental data obtained, 1.0 g/L of the synthesized chitosan was found to achieve maximum adsorption of Pb^{2+}.

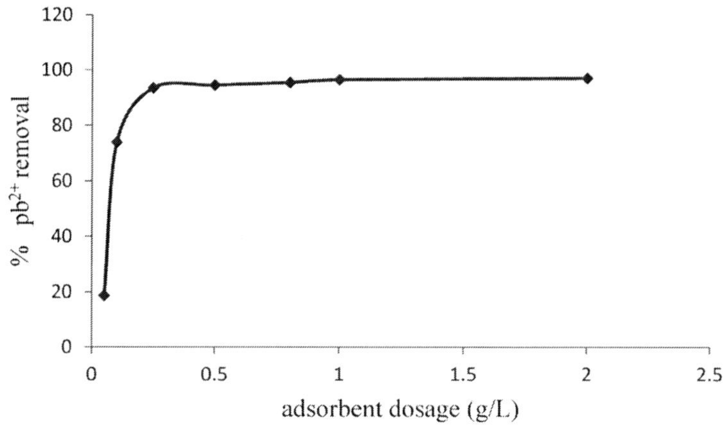

Figure 4. Variation of percent Pb^{2+} removal with adsorbent dosage (initial concentration = 10.7 mg/L, contact time = 40 min, T = 298 K, pH = 4, shaking speed = 250 rpm).

The dependence of adsorption of Pb^{2+} was nearly constant between pH values of 2 to 10, although a slight increase in the uptake of Pb^{2+} was recorded between pH 2 and 4. The optimum pH value for this study is four (Figure 5). This can be attributed to electrostatic balance between the protonated amine sites and Pb^{2+} [27,28]. There was a decrease in the uptake of Pb^{2+} by the synthesized chitosan at pH > 10. The cause for the phenomenon could be due to reduced solubility and precipitation of Pb^{2+} under alkaline condition [26]. Pb^{2+} is soluble in pH < 4 but will precipitate out of solution under neutral and alkaline conditions. The mechanism of the interaction based on the optimum pH is purely by adsorption while that of alkaline pH can be regarded as a mixture of adsorption/precipitation [27,28].

There was a rapid uptake of Pb^{2+} by the synthesized chitosan within 5 min of equilibration (98%) which is due to the availability of vacant binding sites on the surface of the adsorbent. A gradual increase in the uptake of Pb^{2+} was further recorded up to 40 min of equilibration (99%) due to the filling of the remaining binding sites on the adsorbent. Beyond this time (Figure 6), no significant increase in the adsorption capacity of Pb^{2+} was determined, indicating that equilibrium has been reached. This phenomenon can be explained as due to repulsive forces between Pb^{2+} on the surface of the adsorbent and Pb^{2+} in the aqueous phase [25]. Similar optimum time for Pb^{2+} sorption has been reported by Hikmat *et al.* [29], although Santi *et al.* [30] and Tahiruddin and Ab Rahman [31] reported an optimum time of 50 min and 30 min for the adsorption of Pb^{2+} onto activated carbon from rice husk and peanut shells, respectively.

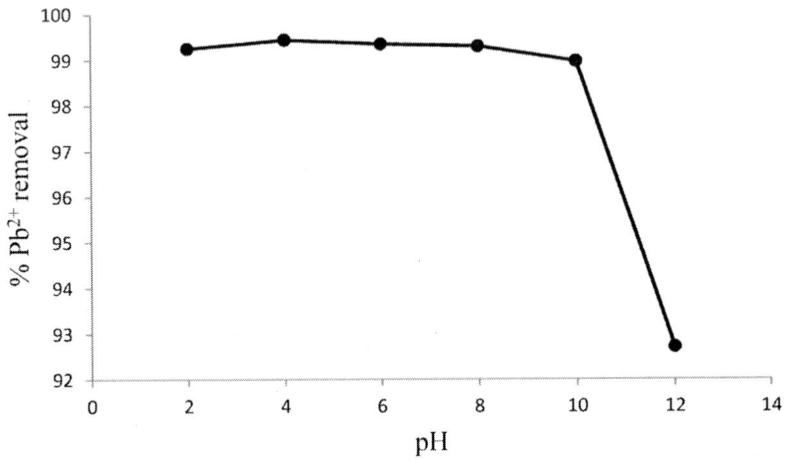

Figure 5. Variation of percent Pb^{2+} ion removal with change in pH value (initial concentration = 10.7 mg/L, contact time = 40 min, T = 298 K, shaking speed = 250 rpm, adsorbent dosage = 1.0 g).

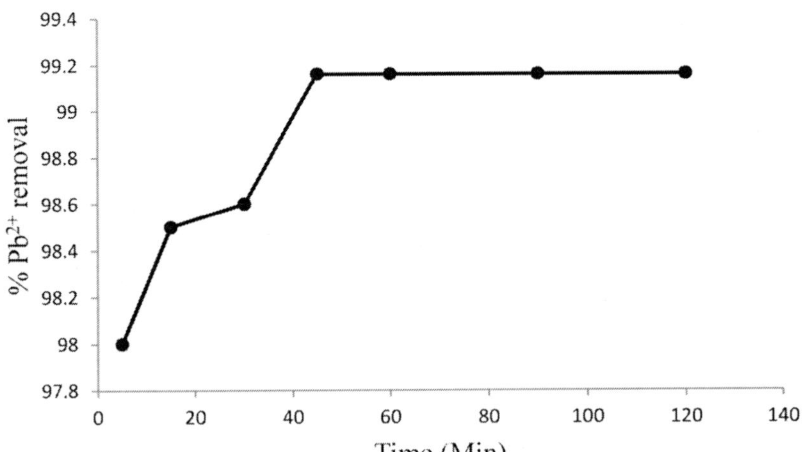

Figure 6. Variation of percent Pb^{2+} ion removal with contact time (initial concentration = 10.7 mg/L, T = 298 K, pH = 4, shaking speed = 250 rpm, adsorbent dosage =1.0 g).

3.3. Adsorption Studies

The adsorption isotherm is an extremely important tool for understanding the distribution of the adsorbate on the adsorbent surface at equilibrium [32,33]. Langmuir and Freundlich isotherm models were used to describe the equilibrium data obtained. Langmuir isotherm applies to the monolayer adsorption process while the Freundlich isotherm is a semi empirical equation based on the adsorption process that occurs on heterogeneous surfaces [34,35]. The linearized equation of Langmuir and Freundlich isotherm models are:

$$\frac{1}{q_e} = \frac{1}{q_{\max}} + \left(\frac{1}{bq_{\max}}\right)\frac{1}{C_e} \tag{1}$$

$$\log q_e = \log K_f + \left(\frac{1}{n}\right)\log C_e \tag{2}$$

A plot of $1/q_e$ *versus* $1/C_e$ gave a straight line with a regression coefficient (R^2) of 0.99 (Figure 7) with $1/q_{max}$ as intercept and $1/bq_{max}$ as slope. Langmuir isotherm can also be expressed in terms of a dimensionless constant called separation factor R_L[34];

$$R_L = \frac{1}{(1+bC_O)} \tag{3}$$

where C_o is the initial metal concentration (mg/L) and b the Langmuir constant (L/mg). $R_L > 1$ indicates an unfavourable monolayer adsorption process, $R_L = 1$ linear, $0 < R_L < 1$ favourable and $R_L = 0$ irreversible [24,36]. The result obtained from this study has an R_L value between zero and one, indicating a favourable adsorption process. This implies that the chemisorption process duly explains the adsorption process.

Figure 8 shows the plot of log q_e against log Ce, which also gave a straight line with a linearized coefficient of 0.89, implying that the adsorption process could also be controlled by physisorption as a result of weak van der Waals forces between the surface of adsorbent and adsorbate. Based on the correlation coefficients obtained from the Langmuir and Freundlich isotherm models, it can be deduced that the adsorption process is more favored by the Langmuir isotherm model. This is consistent with other results reported in literature [37,38].

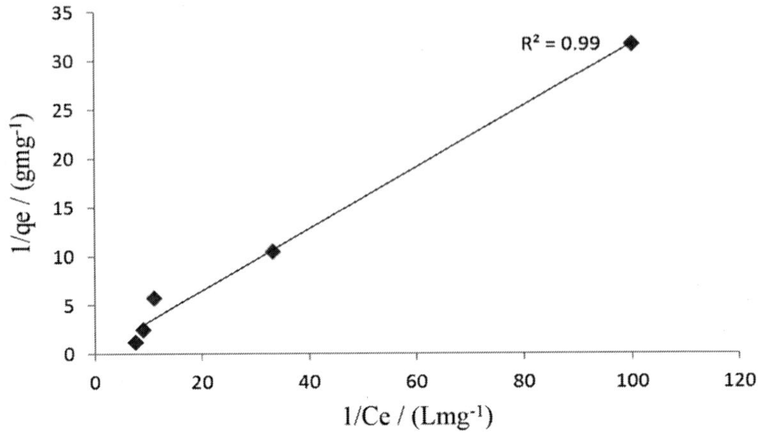

Figure 7. Langmuir plot for Pb^{2+} adsorption onto synthesized chitosan.

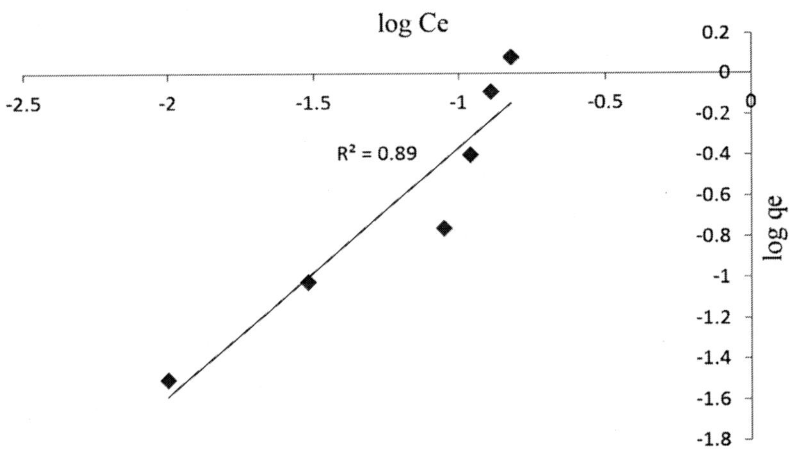

Figure 8. Freundlich plot for Pb^{2+} adsorption onto synthesized chitosan.

3.4. Kinetic Studies

To explain the kinetics of the adsorption process, the experimental data were subjected to the pseudo first order and pseudo second order kinetic models. Equations (4) and (5) show the linearized forms of the pseudo first and pseudo second order kinetic models, respectively.

$$\log(q_e - q_t) = \log q_e - (\frac{k_1}{2.303}) \quad (4)$$

$$\frac{t}{q_t} = (\frac{1}{k_2 q_e^2}) + \frac{t}{q_e} \quad (5)$$

where q_e and q_t are the amounts of Pb^{2+} adsorbed at equilibrium and at time t, k_1 and k_2 are the rate constants of the pseudo first and pseudo second order model. The correlation coefficients obtained from the plots of the pseudo first order and pseudo second order models are 0.65 and 1.0, respectively (Figure 9 and Figure 10). The pseudo second order best describes the kinetics of the adsorption process which corroborate with the findings of Ahmad *et al.* [34], Moyo and Chikazaza [39] and Chen *et al.* [40] in the adsorption of Pb^{2+} by macrocylic calyx[4]naphthalene, acid treated maize tassels and thiacalix[4]arene composites, respectively.

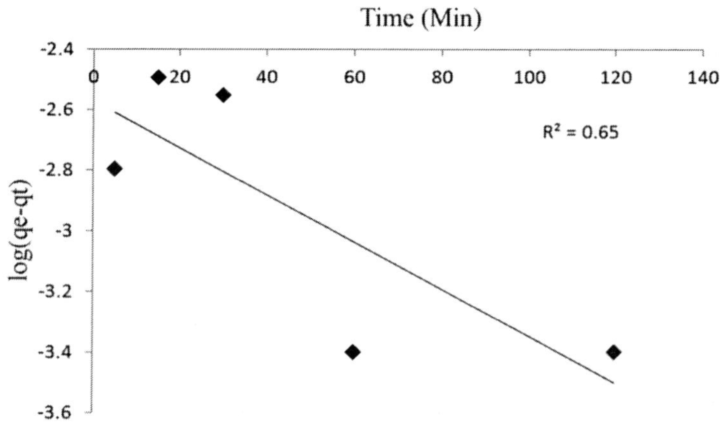

Figure 9. Pseudo first order kinetics for Pb^{2+} adsorption onto synthesized chitosan.

3.5. Desorption

De-ionized water and 0.1 M HCl were very poor desorbing agents, removing less than 5% of the adsorbed Pb^{2+} from the surface of the adsorbent while 0.1 M NaOH was a better desorbing agent with about 30% removal. This also suggests the formation of a bond between the surface of the adsorbent and Pb^{2+}.

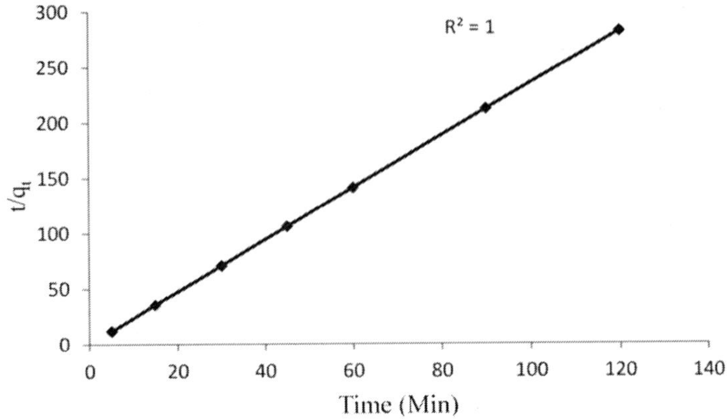

Figure 10. Pseudo second order kinetics for Pb^{2+}adsorption onto synthesized chitosan.

4. CONCLUSIONS

This study has shown that synthesized chitosan from land snail shells can be employed for the adsorption of Pb^{2+} from aqueous solution and wastewater. The mechanism of the adsorption of Pb^{2+} onto the adsorbent is believed to be through the adsorption of the metal onto the amine functional group of the adsorbent. The equilibrium data obtained fit better into the Langmuir isotherm model than the Freundlich isotherm model, implying a chemisorption mechanism.

ACKNOWLEDGMENTS

We appreciate The Research and Publication Committee (RPC) of the University of Venda for funding this research (grant number: S357).

AUTHOR CONTRIBUTIONS

Oluwagbemiga S. Alayande prepared the synthesized chitosan from land snail shells, Joshua N. Edokpayi performed the characterization and batch adsorption experiments. John O. Odiyo, Titus A. M. Msagati and Elizabeth

O. Popoola contributed immensely to the writing and discussion of the manuscript. All authors approved the final manuscript for publication.

REFERENCES

1. Hernandez-Montoya, V.; Perez-Cruz, M.A.; Mendoza-Casillo, D.I.; Moreno-Virgen, M.R.; Bonilla-Petriciolet, A. Competitive adsorption of dyes and heavy metals on zeolitic structure. *J. Environ. Manag.* **2013**, *116*, 213–221.

2. Das, N.; Karthiica, P.; Vimala, R.; Vinidhini, V. Use of natural products as biosorbent of heavy metals, An overview.*Nat. Prod. Radiance* **2008**, *7*, 133–138.

3. Cataldo, S.; Muratore, N.; Orecchio, S.; Pettignano, A. Enhancement of adsorption ability of calcium alginate gel beads towards Pd(II) ion. A kinetic and equilibrium study on hybrid Laponite and Montmorillonite-alginate gel beads. *Appl. Clay Sci.* **2015**, *118*, 162–170.

4. Koh, L.L.; Wong, M.K.; Gan, L.M. Factors affecting the leaching of lead from UPVC pipes. *Environ. Monit. Asses.* **1991**,*19*, 203–214.

5. Sublet, R.; Simonnot, M.; Boireau, A.; Sardin, M. Selection of an adsorbent for lead removal from drinking water by a point-of-use treatment device. *Water Res.* **2003**, *37*, 4904–4912.

6. Minnesota Department of Health. Point-of-Use Water Treatment Units for Lead Reduction. 141–0717. August 2010. Available online: http://www.health.state.mn.us/divs/eh/water/factsheet/com/poulead .html (accessed on 15 December 2014).

7. Edokpayi, J.N.; Odiyo, J.O.; Olasoji, S.O. Assessment of heavy metal contamination of Dzindi River, in Limpopo Province, South Africa. *Int. J. Nat. Sci. Res.* **2014**, *2*, 185–194.

8. Zvinowanda, C.M.; Okonkwo, J.O.; Agyei, N.M.; Shabalala, P.N. Physicochemical characterization of maize tassel as an adsorbent. I. surface texture, microstructure and thermal stability. *J. Appl. Polym. Sci.* **2009**, *111*, 1923–1930.

9. Mihajlovic, M.T.; Lazarevic, S.S.; Jankovic-Castvan, I.M.; Kovac, J.; Jokic, B.M.; Janackovic, D.T.; Petrovic, R.D. Kinetics, thermodynamics, and structural investigations on the removal of Pb^{2+}, Cd^{2+}, and Zn^{2+} from multicomponent solutions onto natural and Fe(III)-modified zeolites. *Clean Technol. Environ. Policy* **2015**, *14*, 407–419.

10. Rorrer, G. Heavy metal ions. In *Removal from Wastewater: Encyclopaedia of Environmental Analysis and Remediation*; Meyers, R.A., Ed.; Wiley: New York, NY, USA, 1998; Volume 4, pp. 2101–2125.

11. Johnson, P.D.; Padmanabhan, G.P.; Ohlinger, K.N.; Ritchie, S.; Teuber, L.; Kirby, J. Enhanced removal of heavy metals in primary treatment using coagulation and flocculation. *Water Environ. Res.* **2008**, *80*, 472–479.

12. Freeman, H.F. *Standard Handbook of Hazardous Waste Treatment and Disposal*; McGraw-Hill: New York, NY, USA, 1989.

13. Cheremisinoff, N.P. *Handbook of Water and Wastewater Treatment Technologies*; Butterworth-Heinemann: Woburn, MA, USA, 2002.

14. Geselbarcht, J. Micro filtration/reverse osmosis pilot trials for Livermore, California, advanced water reclamation. In Proceedings of the AWWA 1996 Water reuse conference, San Diego, CA, USA, 25–28 February 1996; p. 187.

15. Ilium, L. Chitosan and its use as a pharmaceutical excipient. *Pharm. Res.* **1998**, *15*, 1326–1331.

16. Sewvandi, G.A.; Adikary, S.U. Removal of Heavy Metals from Wastewater Using Chitosan. *Society for Social Management Systems Internet Journal*, 2011. Available online: http://hdl.handle.net/10173/836 (accessed on 18 December 2014).

17. Mohanasrinivasan, V.; Mishra, M.; Paliwal, J.S.; Singh, S.K.; Selvarajan, E.; Suganthi, V.; Devi, C.S. Studies on heavy metal removal efficiency and antibacterial activity of chitosan prepared from shrimp shell waste. *3 Biotech.* **2014**, *4*, 167–175.

18. Jatto, E.O.; Asia, I.O.; Egbon, E.E.; Otutu, J.O.; Chukwuedo, M.E.; Ewansiha, C.J. Treatment of waste water from food industry using snail shell. *Acaedmia Arena* **2010**, *2*, 32–36.

19. Coughlin, R.W.; Deshaies, M.R.; Davis, E.M. Preparation of chitosan for heavy metal removal. *Environ. Prog.* **1990**, *9*, 35–39.

20. Bhumkar, D.R.; Pokharkar, V.B. Studies on effect of pH on cross-linking of chitosan with sodium tripolyphosphate: A technical note. *APS PharmSciTech* **2006**, *7*, 1–6.

21. Ray, M.; Anis, K.P.A.; Banthia, A.K. Development and characterization of chitosan based polymeric hydrogel membranes. *Des. Monomers Polym.* **2010**, *13*, 93–206.

22. Coates, J. Interpretation of Infrared Spectra, a practical approach. In *Encyclopedia of Analytical Chemistry*; Meyers, R.A., Ed.; John Wiley & Sons Ltd: Chichester, UK, 2000; pp. 10815–10837.

23. Özbay, İ.; Özdemir, U.; Özbay, B.; Veli, S. Kinetic, thermodynamic, and equilibrium studies for adsorption of azo reactive dye onto a novel waste adsorbent: Charcoal ash. *Desalin. Water Treat.* **2013**, *51*, 6091–6100.

24. Edokpayi, J.N.; Odiyo, J.O.; Msagati, T.A.M.; Popoola, E.O. A Novel Approach for the removal of lead(II) ion from wastewater using mucilaginous leaves of diceriocaryum eriocarpum plant. *Sustainability* **2015**, *7*, 14026–14041.

25. Lingamdinne, L.P.; Koduru, J.R.; Jyothi, R.K.; Chang, Y.; Yang, J. Factors affect on bioremediation of Co(II) and Pb(II) onto Lonicera japonica flowers powder. *Desalin. Water Treat.* **2015**.

26. Koduru, J.R.; Chang, Y.; Yang, J.; Kim, I. Iron oxide impregnated morus alba l. fruit peel for biosorption of Co(II): Biosorption properties and mechanism. *Sci. World J.* **2013**, *2013*, 1–14.

27. Dambies, L.; Guimon, C.; Yiacoumi, S.; Guiba, E. Characterization of metal ion interactions with chitosan by X-ray photoelectron spectroscopy. *Coll. Surf. A Physicochem. Eng. Asp.* **2001**, *177*, 203–214.

28. Guibal, E. Interactions of metal ions with chitosan-based sorbents: A review. *Sep. Puri. Technol.* **2004**, *38*, 43–74.

29. Hikmat, N.A.; Qassim, B.B.; Khethi, M.T. Thermodynamic and kinetic studies of lead adsorption from aqueous solution onto petiole and fibre of palm tree. *Am. J. Chem.* **2014**, *4*, 116–124.

30. Raya, I.; Zakir, M. The adsorption of Pb (II) ions on activated carbon from rice husk, irradiated by ultrasonic waves: Kinetic and thermodynamics studies. *J. Nat. Sci. Res.* **2014**, *4*, 18–24.

31. Tahiruddin, N.S.M.; Ab Rahman, S.Z. Adsorption of lead in aqueous solution by a mixture of activated charcoal and peanut shell. *World J. Sci. Technol. Res.* **2013**, *1*, 102–109.

32. Kannamba, B.; Reddy, K.L.; AppaRao, B.V. Removal of Cu (II) from aqueous solutions using chemically modified chitosan. *J. Hazard. Mater.* **2010**, *175*, 939–948.

33. Rabelo, R.B.; Vieira, R.S.; Luna, F.M.T.; Guibal, E.; Beppu, M.M. Adsorption of copper(II) and mercury(II) ions onto chemically-modified chitosan membranes: Equilibrium and kinetic properties. *Adsorpt. Sci. Technol.* **2012**, *30*, 1–12.

34. Ahmad, R.; Kumar, R.; Laskar, M.A. Adsorptive removal of Pb^{2+} form aqueous solution by macrocyclic calyx[4]naphthalene: Kinetic, thermodynamic, and isotherm analysis. *Environ. Sci. Pollut. Res.* **2013**, *20*, 219–226.

35. Gupta, V K.; Ali, I. Removal of endosulfan and methoxychlor from water on carbon slurry. *Environ. Sci. Technol.* **2008**,*42*, 766–770.

36. Gupta, V.K.; Gupta, M.; Shamar, S. Process development for the removal of lead and chromium from aqueous solutions using red mud—An aluminium industry waste. *Water Res.* **2001**, *35*, 1125–1134.

37. Asandei, D.; Bulgariu, L.; Bobu, E. Lead(II) removal from aqueous solutions by adsorption onto chitosan. *Cellul. Chem. Technol.* **2009**, *43*, 211–216.

38. Ng, J.C.Y.; Cheung, W.H.; Mckay, G. Equilibrium studies of the sorption of Cu(II) ions onto chitosan. *J. Colloid Interface Sci.* **2002**, *255*, 64–74.

39. Moyo, M.; Chikazaza, L. Bioremediation of lead(II) from polluted wastewaters employing sulphuric acid treated maize tassel biomass. *Am. J. Anal. Chem.* **2013**, *4*, 689–695.

40. Chen, D.; Hu, B.; He, M.; Haung, C. Micro-column preconcentration/separation using thiacalix[4]arene tetracarboxylate derivative modified mesoporous TiO_2 as packing materials on-line coupled to inductively coupled plasma optical emission spectrometry for the determination of trace heavy metals in environmental water samples.*Microchem. J.* **2010**, *95*, 90–95

Chapter 3

Extraction and characterization of keratin from bovine hoof: A potential material for biomedical applications

Prachi Kakkar, Balaraman Madhan and Ganesh Shanmugam

Central Leather Research Institute, Council of Scientific and Industrial Research

ABSTRACT

Keratin from the hoof is a less explored source for making valuable products. In this paper we present the extraction of pure keratin from bovine hooves and characterized them to better address the possible exploitation of this bio-resource as an alternative material for tissue engineering applications. The keratin protein from the pulverized hooves was extracted by reduction, which was observed to be pure, and two polypeptide chains of molecular weight in the range of 45–50 and 55–60 KDa were determined using SDS-PAGE assay. FTIR analysis complementing circular dichroism (CD) data, established that hoof keratin predominantly adopted α-helical conformation with admixture of β-sheet. The keratin was shown to have appreciably high denaturation temperature (215°C) as indicated by differential scanning calorimetric (DSC) analysis. Thermogravimetric analysis (TGA) also showed the retention of 50% of the original weight of the sample even at a temperature of 346°C. The keratin from the hoof had been observed to be biocompatible when analyzed with MTT assay using fibroblast cells, showing more than 90% cell viability. Hence, hoof keratin would be useful for high value biomedical applications.

KEYWORDS

Proteins, Keratin, Biocompatibility, α-helix, β-sheet, Biopolymers & renewable polymers

1. INTRODUCTION

A broad category of insoluble proteins that associate as intermediate filaments (IFs), a cytoskeletal element with 8–10 nm diameter, were being referred to the term "keratin". Apart from being the principal constituent of the stratum corneum of human and animal skin, keratins are also found in exoskeletal materials such as horns, hooves, hair, feathers, wool and nails (Aluigi et al.2007; Gupta and Nayak2014). Keratins (mammalian) were categorized into two distinct groups namely hard and soft, based on their structure, function and regulation. Hard keratins form ordered arrays of IFs embedded in a matrix of cystine rich proteins and contribute to the tough structure of epidermal appendages, whereas soft keratins preferentially form loosely-packed bundles of cytoplasmic IFs (Coulombe et al.2000; Fraser et al.1986; Moll et al.1982) and typically contain less sulphur (Zoccola et al.2009). There are also variations in the secondary structure of the keratin from these sources. As observed earlier (Hill et al.2010; Iridag and Kazanci2006; Zoccola et al.2009), mammalian keratins are found to be predominantly of the α-type (contains α-helical structure), whereas birds and reptiles can possess both α- and β-types (contain a mixture of α-helical and β-sheet structures). Fraser and Parry (Fraser and Parry1996) suggested that the inner regions of compact assemblies of keratin molecules are predominantly hydrophobic, and that charged residues are being concentrated at the surface. Characteristically, keratins show high stability and low solubility due to –S–S– cross-linking between cysteine amino acid residues (Brandelli2008; Feughelman2002). The most distinctive feature of keratin at a molecular level is the high concentration of half-cystine residues (7%–20% of the total amino acid residues), most of which are localized at the terminal regions of the proteins (Hearle2000).

Huge amount of keratin byproducts are wasted without ample utility (Korniłłowicz-Kowalska and Bohacz2011). Several attempts have been made for the mechanical (Korol2012) as well as microbial utilization of keratin wastes (Chaudhari et al.2013; Fang et al.2013; Jeong et al.2010). Keratin like other natural biopolymers viz., collagen and chitosan can be used for making biomaterials in tissue repair and regeneration (Alsarra2009; Natarajan et al.2012; Ramshaw et al.2009; Ramadass et al.2013). Such applications would enhance the value of utilization of keratins, as recent reports support the acceptance of keratin as a material for biotechnology and biomedical applications (Hill et al.2010; Rouse and Van Dyke2010; Zhuang et al.2013; Patrucco et al.2011).

Cell adhesion sequences, RGD (Arg-Gly-Asp), and LDV (Leu-Asp-Val), which are found in the extra cellular matrix proteins such as fibronectin, are present in the keratins of wool, silk, and human hair (Feughelmann1985; Marshall et al.1991). Moreover, keratin also contains cellular-binding motifs (i.e. super secondary structure with binding capacity) which mimic the sites of cellular attachment found in the native extra cellular matrix because of which, keratin could be used for the development of tissue engineering constructs. Recently, Srinivasan et al. (Balaji et al.2012) reported the use of keratin from horn for biomaterial development. Although keratin has been extracted from many sources such as wool, hair, feather, horn etc., the source of bovine hoof, which is a solid waste from the slaughter houses, is less explored. Hence, this paper deals with the extraction of high value keratin from bovine hoof. The extracted keratin from hoof has been characterized to establish its structure and biocompatibility.

2. MATERIALS AND METHODS

Pulverized raw hooves were obtained from the local slaughter house at Perambur, Chennai (India). This material was washed thrice with distilled water to remove all dirt and then drained. After drying it completely in an oven, it was used as raw material for further studies. All other chemicals used for analytical purpose were purchased from Sigma-Aldrich.

2.1 Extraction of keratin

The first step of extraction is defatting i.e. removal of fats from the raw material. Soxhlet's apparatus was used to carry out defatting/delipidization of pulverized hoof sample for about two days. Mixture of hexane and dichloromethane (1:1, v/v) was used for refluxing. Ten gram of defatted hoof sample was mixed with 7 M urea, 6 g SDS and 15 ml of 2-mercaptoethanol in a 300 mL round-bottom flask and kept in orbital shaker at 60°C for 12 h to extract keratin at pH 7. The resulting solution was then centrifuged for 15 mins at 6,000 rpm and the supernatant was dialyzed against degassed water for 5–6 days. Some of the extracted keratin was kept in a deep freezer at -80°C for 5 h and lyophilized to make it into powder. Hereafter, the dialyzed keratin i.e. prior to lyophilization will be mentioned as liquid keratin and the powdered one will be mentioned as lyophilized keratin.

2.2 Quantification of protein

The protein-content was determined through CHNS analysis. Carbon, hydrogen, nitrogen and sulfur contents of the lyophilized keratin sample were analyzed using varioMICRO CHNSO Sr no: 15091002. Based on this analysis along with the available literature of amino acids present in hoof keratin, protein content in the sample was quantified.

To determine the molecular weight, the liquid keratin was denatured prior to subjecting it to one-dimensional slab SDS-PAGE (10% gel) by heating the sample mixed with an equal amount of sample buffer containing SDS and β-mercaptoethanol. In the beginning, it was at kept at 50 V and then switched to 100 V once the protein (sample) just reached the separating gel. The concentration of liquid keratin used was about ~30 µg per well. The gel was stained using coomassie brilliant blue stain (Neuhoff et al.1985).

2.3 CD Spectroscopy

CD spectrum of keratin was recorded at 25°C on a JASCO J-715 spectropolarimeter using 1 mm rectangular quartz cell. The sample was prepared by dissolving lyophilized keratin in 50 mM phosphate buffer (pH 5). The sample concentration was 0.1 mg/ml. The CD spectrum (θ) was recorded from 190 to 260 nm with standard sensitivity and at the scan rate of 100 nm/min. The other parameters such as band width, response, and data pitch were set at 1 nm, 1 sec, and 0.5 nm, respectively. The CD spectrum represents an average of three individual scans. After subtracting the CD spectrum of solvent (50 mM phosphate buffer, pH 5) from the sample spectrum, it was converted to specific ellipticity $[\Psi]_\lambda$ using the following formula:

$$[\Psi]_\lambda = \theta/100 \times C \times \ell \tag{1}$$

Where, θ is an observed ellipticity, C is the concentration of sample in g/ml and ℓ is the cell path length in cm.

2.4 FTIR Spectroscopy

FTIR spectra of keratin were recorded from 400 to 4000 cm^{-1} using a Nicolet 20 DXB FT-IR spectrophotometer. For solid-state measurement, a pellet was prepared by mixing lyophilized keratin with potassium bromide. For film measurement, an aqueous buffer solution (pH 5) containing keratin (5 mg/ml) was cast on ZnSe plate. The sample was kept at room temperature until a dry thin film formed on the surface of ZnSe. All spectra were measured at a resolution of 8 cm^{-1}.

2.5 Thermal analysis

DSC analysis of lyophilized keratin after conditioning (for 24 hrs) the samples at 24°C, 65% relative humidity was performed from 30°C to 250°C, at 10°C/min using universal V4.4A TA instruments. The instrument was calibrated by an indium standard, and the calorimeter cell was flushed with 100 mL/min liquid nitrogen.

TGA of lyophilized keratin was performed on universal V4.4A TA instruments. The sample was heated at 10°C/min at a temperature range of 30°C-600°C using Al_2O_3 crucibles.

2.6 MTT Assay for cell viability

3T3-Li fibroblast cells were used for biocompatibility analysis. The cells were cultured in flask containing Dulbecco's Modified Eagle's Medium (DMEM; Sigma) with 10% Fetal Bovine Serum (FBS; Invitrogen) supplemented with 50 U/ml penicillin, 50 µg/ml streptomycin and

2.5 µg/ml amphoterecin B. To determine the biocompatibility of the extracted keratin, the cells were cultured as monolayer in polystyrene cell culture 96-well plate with and without keratin at 37°C under a humidified atmosphere of 5% CO_2 in air. Different concentrations of the keratin used for the biocompatibility assay were 10, 25 and 50 µg/well.

3. RESULTS AND DISCUSSION

3.1 Extraction of keratin

While a significant amount of research carried out on several sources of keratin, there are scanty reports available on the extraction, characterization and utilization of keratin from bovine hooves. Keratin from hooves is a significant source along with other sources of keratin. Here we report the extraction and characterization of keratin from the much un-explored raw material- bovine hoof in the reduced form by adopting a well established method by Yamauchi et al. (Yamauchi et al.1996). Keratin was obtained in the form of an aqueous solution from raw hooves using urea (to break non-covalent bonds), sodium dodecyl sulfate (for disruption of strong intermolecular interactions), and mercaptoethanol (to cleave the disulfide bonds in keratin), at 60°C. The process followed for the extraction of keratin is presented in the flow chart given in Figure 1A. The optimum pH range adopted for extraction was 6–8, as keratin could not be extracted at acidic pH and is likely to undergo decomposition at alkaline pH. This solution is expected to be stable for 1 year when stored at ambient temperature (20°C-24°C) (Srinivasan et al.2010). The lyophilized powder of this keratin solution is shown in Figure 1B.

3.2 Quantitative estimation of keratin

The quantitative estimation of keratin was initially determined by both Lowry's and Bradford's method of protein estimation. But, the results shown by both these analyses were not accurate and constant. This might be due to the presence of some traces of mercaptoethanol in the final dialysate. Hence, nitrogen content was determined using CHNS analyzer. The lyophilized keratin sample contained 13.3% of nitrogen, 45.3% of carbon, and 6.84% of hydrogen. Based on the amino acid analysis of the horn-hoof reported earlier (Zoccola et al.2009), the nitrogen content of the keratin is 16.7%. Higher amount of hydrogen in our sample does not rule out the possibility of the presence of bound water attributing towards the constituent of the sample. Hence the purity of the keratin sample is observed to be greater than 80%. Based on the initial weight of the hooves, 44% yield of keratin was obtained.

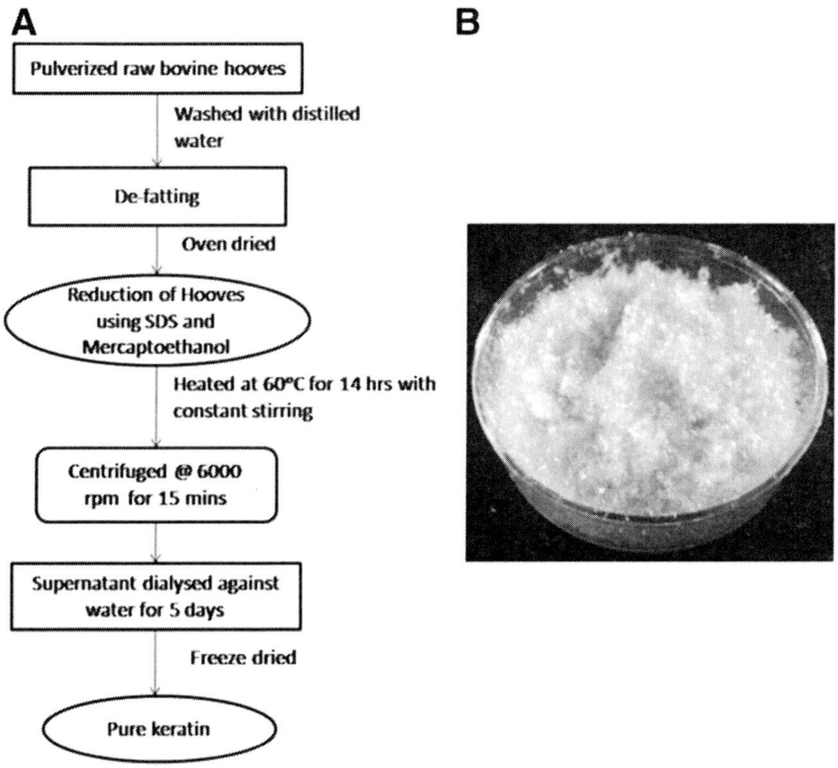

Figure 1. Extraction of Hoof Keratin. **(A)** Flow chart describing the extraction procedure. **(B)** Lyophilized pure keratin.

Molecular weight of the keratin had been estimated by SDS-PAGE using 10% polyacrylamide gel, which showed two clear protein fractions with equal intensities between 45–50 and 55–60 kDa as shown in Figure 2. These bands are similar to the low-sulphur keratin observed earlier (Fraser and Parry1996).

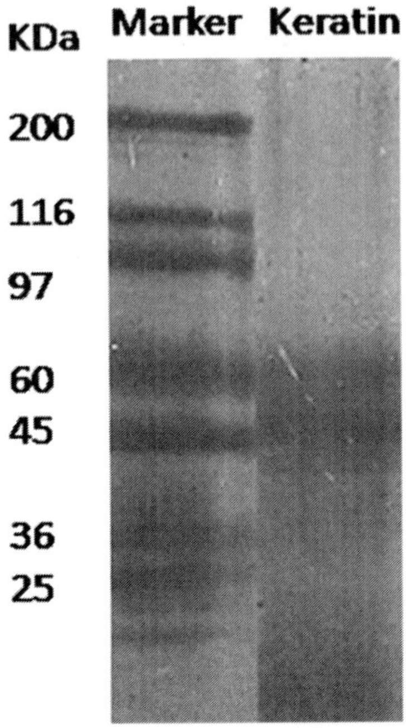

Figure 2. SDS-PAGE of standard protein molecular weight markers (left lane) and hoof keratin (right lane).

3.3 CD analysis of keratin

CD is a quick method to elucidate the secondary structure of protein and peptide in solution. By observing the wavelength and sign of the emerging bands, the kind of secondary structure (α-helix, β-sheet, or random coil etc.) can be evaluated. Generally, an α-helical structure shows two negative minimums of almost same degree of ellipticity at 222 nm and 208 nm followed by a positive maximum at 190 nm while a negative minimum near 218 nm followed by a positive maximum at 195 nm depicts the presence of β-sheet structure (Kelly et al.2005). In the present case (as shown in Figure 3), the CD spectrum shows the negative bands at 208 nm and shoulder at 222 nm, indicating that hoof keratin adopts predominantly α-helical conformation. However, the intensities of these two negative bands are not equal and also CD spectrum has broad band between 200 and 235 nm, suggesting that keratin also adopts β-sheet conformation (Amiya et al.1982). Hence, hoof keratin is comprised of predominantly α-helical conformation with an admixture of β-sheet in liquid state.

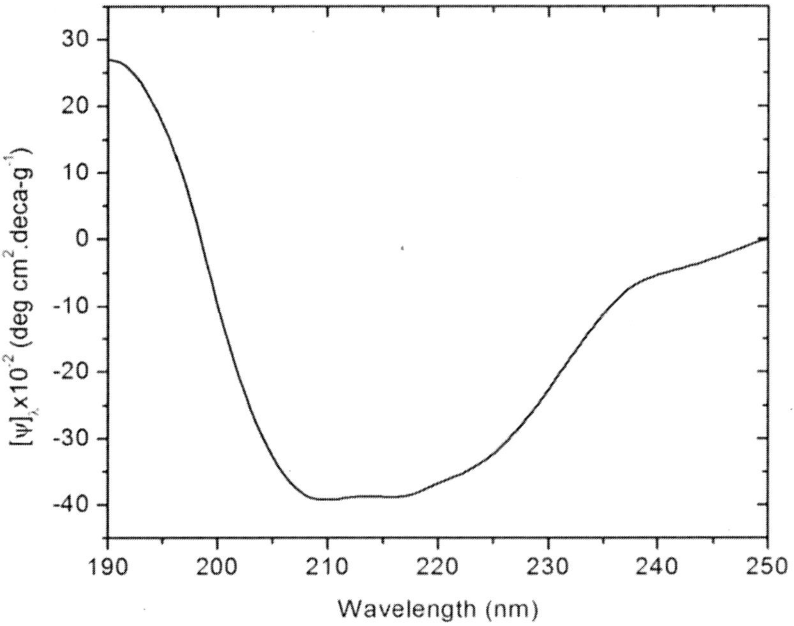

Figure 3. CD spectrum of hoof keratin.

3.4 FTIR analysis of keratin

The secondary structure of keratin was also analyzed using FTIR spectroscopy. Similar to CD, FTIR provides the information about the secondary structure of protein in solution. However, FITR can be directly used in solid and film states compared to CD where special modification requires in the instrument to avoid artifacts (Kuroda et al.2001). Consequently, the secondary structure of hoof keratin was examined in both film, derived from aqueous solution, and solid states. Similar to CD, FTIR exhibits characteristic bands for α-helix, β-sheet, β-turn, and random coil conformations in the amide I (1700–1600 cm^{-1}) and amide II (1560–1500 cm^{-1}) regions (Kong and Yu2007). Among these regions, amide I (due to the C = O stretch vibrations of the peptide linkages) is more sensitive to protein secondary structures. It is well established that α-helical conformation has an amide I and II (mainly from in-plane N-H bending and from the C-N stretching vibration) bands between 1657 and 1650 cm^{-1} and between 1550 and 1540 cm^{-1}, respectively (Lyman et al.2001; Wojciechowska et al.2000), while the β-sheet has an amide I and II bands between 1635 and 1615 cm^{-1} and between 1535 and 1520 cm^{-1}, respectively (Goormaghtigh et al.2006).

The FTIR spectrum of keratin in solid (lyophilized) and supported film (liquid) states are shown in Figure 4. The keratin film, derived from aqueous buffer solution (pH 5), shows an amide I and II bands at 1655 and 1543 cm^{-1} (Figure 4A), respectively, which are characteristic of α-helical conformation. However, the expanded region clearly shows an additional shoulder band at 1620 cm^{-1} in the amide I region and shoulder band at 1520 cm^{-1} in the amide II regions, which are the signature of β-sheet conformation (Figure 4C). The FTIR result thus clearly suggests that keratin adopts predominantly α-helical conformation with admixture of β-sheet conformation in the supported film state. Similar to supported film, the appearance of amide I and II bands at 1655 and 1545 cm^{-1} (Figure 4B), respectively, indicates that keratin adopts predominantly α-helical conformation in the solid state. Again, the broadness of amide I band and the appearance of weak amide II bands at 1516 cm^{-1} (Figure 4D) suggest that keratin also adopts admixture of β-sheet conformation in the solid state. These results reveal that the secondary structure of keratin in the film and solid states are similar. Further, the FTIR result, along with solution CD, demonstrates that the secondary structure of hoof keratin in solid and film states are similar to that of solution state.

Unlike amide I and II, the N-H stretching vibration (3310–3270 cm^{-1}) of peptide bond (-CO-NH-) is insensitive to the conformation of the polypeptide backbone as it is exclusively localized on the NH group. But the frequency of N-H stretching depends on the strength of the hydrogen bond. In the present case, the characteristic frequency of hydrogen bonded N-H has been observed at 3288, and 3308 cm^{-1} for keratin in both film and solid states, respectively, confirming the folded conformation of keratin. Hence both CD and FTIR confirm that the extracted keratin contains ordered structure and is not in denatured form. This keratin having an admixture of α helix and β sheets, seems to have structural similarity with the keratins in birds and reptiles.

3.5 Thermal analysis

DSC is the most popular thermal technique, which measures the heat absorbed or liberated during various transitions in the sample due to temperature treatment. The DSC thermogram of keratin is shown in Figure 5A.

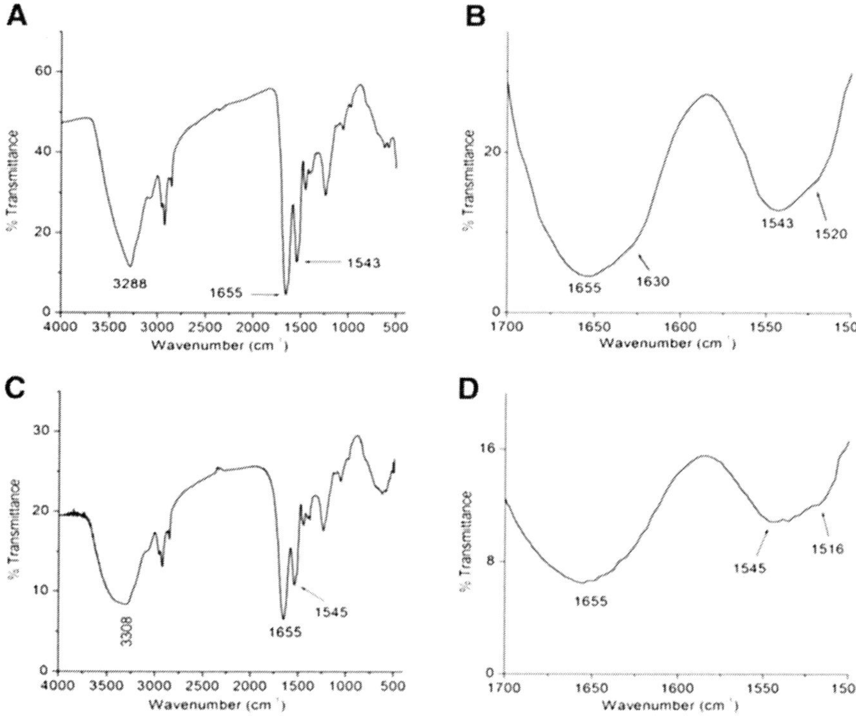

Figure 4. FTIR spectra of hoof keratin in the film **(A)** and solid **(C)** states. **(B)** and **(D)** represent the expanded region (amide I and II) of the corresponding FTIR spectra.

The denaturation temperature observed around 111°C is due to water evaporation [38]. The endothermic peak observed above 200°C is reported for helix denaturation in keratin [39]. The second endothermic peak for the hoof keratin is observed around 215°C. Appreciably high thermal stability of the biomaterial would be an advantage for applications where thermal stability is important. Hence, keratin has an obvious advantage in relation to its thermal stability compared to some of the commonly used biopolymers viz., collagen.

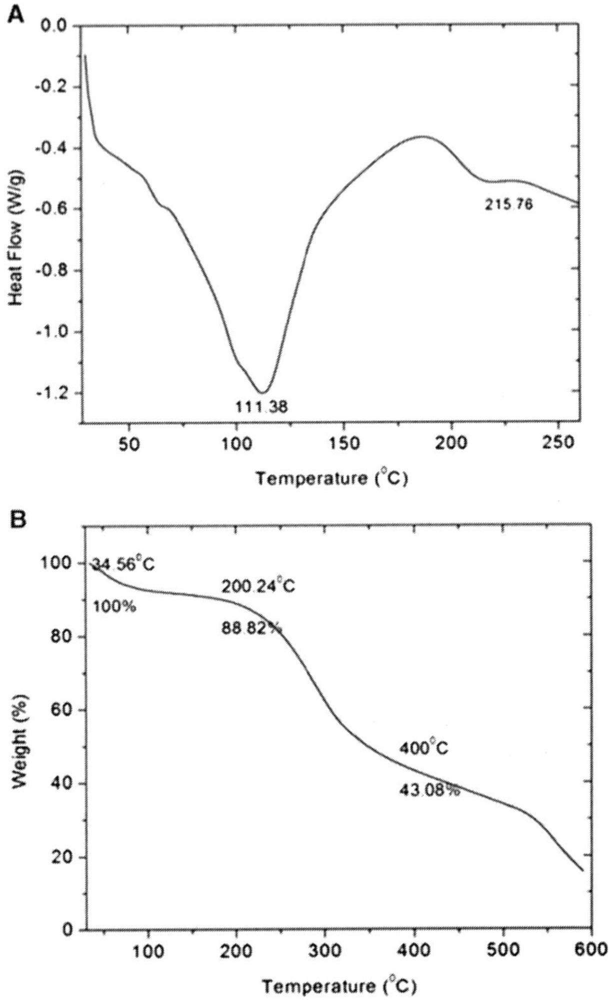

Figure 5. Thermal analysis of hoof keratin (A) DSC profile (B) TGA profile.

Thermogravimetric analysis was carried out to quantify the thermal degradation of keratin in a controlled atmosphere. The TGA curve for keratin sample is shown in Figure 5B. The initial weight loss of around 12% observed between 100°C and 150°C is due to the loss of bound water associated with keratin. Presence of 12% moisture substantiates that the purity of keratin sample could be greater than 90% based on the composition of nitrogen and water. The weight loss between 200 and 400°C is due to the breakdown of polypeptides and associated degradation. More than 50% weight loss was observed at 400°C.

3.6 Biocompatibity assay

The measurement of cell viability and growth is essential to establish the biocompatibility of biomaterials. The MTT assay, a colorimetric assay is employed for measuring the cell viability, involving the reduction of MTT reagent, giving a purple color formazan product in the presence of viable cells. Thus, the absorbance of formazan product proportionally reflects the measure of viable cells. To analyze the biocompatibility of hoof keratin, MTT assay was carried out at different concentrations of keratin (μg/well). It showed good cell viability with reference to control (untreated) as depicted by Figure 6.

The figure shows the images of immature fibroblast cells. We can observe normal aggregation and formation of actin filaments in control (untreated), where cells were allowed to grow on polystyrene plate. When the concentration of keratin was increased from 10 to 25 μg/well, cell membranes are observed to be disctinct and internal boundaries are segregated. Even at concentration of 50 μg/well of keratin, normal cell attachement can be observed. From Figure 6, it is clear that hoof keratin is a biocompatible material and hence it can find wide-spread applications as one of the alternative materials for tissue engineering applications.

4. CONCLUSIONS

In this work we show that pure keratin biopolymer can be extracted from bovine hooves. Characterization of the hoof keratin suggests that, structurally, it belongs to the category of birds and reptiles keratins as it contains a mixture of α- helical and β-sheet structure. Hoof keratin possesses high thermal stability. MTT assay using fibroblast cells shows that the hoof keratin is a biocompatible material for promoting cellular attachment. Hoof keratin appears to be new and promising source of biomaterial for biomedical and tissue engineering applications.

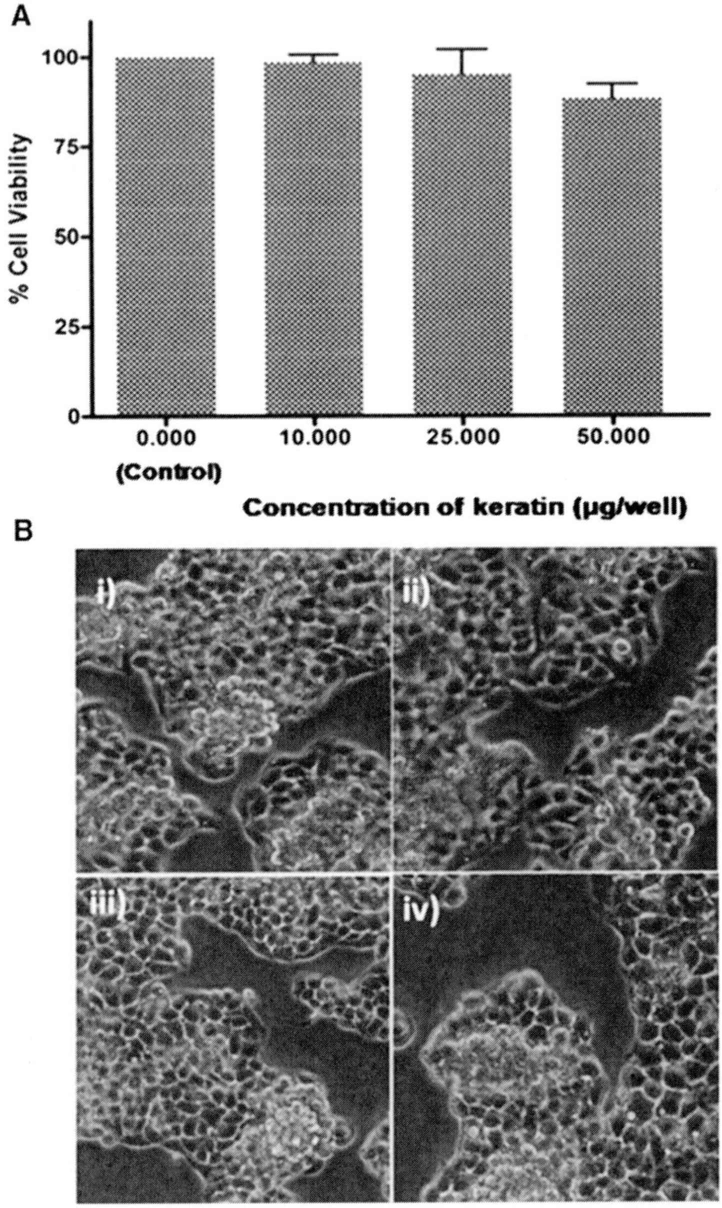

Figure 6. Biocompatibility of the extracted hoof keratin (A) MTT assay using 3 T3 fibroblast cells and (B) the growth of fibroblasts cell shown in the optical micrographs for (i) Control cells- 3 T3-fibroblastic untreated cells with normal aggregation (ii) 10 µg/well treated cells showing syncytium formation (iii) 25 µg/well treated cells with distinct cell membranes eiv) 50 µg/well treated cells.

ACKNOWLEDGEMENTS

Authors thank Council of Scientific and Industrial Research (CSIR) for funding under the CSIR-Young Scientist Award of BM. Author PK would like to thank CSIR for the award of Senior Research Fellowship.

AUTHOR'S CONTRIBUTIONS

BM and PK conceived and designed this study. PK carried out the designed laboratory analyses. BM, GS and PK analyzed the data and drafted the manuscript. All authors read and approved the final manuscript.

REFERENCES

1. Alsarra IA: Chitosan topical gel formulation in the management of burn wounds. *Int J Biol Macromol* 2009, 45(1):16-21.

2. Aluigi A, Varesano A, Montarsolo A, Vineis C, Ferrero F, Mazzuchetti G, Tonin C: Electrospinning of keratin/poly (ethylene oxide) blend nanofibers. *J Appl Polymer Sci* 2007, 104(2):863-870.

3. Amiya T, Kajiwara K, Miyamoto T, Inagaki H: Conformational studies of the α-helical proteins from wool keratin by c.d. *Intl J Biol Macromol*1982, 4(3):165-172.

4. Balaji S, Kumar R, Sripriya R, Kakkar P, Ramesh DV, Reddy PNK, Sehgal PK: Preparation and comparative characterization of keratin–chitosan and keratin–gelatin composite scaffolds for tissue engineering applications. *Mater Sci Eng* 2012, C 32(4):975-982.

5. Brandelli A: Bacterial keratinases: useful enzymes for bioprocessing agroindustrial wastes and beyond. *Food Bioprocess Technol* 2008, 1(2):105-116.

6. Chaudhari PN, Chincholkar SB, Chaudhari BL: Biodegradation of feather keratin with a PEGylated protease of Chryseobacterium gleum.*Process Biochem* 2013, 48(12):1952-1963.

7. Coulombe PA, Bousquet O, Ma L, Yamada S, Wirtz D: The 'ins' and 'outs' of intermediate filament organization. *Trends Cell Biol* 2000, 10(10):420-428.

8. Fang Z, Zhang J, Liu B, Du G, Chen J: Biodegradation of wool waste and keratinase production in scale-up fermenter with different strategies by Stenotrophomonas maltophilia BBE11-1. *Bioresour Technol* 2013, 140: 286-291.

9. Feughelman M: Natural protein fibers. *J Appl Polymer Sci* 2002, 83(3):489-507.

10. Feughelmann M: *Keratin. Encyclopedia of Polymer Science and Engineering*. Wiley, New York; 1985.

11. Fraser RDB, Parry DAD: The molecular structure of reptilian keratin. *Int J Biol Macromol* 1996, 19(3):207-211. -4

12. Fraser RD, MacRae TP, Parry DA, Suzuki E: Intermediate filaments in alpha-keratins. *Proceed Nat Acad Sci* 1986, 83(5):1179-1183.

13. Goormaghtigh E, Ruysschaert J-M, Raussens V: Evaluation of the information content in infrared spectra for protein secondary structure determination. *Biophys J* 2006, 90(8):2946-2957.

14. Gupta P, Nayak KK: Characteristics of protein-based biopolymer and its application. *Polymer Eng Sci* 2014.

15. Hearle JWS: A critical review of the structural mechanics of wool and hair fibres. *Int J Biol Macromol* 2000, 27(2):123-138. -1

16. Hill P, Brantley H, Van Dyke M: Some properties of keratin biomaterials: Kerateines. *Biomaterials* 2010, 31(4):585-593.

17. Iridag Y, Kazanci M: Preparation and characterization of Bombyx mori silk fibroin and wool keratin. *J Appl Polymer Sci* 2006, 100(5):4260-4264.

18. Jeong J-H, Lee OM, Jeon Y-D, Kim J-D, Lee N-R, Lee C-Y, Son H-J: Production of keratinolytic enzyme by a newly isolated feather-degrading Stenotrophomonas maltophilia that produces plant growth-promoting activity. *Process Biochem* 2010, 45(10):1738-1745.

19. Kelly SM, Jess TJ, Price NC: How to study proteins by circular dichroism. *Biochimica et Biophysica Acta (BBA) Proteins Proteomics* 2005, 1751(2):119-139.

20. Kong J, Yu S: Fourier transform infrared spectroscopic analysis of protein secondary structures. *Acta Biochimica et Biophysica Sinica* 2007, 39(8):549-559.

21. Korniłłowicz-Kowalska T, Bohacz J: Biodegradation of keratin waste: Theory and practical aspects. *Waste Manag* 2011, 31(8):1689-1701.

22. Korol J: Polyethylene matrix composites reinforced with keratin fibers obtained from waste chicken feathers. *J Biobased Mater Bioenerg*2012, 6(4):355-360.

23. Kuroda R, Harada T, Shindo Y: A solid-state dedicated circular dichroism spectrophotometer: Development and application. *Rev Sci Inst*2001, 72(10):3802-3810.

24. Lyman DJ, Murray-Wijelath J, Feughelman M: Effect of temperature on the conformation of extended ?-Keratin. *Appl Spectrosc* 2001, 55(5): 552-554.

25. Marshall RC, Orwin DFG, Gillespie JM: Structure and biochemistry of mammalian hard keratin. *Electr Microsc Rev* 1991, 4(1):47-83.

26. Moll R, Franke WW, Schiller DL, Geiger B, Krepler R: The catalog of human cytokeratins: Patterns of expression in normal epithelia, tumors and cultured cells. *Cell* 1982, 31(1):11-24. -7

27. Natarajan V, Saravanakumar P, Madhan B: Collagen adsorption on quercetin loaded polycaprolactone microspheres: Approach for "stealth" implant. *Int J Biol Macromol* 2012, 50(4):1091-1094.

28. Neuhoff V, Stamm R, Eibl H: Clear background and highly sensitive protein staining with Coomassie Blue dyes in polyacrylamide gels: A systematic analysis. *ELECTROPHORESIS* 1985, 6(9):427-448.

29. Patrucco A, Aluigi A, Vineis C, Tonin C: Bio-composite keratin films from wool fibrillation. *J Biobased Mater Bioener* 2011, 5(1):124-131.

30. Ramadass SK, Perumal S, Jabaris SL, Madhan B: Preparation and evaluation of mesalamine collagen in situ rectal gel: A novel

therapeutic approach for treating ulcerative colitis. *Eur J Pharm Sci* 2013, 48(1–2):104-110.

31. Ramshaw JM, Peng Y, Glattauer V, Werkmeister J: Collagens as biomaterials. *J Mater Sci* 2009, 20(1):3-8.

32. Rouse JG, Van Dyke ME: A review of keratin-based biomaterials for biomedical applications. *Materials* 2010, 3(2):999-1014.

33. Srinivasan B, Kumar R, Shanmugam K, Sivagnam UT, Reddy NP, Sehgal PK: Porous keratin scaffold–promising biomaterial for tissue engineering and drug delivery. *J Biomed Mater Res Part B* 2010, 92B(1):5-12.

34. Wojciechowska E, Włochowicz A, Wesełucha-Birczyńska A: Application of Fourier-transform infrared and Raman spectroscopy to the study of the influence of orthosilicic acid on the structure of wool fibre. *J Mol Struct* 2000, 555(1–3):397-406. -8

35. Yamauchi K, Yamauchi A, Kusunoki T, Kohda A, Konishi Y: Preparation of stable aqueous solution of keratins, and physiochemical and biodegradational properties of films. *J Biomed Mater Res* 1996, 31(4):439-444.

36. Zhuang Y, Wu X, Cao Z, Zhao X, Zhou M, Gao P: Preparation and characterization of sponge film made from feathers. *Mater Sci Eng*2013, C 33(8):4732-4738.

37. Zoccola M, Aluigi A, Tonin C: Characterisation of keratin biomass from butchery and wool industry wastes. *J Mol Struct* 2009, 938(1–3):35-40

Chapter 4

Introduction of Microbial Biopolymers in Soil Treatment for Future Environmentally-Friendly and Sustainable Geotechnical Engineering

Ilhan Chang [1,†], Jooyoung Im [2,†] and Gye-Chun Cho [2,*,†]

[1]Geotechnical Engineering Research Institute, Korea Institute of Civil Engineering and Building Technology (KICT), Goyang 10223, Korea

[2]Department of Civil and Environmental Engineering, Korea Advanced Institute of Science and Technology (KAIST), Daejeon 34141, Korea

ABSTRACT

Soil treatment and improvement is commonly performed in the field of geotechnical engineering. Methods and materials to achieve this such as soil stabilization and mixing with cementitious binders have been utilized in engineered soil applications since the beginning of human civilization. Demand for environment-friendly and sustainable alternatives is currently rising. Since cement, the most commonly applied and effective soil treatment material, is responsible for heavy greenhouse gas emissions, alternatives such as geosynthetics, chemical polymers, geopolymers, microbial induction, and biopolymers are being actively studied. This study provides an overall review of the recent applications of biopolymers in geotechnical engineering. Biopolymers are microbially induced polymers that are high-tensile, innocuous, and eco-friendly. Soil–biopolymer interactions and related soil strengthening mechanisms are discussed in the context of recent experimental and microscopic studies. In addition, the economic feasibility of biopolymer implementation in the field is analyzed in comparison to ordinary cement, from environmental perspectives. Findings from this study demonstrate that biopolymers have strong potential to replace cement as a soil treatment material within the context of environment-friendly construction and development. Moreover, continuing research is suggested to ensure performance in terms of practical implementation, reliability, and durability of *in situ*biopolymer applications for geotechnical engineering purposes.

KEYWORDS

soil treatment; engineered soil; geotechnical engineering; biopolymer; environmentally-friendly; CO_2emission; soil strengthening

1. INTRODUCTION

Geotechnical engineering, especially the treatment and usage of soil (or earth) in construction, is a long-standing technical field, dating to the beginning of human civilization. In ancient Mesopotamia and Babylonia, mud was used as a common construction material for various forms of bricks (*i.e.*, sun-dried and kiln-fired) to create the architecture of the city. Sumerians widely used bitumen as a binder to enhance the strength and durability of earth walls [1]. In ancient Egypt, earth engineering methods were developed to create river embankments to control the annual flooding of the Nile river [2]. Engineers in ancient China used a sticky rice mortar that contained amylopectin as a construction binder for the Great Wall [3].

As civilization has advanced, the durability and strength of construction materials has improved as well. The discovery of natural pozzolan materials such as volcanic ash improved construction materials, including the well-known concrete of ancient Rome. The representative formula for Roman concrete was a mixture of volcanic dust with a binder such as gypsum or lime, with added aggregates, and it was used to create highly durable structures including arches and domes [4]. After the Industrial Revolution, ordinary cement (*i.e.*, Portland cement) became the most widely used construction and building material, not only for structural purposes, but also for soil stabilization and strengthening purposes (Figure 1).

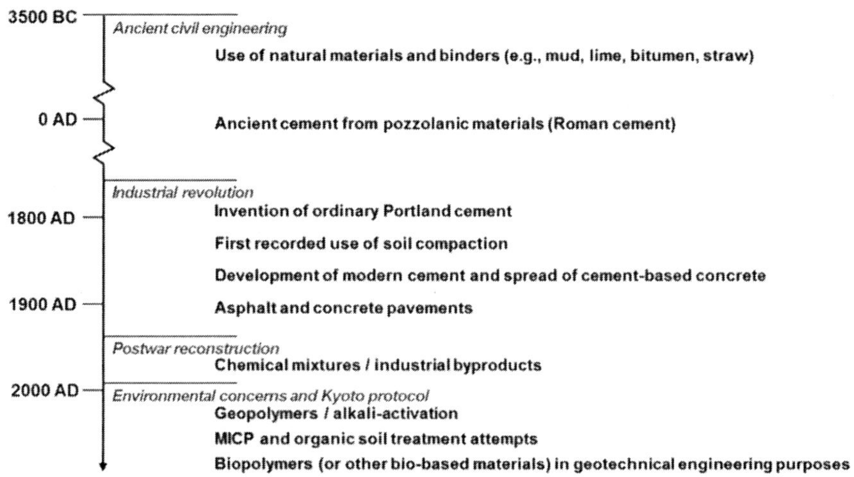

Figure 1. Advancement of engineered soil materials in geotechnical engineering.

The main purpose of soil treatment and improvement (*i.e.*, engineered soil) is to enhance the engineering characteristics of a particular soil, including its strength (*i.e.*, resistance), hydraulic conductivity, and durability against repeating wetting and drying, as well as for environmental revitalization [5]. Two primary methods are conventionally applied to produce engineered soil: mechanical improvement and chemical treatment. Mechanical improvement is a process of reinforcing the strength of the soil through physical processes such as compaction, drainage, external loading (e.g., surcharge), consolidation, or other means. Chemical treatment involves chemical reactions such as hydration or pozzolanic reactions inside the soil to create artificial binding, such as the use of calcium silicate hydrate (C-S-H) between soil particles [5].

As an alternative to such traditional soil treatment and improvement techniques, biological approaches are now being actively investigated in the field of geotechnical engineering, including microbe injection and byproduct precipitation. In particular, microbial induced polymers—or biopolymers—have been introduced as a new type of construction binder, especially for soil treatment and improvement.

To date, most studies on these applications of biopolymers have been experimental efforts that have produced preliminary findings and analyses, and the number of theoretical explanations and case studies of practical implementation in the literature are still limited. In response, this paper provides a detailed review of biopolymer applications in geotechnical engineering including the most recent studies. In this review, strengthening mechanisms between typical biopolymers and soils based on microscopic inter-particle interactions are summarized. The advantages and disadvantages of biopolymer applications are compared with those of existing soil engineering methods. Finally, the potential for practical implementation is evaluated via an economic feasibility analysis, including environment-friendly considerations.

2. ENVIRONMENTAL CONCERNS AFFECTING SOIL TREATMENT IN GEOTECHNICAL ENGINEERING

2.1. CO_2 Emissions related to Cement Usage in Civil and Geotechnical Engineering

Ordinary Portland cement currently dominates the field of materials used for construction and civil engineering purposes: 5245 million tons of hydraulic (*i.e.*, ordinary Portland) cement were produced worldwide in 2012 [6]. Portland cement has multiple engineering benefits, such as high strength and durability, workability, and hydraulicity, as well as low cost (60~100 USD/ton, in developed countries), and these factors have contributed to its widespread use in various applications for ground improvement, concrete structures, and pavement in civil and construction engineering [7].

In geotechnical engineering projects, cement has been used in numerous formats, including deep cement mixing (DCM), grouting, soil nails, and soil stabilization, and is now the most favored material for engineered soil. However, despite cement's benefits and its numerous applications, the overdependence on and overuse of cement has given rise to a number of environmental concerns.

In the production of cement, the calcination of calcium carbonate (*i.e.*, $5CaCO_3 + 2SiO_2 \rightarrow 3CaO{\cdot}SiO_2 + 2CaO{\cdot}SiO_2 + 5CO_2$) liberates 0.55 tons of CO_2 per 1 ton of Portland cement, which is then emitted to the atmosphere. In addition, approximately 0.4 tons of CO_2 are emitted due to the burning of carbon fuel during the production of 1 ton of cement, since calcination requires heat as high as 1450°C. The total amount of CO_2 emitted per each ton of cement produced is 0.95 tons [8].

In 1995, for example, 1453 M·tons of cement were produced worldwide, accounting for approximately 5% of the global annual CO_2 emissions [9]. By 2003, the amount of CO_2 emissions from cement production had increased to 7% of the global CO_2 emissions [10]. This means that the rate of CO_2 emissions related to cement production (Figure 2a) has doubled during the past 30 years, from 4.2% in 1980 to 9.0% in 2012. Moreover, the annual growth rate of cement related CO_2 emission continues to increase annually—it reached 10% in 2010—in spite of global efforts to reduce CO_2 emissions by limiting its annual growth rate to less than 3% (Figure 2b).

CO_2 emissions related to the usage of cement in geotechnical applications (e.g., mixing, grouting, soil stabilization) amount to approximately 2% of the total CO_2 emissions by cement (*i.e.*, 3.05 B·tons in 2011) [11,13]. Thus, replacing 10% of cement usage with low-carbon materials in geotechnical engineering implementations leads to a reduction of 6.1 M·tons of CO_2, which is close to 10% of the annual CO_2 emissions from Austria (66.68 M·tons) in 2012 [14].

2.2. Other Environmental Concerns from the Use of Cement

The presence of cement inside soil (e.g., mixed into soil) quickly raises the pH of the soil up to 12–13, due to the release of alkaline hydroxide (OH-) ions as a byproduct of hydration [15]. This increase in pH can affect biological organisms, and could thereby have detrimental effects [16].

The widespread use of cement-based construction also gives rise to several related issues in the urban environment, such as an increase of urban water runoff, heat islands, and prevention of vegetation growth. Moreover, as the hydration reaction of cement is irreversible, it is difficult to return soil-cement mixtures to their original state (*i.e.*, untreated soil), and furthermore demolition problems arise [17].

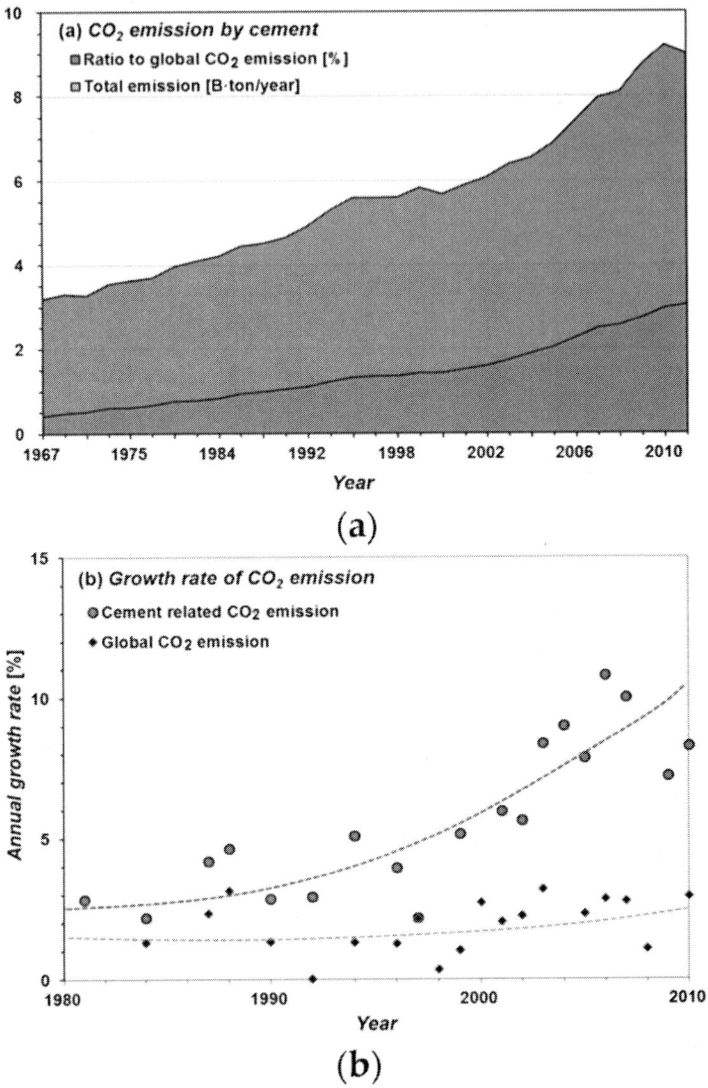

Figure 2. CO_2 emissions of cement through the years (after [11] (**a**) and [12] (**b**)).

Urban runoff is known to be amplified due to the massive use of impervious cement-based concrete with increasing urbanization. The absence of ground infiltration and severe surface runoff can produce floods, and they are strongly related to economic and ecological losses such as residential or infrastructural damage [18,19]. Moreover, surface runoff also degrades overall water quality by flushing various pollutants and contaminants into waterways from urban sources [20,21].

The urban heat island effect is mainly caused by concrete or asphalt materials, whose specific heats are lower than that of soil [22]. Urban heat islands can have a number of environmental impacts such as affecting local wind patterns and humidity and causing the development of urban smog and localized heavy precipitation [23,24].

Moreover, the presence of cementitious covers also prevents the growth of surface vegetation. Vegetation such as trees and grass can lower the surrounding temperature via evapotranspiration and by providing shade. Thus, vegetation can reduce peak summer temperatures in urban areas by 1–5 °C, and can also be used to mitigate urban heat islands [25]. In addition, the presence of vegetation has other benefits, including increasing the efficiency of cooling systems, enhancing rainwater management and water quality by absorbing and filtering precipitation, and improving air quality [26].

Cement demolition waste and its management by disposal or recycling has become a significant environmental concern at the end of the life cycle of cement-based concrete structures [27]. Moreover, cement used for geotechnical purposes such as deep mixing or grouting is difficult to remove from the ground.

Concrete dust can be released into the air through various pathways such as concrete structure demolition or natural disasters (e.g., earthquakes), and can induce lung diseases [28]. For example, high concentrations of concrete dust, 150 $\mu g/m^3$, were monitored after the Great Hanshin earthquake in Japan in 2015, where the alkaline component of the concrete dust in the air was equivalent to an amount sufficient to neutralize the annual acid rainfall in Japan [29].

2.3. Soil Treatment Methods with Environmentally-Friendly Aspects

While the use of cement creates a number of environmental problems, as mentioned in the previous section, the construction industry also accounts for approximately 40% of global energy consumption [30]. In response, environmental-friendly and sustainable approaches involving the use of alternative materials, including geosynthetics, chemical polymers, geopolymers, or biological treatment have recently emerged in the field of geotechnical engineering.

Geosynthetics are artificial polymeric products, including geotextiles, geogrids, geonets, geomembranes, geosynthetic clay liners, geofoam, geocells, and geocomposites, which are implanted into the soil to strengthen or enhance the soil's geotechnical engineering properties. Geosynthetics have high tensile strength, flexibility, and impervious characteristics, and are therefore generally used for soil separation, filtering, reinforcement, and drainage purposes across a large range of applications [31,32,33]. While such materials can provide higher strength, specifically tensile strength, the strengthening is almost completely dependent on the material itself and not the soil, making them undesirable for various engineering practices.

In the late 20th century, geotechnical engineers began introducing chemically synthesized polymers as soil conditioners to replace the use of conventional binders, such as lime and cement, for geotechnical engineering and agricultural purposes. Polyacrylamide (PAM) [34] is a simple linear- or cross-linked polymer synthesized by acrylamide (*i.e.*, $CH_2CHCONH_2$) subunits, which are non-toxic and hydrophilic [35]. Anionic polyacrylamide (PAM) has negative charge density and induces electrostatic bonding with clay particles [36,37], which results in reduced soil erosion and runoff control [38]. PAM consequently has been used in various fields such as agriculture, construction, and military applications (e.g., temporary helicopter landing fields) as a soil erosion reduction agent [38,39]. Sodium silicate ($Na_2(SiO_2)_nO$) and other chemicals, including acrylics, lignosulfonates, and phenolasts, have been widely implemented in soil grouting practices, especially for sandy soils. However, chemically synthesized polymers and solutions have raised concerns about toxicity and water pollution problems [40], and this has restricted the usage of chemical solutions near water conservation regions in particular.

Meanwhile, electroosmotic chemical treatment (ECT) has been attempted in several studies to improve the strength of soft soils [41,42]. However, the ECT method requires massive injection of chemical solutions to produce an increase in osmotic pressure, and has not yet been accepted for practical implementation [43].

Geopolymers are polymers synthesized from either organic or inorganic, such as silicon-based, origins [44,45]. The geopolymerization process involves initial mixing in an alkaline solution, which dissolves silicon and aluminum ions (*i.e.*, alkali-activation) from soil or industrial byproducts such as fly ash or blast furnace slag. Hydroxyl ions ($2OH^-$) remaining in the alkaline solution then react to form one free water molecule (H_2O) and an oxide ion (O^{2-}) [46]. The released oxygens then form monomers with the dissolved silicon and aluminum ions, which undergo condensation. Geopolymers require isolated heating (above 60 °C) for proper geopolymerization (hardening) [47,48], which can limit site application for geotechnical engineering purposes. Studies have shown that geopolymers are highly susceptible to water and show significant degradation in strength when saturated in water, especially in acid solutions, due to depolymerization of the aluminosilicate polymers and removal of silicic acid [49].

3. APPLICATION OF BIOLOGICAL APPROACHES TO GEOTECHNICAL ENGINEERING

3.1. Attempts with Bio-Mineralization

Since the beginning of the 21st century, researchers have been seeking more biogenic alternatives to existing strengthening agents. Recently, environment-friendly approaches that involve the use of biological materials such as microbes and enzymes have emerged in attempts to improve the mechanical properties of soil with a remarkably smaller carbon footprint compared to the use of cement in geotechnical engineering practices. The most common approach is bio-mineralization, which entails mineral precipitation in soil pores via biological organisms [50]. Another approach, based on biofilm formation, starts with a weak attraction force between microorganisms that later develops into a larger attachment that is irreversible [51].

Among bio-mineralization strategies, the most recognized soil treatment method is microbial induced calcite precipitation (MICP). MICP employs microorganisms including *Sporosarcina pasteurii* and *Bacillus pasteurii* to precipitate calcium carbonate crystals in soil voids and thereby provide inter-particle bonding [52,53]. MICP occurs through urea hydrolysis, which primes the soil for calcium carbonate precipitation when calcium is present [54]. Ureolytic bacteria are used to convert urea to ammonium and carbonate. Calcium carbonate precipitates are then formed from the carbonate and calcium ions [55]. The calcium carbonate precipitates bind with the soil grains to increase the strength and stiffness of the soil [56].

Several studies have been performed to increase the strength and stiffness of soils using urease positive microorganisms [53,57]. In these studies, the microorganism *Sporosarcina pasteurii* was injected directly into the soil to allow for calcite precipitation by the microorganisms themselves.

The use of urea, $CaCl_2$, and urease leads to the precipitation of calcite through enzymatic processes without microbial activities, which is suitable for soil strengthening and increasing stiffness [58]. MICP has also been found to be applicable in various other fields, including the remediation of heavy metals [59], CO_2 sequestration [60], and the repair of concrete [61,62].

However, MICP has several shortcomings in terms of field implementation. MICP is most applicable to coarse sediments, and accompanies difficulties when used with fine grained soils due to infiltration problems: the pore sizes of fine grained sediments provide an unsuitable environment for bacteria growth [63]. Moreover, the transport, cultivation, and fixation performance of bacteria is not consistent. Thus, recent studies have adopted the use of urease enzyme instead of using bacteria to promote the hydrolysis of urea, for the chemical precipitation of calcite [64]. Furthermore, the use of MICP results in a highly concentrated ammonium chloride byproduct, which needs to be removed and treated, generally in the form of soil flushing [65].

Even though the limitations of MICP are problematic, the bio-soil method has been shown to have several promising features, including high strengthening, low environmental impact, self-proliferation, and biodegradation. A number of researchers have thus focused on the direct use of biogenic excrement (*i.e.*, biopolymers) instead of attempting to cultivate the microorganisms in the soil.

3.2. Biopolymers: Definition and Usage

Biopolymers are organic polymers that are synthesized by biological organisms. They consist of monomeric units that are bonded into larger formations. The use of biopolymers is, in fact, not an entirely new development in geotechnical engineering. Organic polymers such as natural bitumen, straw, and sticky rice have been used in ancient civilizations and can also be classified as biopolymers in a broad sense. In ancient Chinese civilization, sticky rice mortar was used as a binder. Sticky rice soup mixed with *Actinidia chinensis* cane juice, lime, loess, and river sand produced a mortar with good strength, high toughness, and water resistance [66].

Among the three typical types of biopolymers—polynucleotides (e.g., RNA and DNA), polypeptides (e.g., composed of amino acids), and polysaccharides—polysaccharides have been the most commonly applied contemporary biopolymer type in various practices [67,68].

Polysaccharides are polymeric carbohydrate chains composed of monosaccharide units. Polysaccharides are widely found in nature because they are employed in key biological roles, as substances forming skeletal structures, assimilative reserve substances, and water binding substances [69]. The properties of polysaccharides have led to their widespread use as thickening agents, stabilizers, sweeteners, and gel forming agents in the fields of food production, agriculture, cosmetics, medical treatment, and pharmaceuticals [70,71,72].

3.3. Biopolymers in Geotechnical Engineering

Biopolymers mixed with soil, such as sticky rice mortar, promote strengthening of the soil, including increased cohesion and strength, resistance to erosion, reduced permeability, *etc.*, by acting as a binder. The direct use of biopolymers in soil has several benefits over pre-existing biological soil treatment methods [73]. The direct use of exo-cultivated biopolymers for soil treatment overcomes several shortcomings of other approaches (e.g., microbe injection) such as the need for microbial and nutrient injection, time for cultivation and excrement precipitation, and inappropriateness with clayey soils [74]. Moreover, since biopolymers are readily found in nature and many are known to be harmless and edible, biopolymers can be considered eco-friendly substitutes for soil treatment. Several polysaccharide group biopolymers recently have been examined for use in geotechnical engineering. The characteristics of common biopolymers are summarized in Table 1, and the following sections review notable case studies in geotechnical engineering.

Table 1. Characteristics of common biopolymers (after [35,75,76, 77,78, 79,80,81]).

Biopolymer	Composition	Characteristics
Cellulose	β-(1 → 4)-D-*glucose* linkages	- Hydrophilic - Properties depend on the chain length
Starch	D-*glucose residues* linked by *α-(1, 4) glucosidic bonds*	- Soluble in heated water - Viscous, gelatinization
Chitosan	*P-(1,4)-2-amino-2-deoxy-*D-*glucose*	- Soluble in acidic solvents - Bioadhesive for (-) charged surfaces - Biodegradable
Xanthan	$C_{35}H_{49}O_{29}$	- High viscous rheology
Curdlan	$(C_6H_{10}O_5)_n$	- Gel formation via heating in aqueous solutions
Beta-Glucan	D-glucose monomers linked by β-glycosidic bonds	- Capable of immune activation in humans - Cholesterol absorption capabilities
Polyacrylamide	$(C_3H_5NO)_n$	- Water absorbent, gel formation - Thickener

3.3.1. Cellulose and Starch

Cellulose $(C_6H_{10}O_5)_n$ is abundantly present in nature. It consists of *β-(1 → 4)-D-glucose* linkages, and is the main component in the cell walls of plants, and has the longest history of utilization in engineering (e.g., straw fibers in adobe bricks). Its most notable current uses are in the food industry where it is used for packaging materials and for thickeners or stabilizers for food [82]. Cellulose has various potential uses in geotechnical engineering applications due to its gelation properties, which can be exploited to produce thickeners and stabilizing agents [75]. Natural cellulose-group fibers significantly increase the tensile and flexural strength of soils [83,84]. Furthermore, cellulose has been used in open graded friction coarse (OGFC) pavement. OGFC consists of single size coarse aggregates with large voids, resulting in high asphalt content, and cellulose improves its performance by limiting coarse aggregate pop-out and reflective cracking [80]. In addition, organic compost containing high cellulose content has been reported to induce higher soil aggregate stability by increasing water repellency [85].

Starch is a polysaccharide that consists of *D-glucose residues* linked by *α-(1, 4) glucosidic* bonds [86]. It can be naturally found in a variety of plants, such as maize, rice, wheat, potatoes, and cassava [87], and it has been applied in numerous areas outside of food and agriculture. The major area of starch application in geotechnical engineering has been in adhesives for drilling fluids [88]. The use of starch in thermoplastic materials provides two major benefits: (i) it allows for continuously tunable processed materials; and (ii) because it is an abundant and inexpensive raw material, large scale production can be economically achieved [89].

3.3.2. Chitosan

Chitosan (*P-(1,4)-2-amino-2-deoxy-D-glucose*) is the degraded form of chitin, found in the exoskeletons of crustaceans and the cell walls of fungi. It is typically extracted by alkali sodium hydroxide treatment. This hydrophilic biopolymer has been used in a variety of applications. As a biorenewable, biocompatible, biodegradable, and biofunctional biopolymer, chitosan has been used and researched in the fields of food/nutrition, material science, medical science, microbiology, and immunology, as well as other fields such as agriculture, wastewater treatment, and degradation products [81,90,91]. In geotechnical and geoenvironmental contexts, chitosan has been shown to be capable of decontaminating groundwater containing contaminants such as copper (II) (Cu^{2+}) and phosphorus (P-) [92,93]. Chitosan coated sand particles have practical implications in filters for remediation of contaminated groundwater [81] and can create a suitable plugging effect by reducing the hydraulic conductivity of sandy soils [94].

3.3.3. Curdlan

Curdlan is a high molecular weight linear β-1,3-glucan biopolymer produced by pathogenic bacteria (e.g., *Agrobacterium biobar*, *Alcaligenes faecalis*) that forms an elastic gel when heated in an aqueous solution [79]. Curdlan has been used as an additive in concrete mixtures as a superplasticizer to improve workability and prevent cement-aggregate separation [95]. From a geotechnical perspective, curdlan has been suggested for use in grouting due to its large scale soil clogging potential [96].

3.3.4. Beta-Glucan

Beta-glucans are biopolymers that consist of D-glucose monomers that are linked by glycosidic bonds [97]. *β*-glucan is naturally found in various formations, such as in cellulose, bran, and the cell walls of yeast, fungi, and bacteria [77].

A small amount (0.25%) of beta-glucan biopolymer (*i.e.*, mass ratio to soil) in soil provides a compressive strength value of 2650 kPa, comparable to that of a 10% cement mix (2170 kPa), while 0.5% beta-glucan mixing yields soil strength up to 4310 kPa [98]. From a geotechnical perspective, the presence of beta-glucan in soil increases the Atterberg limit values as well as shear stiffness (G), while it seems to have minor or even no effects on the constrained modulus (M) of soil [99]. A recent study showed that beta-glucan treatment promotes vegetation growth in treated soils as well as high strength, and in particular it promoted vegetation growth on barren soil, indicating its potential for use as a countermeasure for desertification [100].

3.3.5. Xanthan Gum

Xanthan gum is a polysaccharide that is made by the *Xanthomonas campestris* bacterium, and is generally used as a viscosity thickener due to its hydrocolloid rheology [76]. Xanthan gum has been introduced to geotechnical engineering to reduce the hydraulic conductivity of silty sand via pore filling [94,101] as well as to increase the undrained shear strength of soil by increasing the liquid limit [102]. Another recent study has studied possibilities for using xanthan gum as a soil strengthener, and showed that xanthan gum preferentially forms firm xanthan gum-clayey soil matrices via hydrogen bonding [103].

3.3.6. Agar Gum

Agar gum is a polysaccharide that is composed of linked galactose molecules, and is extracted from *rhodophyceae* [96]. Agar gum has rheological properties that make it useful as a thickener, stabilizer, and emulsifier. Moreover, agar gum has a thermogelation property that allows the formation of strong gels when cooled back to room temperature after being dissolved in boiling water [104]. Agar gum effectively improves the strength of sand (especially cohesion) without environmental risks [105] Recently, researchers attempted to use agar gum to enhance the strength of soils, showing that the use of 3% (to soil mass) thermal treated agar gum enhanced the unconfined compressive strength of soil up to 10 MPa under a dried condition [106].

3.3.7. Gellan Gum

Gellan gum is a high molecular weight polysaccharide fermented from *Spingomonas elodea* (formerly known as*Pseudomonas elodea*) microbes [107]. Gellan gum is usually used as a substitute for agar gum, with which it shares many properties, including thermogelation properties [108,109]. Gellan gum shows great promise for application as an engineered soil material due to its high strengthening effect [110]. A recent study showed that 3% of gellan gum dispersed within a clayey soil exhibited a maximum unconfined compressive strength of 12.6 MPa [106]. Although gellan gum has a similar gelation rheology and strengthening effect to agar gum, gellan gum is preferred for future mass commercialization because it can be easily produced via microbial fermentation, whereas agar gum has to be extracted from seaweed (*i.e.*, algae).

4. MECHANISMS OF SOIL–BIOPOLYMER INTERACTION

4.1. How Biopolymers Strengthen Soil

A small quantity (*i.e.*, 0.5%–1% to the soil weight) of biopolymer in soil can produce significant strengthening effects (Figure 3). In general, biopolymers have high specific surfaces with electrical charges, which enable direct interactions between the biopolymers and fine soil particles, thereby providing firm biopolymer-soil matrices with high strength. SEM images of sand-clay-biopolymer mixtures show that biopolymers directly bond with kaolinite particles, producing accumulated face-to-face clay layers, while sand surfaces remain clean or only film-type coats form around particles (Figure 4). Thus, the strengthening is maximized in the presence of clayey particles due to the hydrogen and ionic bonding between the biopolymers and clay particles, which have electrical charges [103,106].

However, this does not mean that biopolymers have a negligible impact in terms of their use with sand particles, because well-graded soil with coarse particles treated with biopolymers shows higher strength than that obtained with pure clay, such as kaolinite [103].

Figure 5 shows the inter-particle cohesion and friction angle variation of 1% gellan gum biopolymer mixed soils, evaluated via direct shear tests. Although the unconfined compressive strength (Figure 3) and cohesion of biopolymer mixed clay is much higher than that produced in pure sand, the friction characteristics of sandy soils (*i.e.*, sand-clay mixtures) are more appropriate for practical applications than those of pure clay. Thus, the strengthening mechanism of biopolymers in ordinary soils (*i.e.*, those containing both sand and clay) is believed to be a combination of the formation of biopolymer-clayey soil matrices (*i.e.*, cohesion enhancement) and friction improvement via coarse particles acting as aggregates (Figure 6). Inter-particle cohesion enhancement depends on the strength of the biopolymer-clayey soil matrices, which is higher with lower water content. The friction angles of the soil increase with biopolymer treatment due to improved particle contact. The stiffer dried gels add a substantial amount of strength with a small increase in the contact radius, especially at lower water content, as opposed to the softer wet gels, which exhibit almost no increase in strength with a small increase in the contact radius.

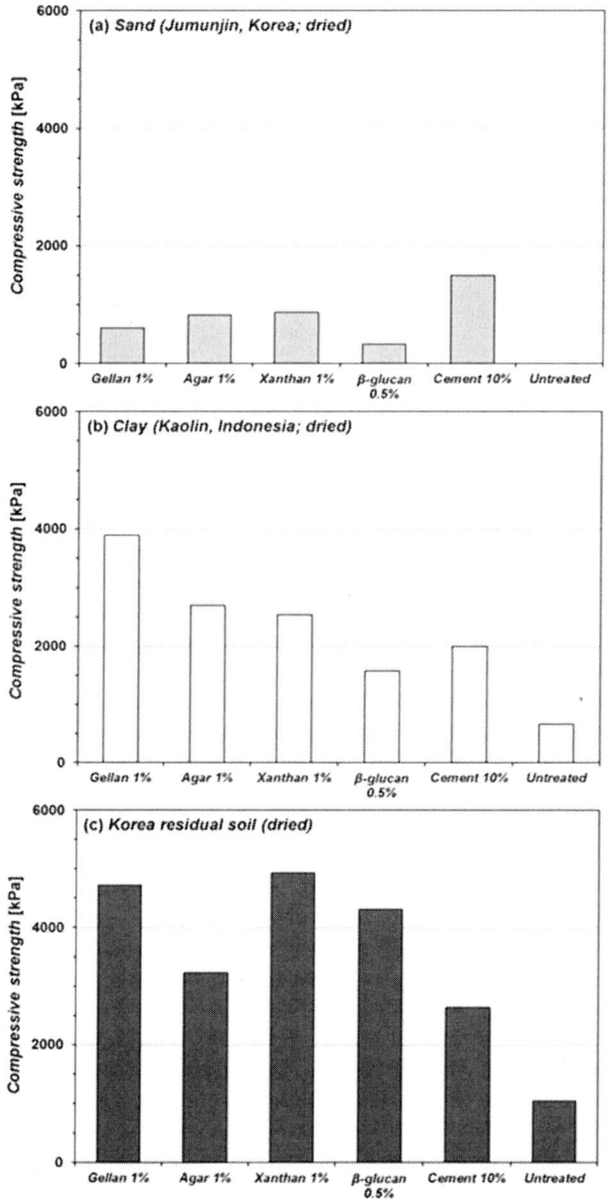

Figure 3. Compressive strengths of biopolymer-treated soils.

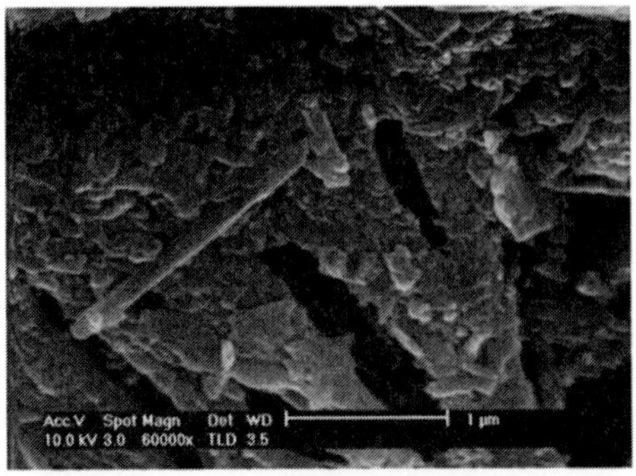
Acc.V Spot Magn Det WD 1 µm
10.0 kV 3.0 60000x TLD 3.5

(a)

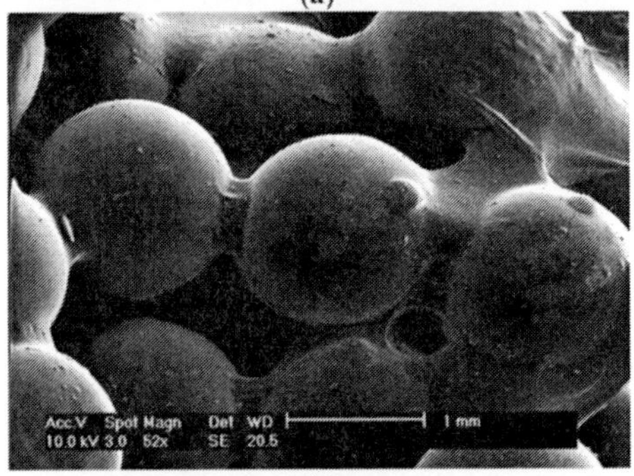
Acc.V Spot Magn Det WD 1 mm
10.0 kV 3.0 52x SE 20.5

(b)

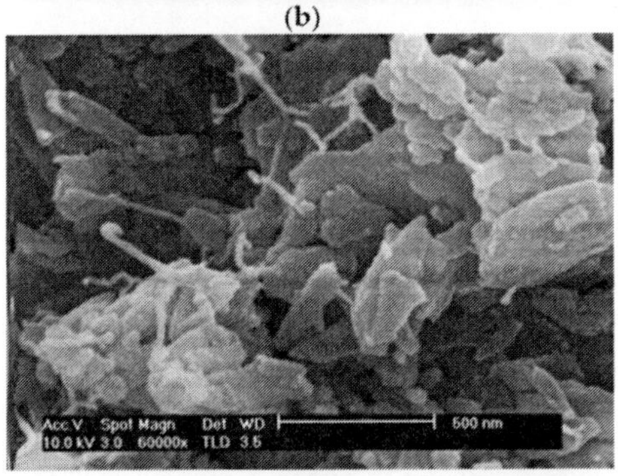
Acc.V Spot Magn Det WD 500 nm
10.0 kV 3.0 60000x TLD 3.5

(c)

Acc.V Spot Magn Det WD 1 mm
15.0 kV 3.0 30x SE 13.7

(d)

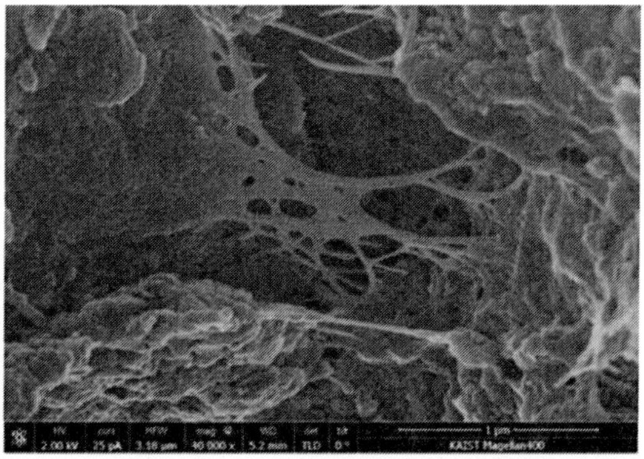

(e)

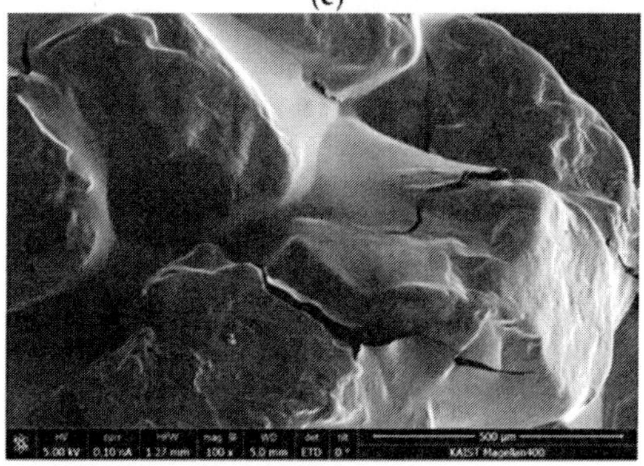

(f)

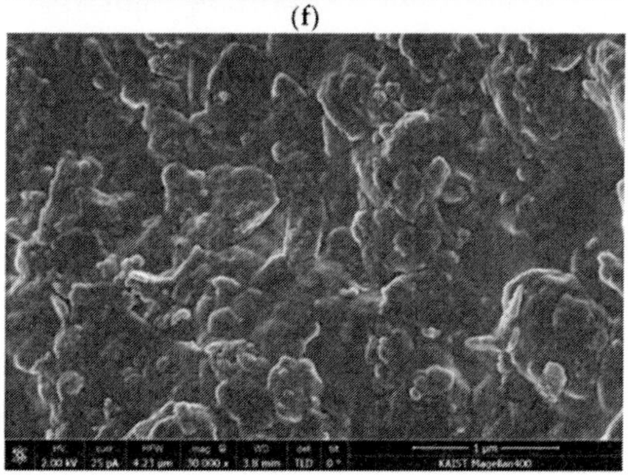

(g)

Figure 4. SEM images of biopolymer-soil mixtures: (**a**) Beta-glucan and Korea residual soil; (**b**) Beta-glucan and glassbeads; (**c**) Xanthan gum and kaolin clay; (**d**) Xanthan gum and sand (*Jumunjin*, Korea); (**e**) Gellan gum and kaolin clay; (**f**) Gellan gum and sand (*Jumunjin*, Korea); and (**g**) Agar gum and kaolin clay (images from [98,103,106]).

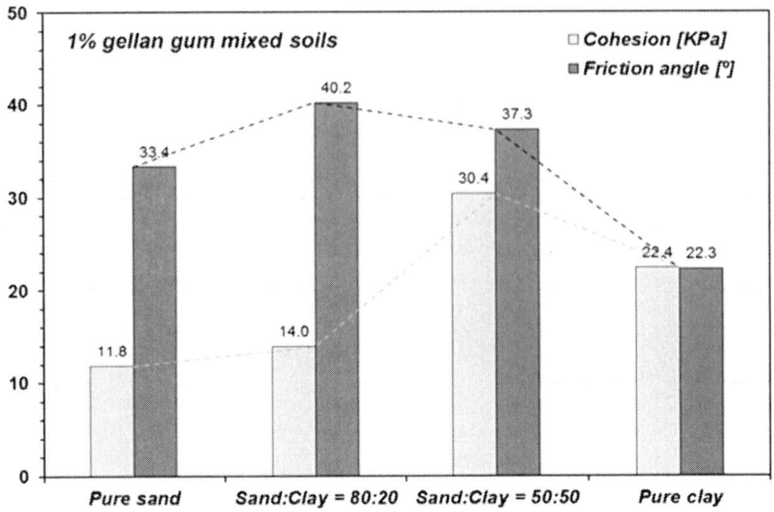

Figure 5. Direct shear test result of 1% gellan gum biopolymer mixed soils.

Therefore, to employ biopolymers, specific *in-situ* characterization must be conducted in advance, to understand the particular composition of the targeted soil, such as its coarseness and clay ratio, which is important for determining the proper quantities of biopolymers and the best utilization methods for geotechnical engineering purposes. For instance, biopolymers have the potential to reduce aeolian erosion and promote vegetation growth in arid or semi-arid deserts [100]. Since desert soil mostly consists of coarse grains, biopolymers are expected to show suitable workability in terms of material rheology and soil structure. Moreover, additional clay-type materials or ionic chemicals (e.g., Ca^{2+} and Mg^{2+}) can be applied to sandy soils to enhance the strengthening efficiency of biopolymers.

4.2. Hydro-Dependency and Clogging Effect of Biopolymers

When soils have increasing amounts of water, dehydrated biopolymer gels (Figure 6a,b) or biopolymer-clay matrices (Figure 6c,d) will adsorb and transfer water to hydrogels with a resulting volumetric expansion (*i.e.*, swelling) [111,112]. The elastic properties of biopolymer hydrogels, such as their tensile strength and stiffness, diminish exponentially with increased water content [113], resulting in a remarkable reduction in soil strength, to approximately 1/10th of the strength of the dried state in a fully saturated condition. However, the unconfined compressive strength of re-wetted biopolymer soil mixtures (*i.e.*, clayey soil ≥ 200 kPa; sandy soil ≥ 50 kPa) is much higher than that of untreated soils, and is immeasurable in most cases [106].

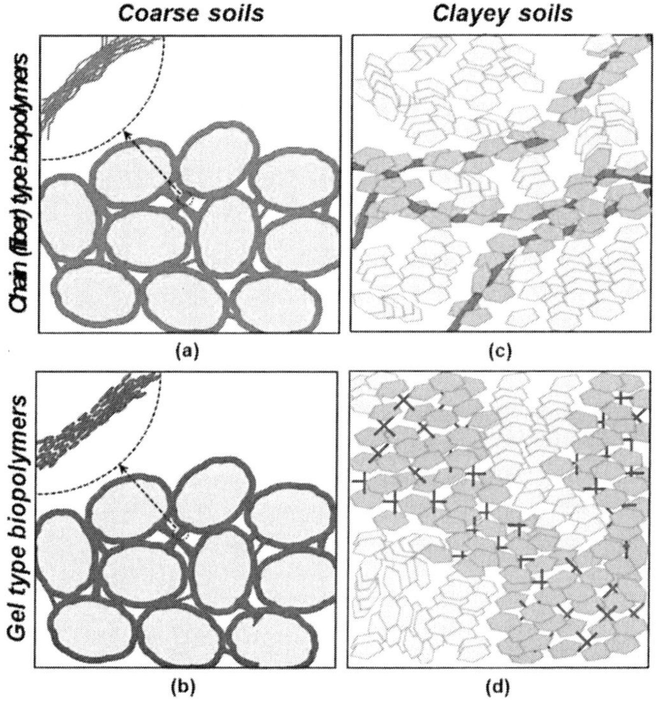

Figure 6. Schematic model for biopolymer treated soils: (**a**) chain type biopolymers with sandy soils; (**b**) gel type biopolymers with sandy soils; (**c**) chain type biopolymer with clayey soils; and (**d**) gel type biopolymers with clayey soils.

Furthermore, swelled viscous biopolymer hydrogels fill the pore spaces of soils (especially sand) and induce pore clogging, which reduces the hydraulic conductivity of soils by more than 3~4 orders of magnitude [94,96,101]. Biopolymers thus have potential to be applied for hydraulic purposes in geotechnical engineering, such as in slurry walls, (temporary) seepage barriers, and grouting.

5. DISCUSSION

5.1. Opportunities for Biopolymers in Geotechnical Engineering

Table 2 summarizes the features and engineering performance of biopolymer treatments and the other aforementioned engineered soil methods used in geotechnical engineering. Although OPC and geopolymers exhibit higher strengthening effects than biopolymer treatments, both OPC and geopolymers require huge amounts of binders (10%–50% binder content to the weight of soil, in general), whereas biopolymers only require 0.5%–1.0% of binder content to achieve strength levels equivalent to those attained by massive cement mixing. Specifically, it has been reported that a 0.5% biopolymer mix is necessary to achieve strength levels that are equivalent to or even higher than those for 10% OPC-soil mixtures [98]. In addition, unlike other ground improvement methods, biopolymers work well in the presence of fine soils [103].

Moreover, biopolymers can be introduced into the soil by various practical modes of application including mixing, injection, spraying, and grouting, and they can be used for building materials, earth pavement, and farmland erosion prevention (Figure 7). Furthermore, biopolymers form a stable gel matrix inside soil that does not damage the local ecosystem. Combined with their water retaining properties in soil, biopolymers are thought to be capable of promoting vegetation growth.

The largest advantage of direct biopolymer implementation compared to other bio-soil methods is that biopolymers can be produced ex-situ (*i.e.*, exo-cultivation) and applied in-situ with a higher degree of quality control, whereas MICP requires time-consuming in-situ cultivation. Moreover, biopolymers can be commercially mass produced, and react with soil particles immediately after mixing, which allows them to be utilized for temporary or rapid supporting purposes.

Table 2. Overview of ground improvement methods (after [7,72,98,103, 114,115,116,117]).

Properties	Cement	Chemicals	Geosynthetics	Geopolymer	MICP	Biopolymer
Soil Competitiveness	Course grained soils Fine grained soils	Course grained soils	Course grained soils	Course grained soils Fine grained soils	Course grained soils	Course grained to Fine grained soils
Method/Construction	Deep mixing, injection, grouting, direct mixing spray	Injection, grouting, spray	*In-situ* installation	Deep mixing, injection grouting, direct mixing Spray	Direct microbe and nutrient injection in to ground	Direct mixing, injection, spraying exo-cultivated biopolymers
Main Materials	Cement Water	Chemically synthesized polymers	Various synthesized materials	Pozzolanic materials, water, cement	Micro-organisms urea, starch	Biopolymers, water
Major Reactions	Hydration	Solidification, chemical bonding	Tensile reinforcement	Pozzolanic reactions	Biological calcite precipitation	Hydrogen-/ionic-bonding, gelation
Advantages	Strong, durable Numerous case studies	Strong, durable Quick isolation Numerous case studies	Quick, Easy to use	Strong, durable Good waste sink Lower carbon footprint than cement	No carbon footprint Possible use with material remediation	Environment-friendly Smaller concentrations Stronger in fine soils
Disadvantages/ Challenges	Large carbon footprint Disruption of local ecosystem. Disposal Problems	High price Environmental impacts. Washout and durability problems	Heterogeneous method Use limited to shallow depths	Similar environmental problems as cement	Ammonia byproduct Limitations in fine soils	High price High sensitivity (hydrophilic) to presence of water
Compressive Strength (MPa) (average)	2 [a]	0.8 (acrylic polymer-treated sand)	-	10 [c]	1.8 (shear strength) [d]	4.3 [f]
Price for Soil Treatment (average)	45.5 USD/ton [a]	200 USD/ton [b]	2 USD/m^2	30 USD/ton [c]	90 USD [e]	28 [g] – 250 [f] USD/ton

[a] Compressive strength of cement mixed soil (10% cement content to the weight of soil); [b] Acrylic grouting cost 4.4 times higher than cement treatment for the same 10% content to the soil mass (Anagnostopoulos and Hadjispyrou 2004); [c] Clay : fly-ash geopolymer = 100 : 30 mixing ratio in mass (Sukmak *et al.* 2013); [d] Microbial induced calcite precipitated sand; [e] Cost of 100 L of nutrient solution with 10 mM urea/min activity; [f] β-glucan biopolymer (0.5% to the mass of soil) mixed residual soil; [g] Xanthan gum (0.5% to the mass of soil) mixed.

Figure 7. Schematic impression of biopolymer applications in geotechnical engineering practices in field.

5.2. Economic Feasibility of Biopolymers: Global market of Biopolymers

Biopolymers are expected to be an effective, environment-friendly, and promising substitute to replace or reduce the use of OPC in soils for geotechnical engineering purposes. However, while biopolymers are effective at smaller concentrations compared to OPC, the current market price of biopolymers is significantly higher than that of OPC. The introduction of mass production and expanded utilization is expected to produce a significant cost reduction of biopolymers, as was the case with the considerable price reduction of xanthan gum following its commercialization (Figure 8).

The material cost for 0.5% xanthan gum biopolymer for soil treatment has decreased from \$250 to \$28 during the last three decades.

While the majority of biopolymers are currently quite expensive, the actual cost of their use in geotechnical engineering largely depends on the choice of biopolymer. Additionally, the present market prices of most candidate biopolymers are for food grade quality with very high purity, and that standard results in significantly higher production cost. For geotechnical applications, edible grade purity is unnecessary, and thus the price of biopolymers can be reasonably expected to be lower when they are produced specifically for construction and geotechnical engineering applications.

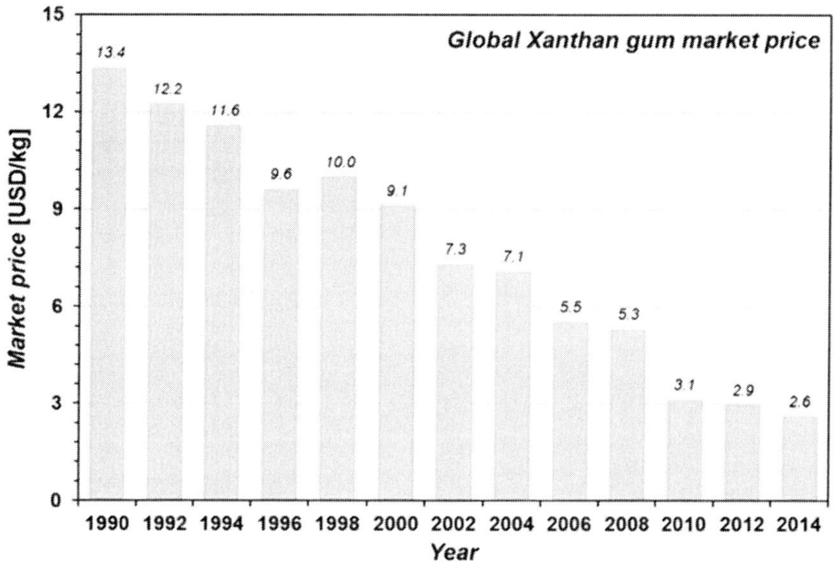

Figure 8. Global market price trend of xanthan gum.

5.3. Biopolymer Competitiveness in Global Carbon Emission Trading

Various raw materials are available for the production of biopolymers. For the first generation in the 1960s to the 1980s, edible biomass such as corn and sugarcane was widely used. Non-edible sources including food wastes and agricultural wastes were employed for the 2nd generation of biopolymer production in the 1990s. CO_2 is now effectively being used as an alternative source in various biopolymer production strategies [118]. For instance, the production of xanthan gum is known to consume approximately 4.97 kg of CO_2 for 1 kg of biopolymer [119,120].

Meanwhile, cement production emits 1.25 tons of CO_2 per 1 ton of cement, as a combined result of chemical reactions (52%) and fossil fuel burning (48%) during the calcination process [9] . For soil improvement in geotechnical engineering, treatment with a 0.5% ratio of xanthan gum to the weight of soil provides an equivalent or greater strengthening effect than that of a 10% cement mix [103]. However, for an equal unit (1 ton) of soil treatment, the material cost at current prices of employing xanthan gum becomes uneconomical compared to conventional cement mixing, being 37% more expensive (xanthan gum treatment: 13.5 USD/ton of soil; cement: 9.85 USD/ton of soil), as shown in Table 3.

Table 3. Economic feasibility of biopolymer (xanthan gum) implementation in geotechnical engineering applications with environmental considerations.

Soil Treatment Material	Cement	Xanthan Gum
Market price of material	98.5 USD/ton [a]	2,700 USD/ton [b]
Required amount for 1 ton soil treatment ($\geq$ 2.5 MPa)	100 kg (10% to soil 1 ton)	5 kg (0.5% to soil 1 ton)
Material price for 1 ton soil treatment	9.85 USD	13.50 USD
CO_2 emission per 1 kg material production (kg CO_2)	+1.25	−4.97
CO_2 emission related to 1 ton soil treatment (kg CO_2)	+125	−24.85
CO_2 emission trade related to 1 ton soil treatment [c]	+2.75 USD	−0.55 USD
Total cost for 1 ton soil treatment (with carbon trade exchange considerations)	12.50 USD	12.95 USD

[a] Depicts the price of cement in the United States, in 2014 (www.statista.com); [b] Depicts the average price of xanthan gum (market price: www.alibaba.com), in 2014; [c] EU ETS (European Union Emission Trading Scheme) carbon emission trade: 22 USD/ton of CO_2, in 2012.

With global greenhouse gas reduction efforts and carbon emission trading (considering a rate of 22 USD/ton CO_2) [121], the CO_2 emission trade imposed on soil-cement mixtures (with 10% cement) has been estimated to be 2.75 USD, while 0.5% xanthan gum treatment would secure 0.55 USD CERs (Certified Emission Reductions) for 1 ton of soil treatment, respectively. When this CO_2 emission trading is considered, in economic terms xanthan gum soil treatment becomes remarkably competitive, being only 3.6% more expensive than cement, as seen in Table 3. Thus, especially when considering the aim of environment-friendly construction and development, biopolymers have high potential to replace high carbon emitting soil treatment materials.

5.4. Further Challenges for Practical Implementation of Biopolymer Technology

In addition to the challenges related to market price, the use of biopolymers in soil engineering also raises concerns about durability. First, since biopolymers are organic materials, there are concerns about whether these biopolymers will biodegrade with time. Moreover, biopolymers are highly sensitive to the presence of water, and therefore durability concerns involving wetting and drying cycles, as well as overall soil strength in the presence of water, must be addressed. However, a recent study shows consistent strengthening performance of xanthan gum biopolymer-treated soils up to two years after mixing [103]. Thus, it is expected that the degradation characteristics of biopolymers would be acceptable when they are used for temporary or short-term practices. They would also be advantageous in certain applications because they would make further remediation processes or removal unnecessary, since the biopolymers are expected to naturally decompose with time.

To address other durability concerns, various types of biopolymers and/or practical implementation methods can be utilized together to improve the overall quality and reliability of biopolymer treated soils. Because most biopolymers exhibit high sensitivity to water, viable methods for increasing the effectiveness of biopolymer treated soils in water will greatly improve the reliability of these soils for ground improvement and stabilization. Several methods have been studied and conceived to resolve this issue, such as the use of biopolymers with thermo-gelation properties. As previously noted, gellan gum and agar gum are among the biopolymers that possess such thermo-gelation properties. These biopolymers become highly soluble in water at around 85–90 °C, and hardening of the gel solution occurs as the gel is cooled to room temperature [108,109]. Such thermo-gelating biopolymers retain considerable strength even when subjected to water.

Another plausible solution is the use of protein based biopolymers. Protein based biopolymers, such as casein proteins, have been employed in a wide variety of fields such as in the manufacturing of adhesives. Since such proteins have lower hydrophilic properties than their polysaccharide biopolymer counterparts, soils that are stabilized by such proteins may provide higher resistance to water. A recent study showed that 6% casein mixed with lime and loam was suitable for use in ceramic manufacturing [122].

In addition to these methods, the use of cross-linking for biopolymers may provide a more powerful soil stabilizing method. Cross-linking is a technique used to greatly improve the properties of a specific material by introducing an agent that promotes interactions between separate polymer chains, thereby enhancing their overall strength. One common application of cross-linking is the use of sulfur in rubber polymer to greatly enhance the strength and stiffness of the rubber, a process also known as vulcanization [123]. Cross-linking has also been used to enhance the properties of gel solutions. A double network hydrogel composed of two different biopolymers with a cross-linking agent was used to create a hydrogel with high mechanical strength. The resulting hydrogels showed extremely high strength and durability, and one case study showed that the hydrogel, which could easily be sliced at 0.2 MPa prior to cross-linking, exhibited a strength of 25 MPa at a strain of over 80% after cross-linking [124,125].

Overall, a wide variety of methods can be applied to improve the mechanical properties of biopolymer treated soils, and as such, the use of biopolymers is emerging as a possible eco-friendly, sustainable method for soil improvement and stabilization. However, various further studies focusing on practical implementation methods and the development of suitable equipment are required to ensure desired construction performance and reliability of biopolymer applications for *in situ* geotechnical engineering purposes.

6. CONCLUSIONS

Ground enhancement methods have been intensively studied and developed over centuries. The use of cement in geotechnical engineering could be considered the beginning of modern ground enhancement methods, and numerous studies have since been performed to enhance the properties of cement. However, in the late 20th century, environmental concerns gave rise to increasing demand for environment-friendly construction methods. The development of green cement and several possible cement substitutes, including the direct use of biopolymers in the soil layer, have been proposed.

Recent studies have shown that biopolymers can strengthen soils, and they offer several advantages in such applications, including being environment-friendly and effective at low concentrations. Several studies have shown that the strengthening induced by biopolymer treatment is maximized in the presence of fines, especially clay particles. For polysaccharide-type biopolymers, hydroxyl groups on the polymer surfaces induce hydrogen bonding with water molecules, making them hydrophilic, and enabling the formation of firm (*i.e.*, viscous) hydrocolloids or hydrogels.

Meanwhile, when water is scarce, as in conditions of drought or dry soils, biopolymers can form direct hydrogen bonds with clay particles, or indirect ionic bonds with these particles, in the presence of intermediate ions such as alkali- or alkali earth- metal ions in the soil. Direct and indirect bonding leads to the formation of a firm biopolymer-clay matrix, which provides a significant increase in soil cohesion. Proper mixing of coarse particles, clay particles, and biopolymers is thus expected to provide optimal strengthening effects, due to the combination of increased mechanical friction between coarse particles, and a cementation effect between biopolymer-clay matrices.

In contrast to cement, biopolymers have strong potential to reduce carbon dioxide emissions. Moreover, some biopolymers show functionality to support vegetation growth and stabilization, which can be applied as countermeasures for farmland preservation and anti-desertification and against other threats to environmental conservation.

Although their benefits are numerous, several challenges related to the use of biopolymers remain to be addressed, including sensitivity to water, market costs, and possible biological degradation. Overall, given the wide variety of available biopolymers, the flexibility of their modification, and the numerous advantageous properties that they possess, the use of biopolymers in geotechnical engineering appears to have a promising future.

ACKNOWLEDGMENTS

The research described in this paper was financially supported by a National Research Foundation of Korea (NRF) grant funded by the Korean government (MSIP) (No. 2015R1A2A2A03006268), by a grant from the Strategic Research Project (Development of Key Excavation Solutions for Expandable Urban Underground Space) funded by the Korea Institute of Civil Engineering and Building Technology (KICT), and by the KAIST End-Run Program (No. N01150661) supported by the Korea Ministry of Science, ICT and Future Planning (MISP). The second author is supported by the "U-City

Master and Doctor Course Grant (Education) Program" under the Korea Ministry of Land, Infrastructure and Transport (MOLIT).

AUTHOR CONTRIBUTIONS

Ilhan Chang and Gye-Chun Cho conceived and constructed the outline and context of the review. Ilhan Chang and Jooyoung Im provided detailed reviews on previous studies and analyzed the data. All authors wrote the paper together with different contributions (Jooyoung Im was responsible for Section 2, Section 3 and Section 4; Ilhan Chang contributed toSection 5 and Section 6; and Gye-Chun Cho provided an overall summary in Section 1 and Section 6).

REFERENCES

1. Potts, D.T. *Mesopotamian Civilization: The Material Foundations*; Cornell University Press: Ithaca, NY, USA, 1997; p. 366.

2. Kemp, B.J. *Ancient Egypt: Anatomy of a Civilization*; Routledge: London, UK, 1989; p. 437.

3. Yang, F.; Zhang, B.; Ma, Q. Study of Sticky Rice-Lime Mortar Technology for the Restoration of Historical Masonry Construction. *Acc. Chem. Res.* **2010**, *43*, 936–944.

4. Delatte, N.J. Lessons from Roman Cement and Concrete. *J. Prof. Issues Eng. Deucation Pract.* **2001**, *127*, 109–115.

5. Sherwood, P.T. *Soil Stabilization with Cement and Lime*; HMSO: London, UK, 1993.

6. Hargreaves, D. *The Global Cement Report 10th Edition*; International Cement Review: Surrey, UK, 2013.

7. Maclaren, D.C.; White, M.A. Cement: Its Chemistry and Properties. *J. Cemical Educ.* **2003**, *80*, 623.

8. Larson, A. *Sustainability, Innovation, and Entrepreneurship*; University of Virginia: Charlottesville, VA, USA, 2011.

9. Worrell, E.; Price, L.; Martin, N.; Hendriks, C.; Meida, L.O. Carbon Dioxide Emissions from the Global Cement Industry. *Ann. Rev. Energy Environ.* **2001**, *26*, 303–329

10. Metz, B.; Davidson, O.; Coninck, H.D.; Loos, M.; Meyer, L. *Carbon Dioxide Capture and Storage*; Intergovernmental Panel on Climate Change: New York, NY, USA, 2005.

11. Oss, H.G.V. *Cement Statistics and Information*; US Geological Survey: Reston, VA, USA, 2014.

12. Rapier, R. *Global Carbon Dioxide Emissions - Facts and Figures*; Consumer Energy Repor, 2012.

13. Plank, J. Applications of biopolymers in construction engineering. In *Biopolymers Online*; Wiley-VCH Verlag GmbH & Co. KGaA: Weinheim, Germany, 2005.

14. U.S. Energy Information Administration. *International Energy Statistics*; U.S. Department of Energy: Washington, DC, USA, 2015.

15. Taylor, H.F.W. *Cement Chemistry*; Thomas Telford Publishing: London, UK, 1997.

16. Hansen, P.J. Effect of high pH on the growth and survival of marine phytoplankton: Implications for species succession. *Aquat. Microb. Ecol.* **2002**, *28*, 279–288.

17. Rao, A.; Jha, K.N.; Misra, S. Use of aggregates from recycled construction and demolition waste in concrete. *Resour. Conserv. Recycl.* **2007**, *50*, 71–81.

18. Water Environment Federation. *Urban Runoff Quality Manaement*; Water Environment Federation: Alexandria, VA, USA, 1998.

19. Parliamentary Office of Science and Technology. *Urban Flooding*; Postnote: London, UK, 2007.

20. Burton, G.A.; Pitt, R. *Stormwater Effects Handbook: A Toolbox for Watershed Managers, Scientists, and Engineers*; Lewis Publishers: Boca Raton, FL, USA, 2002.

21. Schueler, T.R. *Watershed Protection Techniques: A Quarterly Bulletin on Urban Watershed Restoration and Protection Tools*; Center for Watershed Restoration: Silver Spring, MD, USA, 1994.

22. Solecki, W.D.; Rosenzweig, C.; Parshall, L.; Pope, G.; Clark, M.; Cox, J.; Wiencke, M. Mitigation of the heat island effect in urban New Jersey. *Glob. Environ. Chang. B Environ. Hazards* **2005**, *6*, 39–49.

23. Regents, A.B.O. *Urban Climate—Climate Study and UHI via the Internet Wayback Machine*; Arizona State University: Tempe, AZ, USA, 2006.

24. Arnfield, A.J. Two decades of urban climate research: A review of turbulence, exchanges of energy and water, and the urban heat island. *Int. J. Climatol.* **2003**, *23*, 1–26.

25. Kurn, D.; Bretz, S.; Huang, B.; Akbari, H. The Potential for Reducting Urban Air Temperatures and Energy Consumption through Vegetative cooling. In *ACEEE Summer Study on Energy Efficiency in Buildings*; American Council for an Energy Efficient Economy: Pacific Grove, CA, USA, 1994.

26. McPherson, E.G.; Simpson, J.R.; Peper, P.J.; Maco, S.E.; Xiao, Q. Municipal forest benefits and costs in five US cities. *J. For.* **2005**, *103*, 411–416.

27. Torring, M. Management of concrete demolition waste. In *Concrete Technology for a Sustainable Development in the 21st Century*; CRC Press: Boca Raton, FL, USA, 2000; pp. 321–331.

28. Meijer, E.; Kromhout, H.; Heederik, D. Respiratory effects of exposure to low levels of concrete dust containing crystalline silica. *Am. J. Ind. Med.* **2001**, *40*, 133–140.

29. Gotoh, T.; Nishimura, T.; Nakata, M.; Nakaguchi, Y.; Hiraki, K. Air Pollution by Concrete Dust from the Great Hanshin Earthquake. *J. Environ. Qual.* **2002**, *31*, 718–723.

30. Basu, D.; Misra, A. Sustainability in Geotechnical Engineering. In Proceedings of the 18th International Conference on Soil Mechanics and Geotechnical Engineering, Paris, France, 2–6 September 2013.

31. Barrett, R.J. Use of Plastic Filters in Coastal Structures. In Proceedings of the 16th International Conference on Coastal Engineers, Tokyo, Japan, September 1966; pp. 1048–1067.

32. Ward, I.M. The Orientation of Polymers to Produce high Performance Materials. In Proceedings of the Symposium on Polymer Grid Reinforcement in Civil Engineering, London, UK, 22–23 March 1984.

33. Eith, A.W.; Koerner, R.M. Field Evaluation of GEonet Flow Rate (Transmissivity) Under Increasing Load. *J. Geotext. Geomembr.* **1992**, *11*, 153–166.

34. Narjary, B.; Aggarwal, P.; Singh, A.; Chakraborty, D.; Singh, R. Water availability in different soils in relation to hydrogel application. *Geoderma* **2012**, *187–188*, 94–101.

35. Kulicke, W.M.; Kniewske, R.; Klein, J. Preparation, characterization, solution properties and rheological behaviour of polyacrylamide. *Prog. Polym. Sci.* **1982**, *8*, 373–468.

36. Malik, M.; Letey, J. Adsorption of Polyacrylamide and Polysaccharide Polymers on Soil Materials. *Soil Sci. Soc. Am. J.***1991**, *55*, 380–383.

37. Laird, D.A. Bonding Between Polyacrylamide and Clay Mineral Surfaces. *Soil Sci.* **1997**, *162*, 826–832.

38. Fox, D.; Bryan, R.B. Influence of a polyacrylamide soil conditioner on runoff generation and soil erosion: Field tests in Baringo District, Kenya. *Soil Technol.* **1992**, *5*, 101–119.

39. Orts, W.J.; Roa-Espinosa, A.; Sojka, R.E.; Glenn, G.M.; Imam, S.H.; Erlacher, K.; Pedersen, J.S. Use of synthetic polymers and biopolymers for soil stabilization in agricultural, construction, and military applications. *J. Mater. Civ. Eng.* **2007**, *19*, 58–66.

40. Karol, R.H. *Chemical Grouting and Soil Stabilization*, 3rd ed.; CRC Press: New York, NY, USA, 2003.

41. Burnotte, F.; Lefebvre, G.; Grondin, G. A case record of electroosmotic consolidation of soft clay with improved soilelectrode contact. *Can. Geotech. J.* **2004**, *41*, 1038–1053.

42. Lefebvre, G.; Burnotte, F. Improvements of electroosmotic consolidation of soft clays by minimizing power loss at electrodes. *Can. Geotech. J.* **2002**, *39*, 399–408.

43. Ou, C.Y.; Chien, S.C.; Chang, H.H. Soil improvement using electroosmosis with the injection of chemical solutions: Field tests. *Can. Geotech. J.* **2009**, *46*, 727–733.

44. Duxon, P.; Jimenez, A.F.; Provis, J.L.; Lukey, G.C.; Palomo, A.; Deventer, J.S.J.V. Geopolymer technology: The current state of the art. *J. Mater. Sci.* **2007**, *42*, 2917–2933.

45. Kim, D.; Lai, H.T.; Chilingar, G.V.; Yen, T.F. Geopolymer formation and its unique properties. *Environ. Geol.* **2006**, *51*, 103–111.

46. Hench, L.L. *Sol-Gel Silica. Properties, Processing and Technology Transfer*; Noyes Publications: Park Ridge, NJ, USA, 1998.

47. Rovnaník, P. Effect of curing temperature on the development of hard structure of metakaolin-based geopolymer.*Constr. Build. Mater.* **2010**, *24*, 1176–1183.

48. Van Jaarsveld, J.G.S.; van Deventer, J.S.J.; Lukey, G.C. The effect of composition and temperature on the properties of fly ash- and kaolinite-based geopolymers. *Chem. Eng. J.* **2002**, *89*, 63–73.

49. Bakharev, T. Resistance of geopolymer materials to acid attack. *Cem. Concr. Rese.* **2005**, *35*, 658–670.

50. Sigel, A.; Sigel, H.; Sigel, R.K.O. *Biomineralization: From Nature to Application*; Wiley: Hoboken, NJ, USA, 2008; Volume 4.

51. Lear, G.; Lewis, G.D. *Microbial Biofilms: Current Research and Applications*; Caister Academic Press: Norwich, UK, 2012.

52. DeJong, J.; Fritzges, M.; Nüsslein, K. Microbially Induced Cementation to Control Sand Response to Undrained Shear.*J. Geotech. Geoenviron. Eng.* **2006**, *132*, 1381–1392.

53. Whiffin, V.S.; van Paassen, L.A.; Harkes, M.P. Microbial Carbonate Precipitation as a Soil Improvement Technique.*Geomicrobiol. J.* **2007**, *24*, 417–423.

54. Stockes-Fischer, S.; Galinat, J.K.; Ban, S.S. Microbiological precipitation of CaCO3. *Soil Biol. Biochem.* **1999**, *31*, 1563–1571.

55. Fujita, Y.; Ferris, F.G.; Lawson, R.D.; Colwell, F.S.; Smith, R.W. Calcium Carbonate Precipitation by Ureolytic Subsurface Bacteria. *Geomicrobiol. J.* **2000**, *17*, 305–318. Mortensen, B.M.; Haber, M.J.; DeJong, J.T.; Caslake, L.G.; Nelson, D.C. Effects of environmental factors on microbial induced calcium carbonate precipitation. *J. Appl. Microbiol.* **2011**, *111*, 338–349.

56. DeJong, J.T.; Mortensen, B.M.; Martinez, B.C.; Nelson, D.C. Bio-mediated soil improvement. *Ecol. Eng.* **2010**, *36*, 197–210.

57. Neupane, D.; Yasuhara, H.; Kinoshita, N.; Unno, T. Applicability of Enzymatic Calcium Carbonate Precipitation as a Soil-Strengthening Technique. *J. Geotech. Geoenviron. Eng.* **2013**, *139*, 2201–2211.

58. Fujita, Y.; Taylor, J.L.; Gresham, T.L.; Delwiche, M.E.; Colwell, F.S.; Mcling, T.L.; Mcling, L.M.; Petzke, L.M.; Smith, R.W. Stimulation of microbial urea hydrolysis in groundwater to enhance calcite precipitation. *Environ. Sci. Technol.***2008**, *42*, 3025–3032.

59. Renforth, R.; Manning, D.A.C.; Lopez-Capel, E. Carbonate precipitation in artificial soils as a sink for atmospheric carbon dioxide. *Appl. Geochem.* **2009**, *24*, 1757–1764.

60. Ramakrishnan, V.; Bang, S.S.; Deo, K.S. A novel technique for repairing cracks in high performance concrete using bacteria. In Proceedings of the International Conference on High Performance High Strength Concrete, Perth, Australia, 10–12 August 1998; Curtin University of Technology: Perth, Australia, 1998; pp. 597–618.

61. Ramachandran, S.K.; Ramakrishnan, V.; Band, S.S. Remediation of concrete using micro-organisms. *ACI Mater. J.* **2001**,*98*, 3–9.

62. Rong, H.; Qian, C.; Wang, R. A cementation method of loose particles based on microbe-based cement. *Sci. China Technol. Sci.* **2011**, *54*, 1722–1729.

63. Yasuhara, H.; Neupane, D.; Hayashi, K.; Okamura, M. Experiments and predictions of physical properties of sand cemented by enzymatically-induced carbonate precipitation. *Soils Found.* **2012**, *52*, 539–549.

64. Pham, V.; Paassen, L.V.; Nakano, A.; Kanayama, M.; Heimovaara, T. Microbially induced carbonate precipitation (MICP) by denitrification as ground improvement method—Process control in sand column experiments. In Paper presented at the EGU General Assembly 2013, Vienna, Austria, 7–12 April 2013.

65. FuWei, Y.; BingJian, Z.; ChangChu, P.; YuYao, Z. Traditional mortar represented by sticky rice lime mortar—One of the great inventions in ancient China. *Sci. China Ser. E Technol. Sci.* **2009**, *52*, 1641–1647.

66. US National Library of Medicine. Medical Subject Headings (Polynucleotides). Available online: http://www.nlm.nih.gov/ mesh/ 2011/mesh_browser/MBrowser.html (accessed on 31 August 2011).

67. Kalia, S.; Averous, L. *Biopolymers: Biomedical and Environmental Applications*; Wiley: Hoboken, NJ, USA, 2011.

68. Belitz, H.D.; Grosch, W.; Schieberle, P. *Food Chemistry*, 4th revised and extended edition ed.; Springer: Leipzig, Germany, 2009.

69. Saha, D.; Bhattacharya, S. Hydrocolloids as thickening and gelling agents in food: A critical review. *J. Food Sci. Technol.***2010**, *47*, 587–597.

70. Lorenzo, G.; Zaritzky, N.; Califano, A. Rheological analysis of emulsion-filled gels based on high acyl gellan gum.*Food Hydrocoll.* **2012**, *30*, 672–680.

71. Velde, K.V.D.; Kiekens, P. Biopolymers: Overview of several properties and consequences on their applications.*Polym. Test.* **2002**, *21*, 433–442.

72. Cole, D.; Ringelberg, D.; Reynolds, C. Small-scale mechanical properties of biopolymers. *J. Geotech. Geoenviron. Eng.***2012**, *138*, 1063–1074.

73. De Muynck, W.; de Belie, N.; Verstraete, W. Microbial carbonate precipitation in construction materials: A review.*Ecol. Eng.* **2010**, *36*, 118–136.

74. Adibkia, K.; MR, S.S.; Nokhodchi, A.; Javadzedeh, A.; Barzegar-Jalali, M.; Barar, J.; Mohammadi, G.; Omidi, Y. Piroxicam nanoparticles for ocular delivery: Physicochemical characterization and implementation in endotoxin-induced uveitis. *J. Drug Target.* **2007**, *15*, 407–416.

75. Barrére, G.C.; Barber, C.E.; Daniels, M.J. Molecular cloning of genes involved in the production of the extracellular polysaccharide xanthan by Xanthomonas campestris pv. campestris. *Int. J. Biol. Macromol.* **1986**, *8*, 372–374. Chang, Y.J.; Lee, S.; Yoo, M.A.; Lee, H.G. Structural and biological characterization of sulfated-derivatized oat β-glucan. *J. Agric. Food Chem.* **2006**, *54*, 3815–3818.

76. Daniel, J.; Whistler, R.L.; Voragen, A.C.J.; Pilnik, W. Starch and other polysaccharides. In *Ullmann's Encyclopedia of Industrial Chemistry*; VCH Verlagsgesellschaft mbH: Weinheim, Germany, 1994; Volume A25.

77. Harada, T.; Misaki, A.; Saito, H. Curdlan: A bacterial gel-forming β-1, 3-glucan. *Arch. Biochem. Biophys.* **1968**, *124*, 292–298.

78. Hassan, H.; Al-Oraimi, S.; Taha, R. Evaluation of open-graded friction course mixtures containing cellulose fibers and styrene butadiene rubber polymer. *J. Mater. Civ. Eng.* **2005**, *17*, 416–422.

79. Wan, M.W.; Petrisor, I.G.; Lai, H.T.; Kim, D.; Yen, T.F. Copper adsorption through chitosan immobilized on sand to demonstrate the feasibility for *in situ* soil decontamination. *Carbohydr. Polym.* **2004**, *55*, 249–254.

80. Fahnestock, K.J.; Austero, M.S.; Schauer, C.L. Natural Polysaccharides: From Membranes to Active Food Packaging. In*Biopolymers: Biomedical and Environmental Applications*; Wiley: Hoboken, NJ, USA, 2011.

81. Maher, M.; Ho, Y. Mechanical Properties of Kaolinite/Fiber Soil Composite. *J. Geotech. Eng.* **1994**, *120*, 1381–1393.

82. Sivakumar Babu, G.; Vasudevan, A. Strength and Stiffness Response of Coir Fiber-Reinforced Tropical Soil. *J. Mater. Civ. Eng.* **2008**, *20*, 571–577.

83. Annabi, M.; Houot, S.; Francou, C.; Poitrenaud, M.; Bissonnais, Y.L. Soil Aggregate Stability Improvement with Urban Composts of Different Maturities. *Soil Sci. Soc. Am. J.* **2007**, *71*, 413–423.

84. Zobel, H.F. Molecules to granules: A comprehensive starch review. *Starch* **1988**, *40*, 1.

85. Food and Agriculture Organization of the United Nations. *FAO Statistical Yearbook 2009*; FAO Economic and Social Development Department, Ed.; FAO: Rome, Italy, 2009.

86. Francis, H.P.; DeBoer, E.D.; Wermers, V.L. High Temperature Drilling Fluid Component. U.S. Patent 4,652,384, 24 March 1987.

87. Jose, A.; Carvalho, F. Starch as Source of Polymeric Materials. In *Biopolymers: Biomedical and Environmental Applications*; Wiley: Salem, MA, USA, 2011.

88. Sahoo, D.; Nayak, P.L. Chitosan: The Most Valuable Derivative of Chitin. In *Biopolymers: Biomedical and Environmental Applications*; Wiley: Hoboken, NJ, USA, 2011.

89. Nascimento, T.; Rego, C.; Oliveira, H. Potential use of chitosan in the control of grapevine trunk diseases. *Phytopathol. Mediterr.* **2007**, *46*, 218–224.

90. Cheung, W.H.; Ng, J.C.Y.; McKay, G. Kinetic analysis of the sorption of copper(II) ions on chitosan. *J. Chem. Technol. Biotechnol.* **2003**, *78*, 562–571.

91. Renault, F.; Sancey, B.; Badot, P.M.; Crini, G. Chitosan for coagulation/flocculation processes—An eco-friendly approach. *Eur. Polym. J.* **2009**, *45*, 1337–1348.

92. Khachatoorian, R.; Petrisor, I.G.; Kwan, C.C.; Yen, T.F. Biopolymer plugging effect: Laboratory-pressurized pumping flow studies. *J. Petroleum Sci. Eng.* **2003**, *38*, 13–21.

93. McIntosh, M.; Ston, B.A.; Stanisich, V.A. Curdlan and other bacterial (1→3)-[beta]-D-glucans. *Appl. Microbiol. Biotechnol.* **2005**, *68*, 163–173.

94. Ivanov, V.; Chu, J. Applications of microorganisms to geotechnical engineering for bioclogging and biocementation of soil in situ. *Rev. Environ. Sci. Biotechnol.* **2008**, *7*, 139–153.

95. Bacic, A.; Fincher, G.B.; Stone, B.A. *Chemistry Biochemistry and Biology of (1–3)-[beta]-glucans and Related Polysaccharides*, 1st ed.; Academic Press: Amsterdam, The Netherlands, 2009.

96. Chang, I.; Cho, G.C. Strengthening of Korean residual soil with β-1,3/1,6-glucan biopolymer. *Constr. Build. Mater.***2012**, *30*, 30–35.

97. Chang, I.; Cho, G.C. Geotechnical behavior of a beta-1,3/1,6-glucan biopolymer-treated residual soil. *Geomech. Eng.***2014**, *7*, 633–647.

98. Chang, I.; Prasidhi, A.K.; Im, J.; Shin, H.D.; Cho, G.C. Soil treatment using microbial biopolymers for anti-desertification purposes. *Geoderma* **2015**, *253–254*, 39–47.

99. Bouazza, A.; Gates, W.P.; Ranhith, P.G. Hydraulic conductivity of biopolymer-treated silty sand. *Geotechnique* **2009**,*59*, 71–72.

100. Nugent, R.A.; Zhang, G.; Gambrell, R.P. Effect of exopolymers on the liquid limit of clays and its engineering implications. *Transp. Res. Rec. J. Transp. Res. Board* **2009**, *2101*, 34–43.

101. Chang, I.; Im, J.; Prasidhi, A.K.; Cho, G.C. Effects of Xanthan gum biopolymer on soil strengthening. *Constr. Build. Mater.* **2015**, *74*, 65–72.

102. McHugh, D.J. *A Guide to the Seaweed Industry*; Food and Agriculture Organization of The United Nations: Rome, Italy, 2003.

103. Khatami, H.; O'Kelly, B. Improving mechanical properties of sand using biopolymers. *J. Geotech. Geoenviron. Eng.***2012**, *139*, 1402–1406.

104. Chang, I.; Prasidhi, A.K.; Im, J.; Cho, G.C. Soil strengthening using thermo-gelation biopolymers. *Constr. Build. Mater.***2015**, *77*, 430–438.

105. Imeson, A. *Food Stabilisers, Thickeners and Gelling Agents*; Wiley-Blackwell Publishing: Chichester, UK, 2009; p. 368.

106. Huang, M.; Kennedy, J.F.; Li, B.; Xu, X.; Xie, B.J. Characters of rice starch gel modified by gellan, carrageenan, and glucomannan: A texture profile analysis study. *Carbohydr. Polym.* **2007**, *69*, 411–418.

107. Tang, J.; Tung, M.A.; Zeng, Y. Gelling Properties of Gellan Solutions Containing Monovalent and Divalent Cations. *J. Food Sci.* **1997**, *62*, 688–692.

108. Ferruzzi, G.G.; Pan, N.; Casey, W.H. Mechanical properties of gellan and polyacrlamide gels with implications for soil stabilization. *Soil Sci.* **2000**, *165*, 778–792. Chang, C.; Duan, B.; Cai, J.; Zhang, L. Superabsorbent hydrogels based on cellulose for smart swelling and controllable delivery. *Eur. Polym. J.* **2010**, *46*, 92–100.

109. Bueno, V.B.; Bentini, R.; Catalani, L.H.; Petri, D.F.S. Synthesis and swelling behavior of xanthan-based hydrogels.*Carbohydr. Polym.* **2013**, *92*, 1091–1099.

110. Yakimets, I.; Paes, S.S.; Wellner, N.; Smith, A.C.; Wilson, R.H.; Mitchell, J.R. Effect of Water Content on the Structural Reorganization and Elastic Properties of Biopolymer Films: A Comparative Study. *Biomacromolecules* **2007**, *8*, 1710–1722.

111. Davidovits, J. *Geopolymer Chemistry and Applications*; Geopolymer Institute: Saint-Quentin, France, 2008; p. 585.

112. DeJong, J.T.; Fritzges, M.B.; Nusslein, K. Microbially Induced Cementation to Control Sand Response to Undrained Shear. *J. Geotech. Geoenviron. Eng.* **2006**, *132*, 1381–1392.

113. Koerner, R.M. *Designing With Geosynthetics*, 6th ed.; Xlibris Publishers: Dartford, UK, 2012; p. 914.

114. Oh, J.E.; Monteiro, P.J.M.; Jun, S.S.; Choi, S.; Clark, S.M. The evolution of strength and crystalline phases for alkali-activated ground blast furnace slag and fly ash-based geopolymers. *Cem. Concr. Res.* **2010**, *40*, 189–196.

115. Niaounakis, M. *Biopolymers: Reuse, Recycling, And Disposal*, 1st ed.; Elsevier/WA: Oxford, UK; Waltham, MA, USA, 2013; p. 432.

116. Mohanty, A.K.; Misra, M.; Drzal, L.T. Sustainable Bio-Composites from Renewable Resources: Opportunities and Challenges in the Green Materials World. *J. Polym. Environ.* **2002**, *10*, 19–26.

117. Krishna Leela, J.; Sharma, G. Studies on xanthan production from Xanthomonas campestris. *Bioprocess Eng.* **2000**, *23*, 687–689.

118. Guerrero-Lemus, R.; Martinez-Duart, J.M. *Renewable energies and CO_2: Cost Analysis, Environmental Impacts and Technological Trends*, 2012 ed.; Springer: London, UK; New York, NY, USA, 2013; p. 373.

119. Minke, G. Building with earth-30 years of research and development at the University of Kassel. In Proceedings of the International Symposium on Earthen Structures, Bangalore, Bangalore, India, 22–24 August 2007; Interline Publishing: Karnataka, India, 2007.

120. Mark, J.E.; Erman, B.; Eirich, F.R. *Science and Technology of Rubber*; Academic Press: San Diego, CA, USA, 1994.

121. Gong, J.P.; Katsuyama, Y.; Kurokawa, T.; Osada, Y. Double-Network Hydrogels with Extremely High Mechanical Strength. *Adv. Mater.* **2003**, *15*, 1155–1158.

122. Nakayama, A.; Kakugo, A.; Gong, J.P.; Osada, Y.; Takai, M.; Erata, T.; Kawano, S. High Mechanical Strength Double-Network Hydrogel with Bacterial Cellulose. *Adv. Funct. Mater.* **2004**, *14*, 1124–1128.

Chapter 5

Biopolymer Blends Based on Poly (lactic acid): Shear and Elongation Rheology/Structure/Blowing Process Relationships

Racha Al-Itry [1,2], **Khalid Lamnawar** [1,3] **and Abderrahim Maazouz** [1,2,4,*]

[1]Institut National des Sciences Appliquées (INSA), Université de Lyon, Lyon F-69361, France

[2]UMR (Unité Mixte de Recherche) 5223, Ingénierie des Matériaux Polymères IMP, Centre National de la Recherche Scientifique (CNRS), INSA Lyon, Villeurbanne F-69621, France

[3]UMR (Unité Mixte de Recherche) 5259, Laboratoire de Mécanique des Contacts et des Structures LaMCoS, Centre National de la Recherche Scientifique (CNRS), INSA Lyon, Villeurbanne F69621, France

[4]Hassan II Academy of Science and Technology, Rabat, Morocco

ABSTRACT

This study was dedicated to the blown film extrusion of poly(lactic acid), which mainly presents poor shear and elongation viscosities, and its blends. In order to enhance its melt strength, two main routes were selected (i) a structural modification through chain extension and branching mechanisms by adding a reactive multifunctional epoxide (named Joncryl) and (ii) blending with poly(butylene adipate-*co*-terephtalate), named PBAT in presence (or not) of Joncryl. The effects of the reactive agent on the shear and elongation rheology, morphological, and interfacial properties of the blends were systematically investigated. A decrease of the interfacial tension has been also demonstrated according to the deformed drop retraction method (DDRM). Hence, the role of Joncryl as a compatibilizer was highlighted. Consequently, finer morphology of the dispersed phase was obtained. Furthermore, the impact of the two modification routes on the blown film extrusion ability of PLA has been studied. Based on the improved shear and elongational rheological properties, a great enlargement of the blowing processing window of PLA modified with Joncryl was demonstrated. Indeed, with the addition of Joncryl into PLA–PBAT blends, a reduction of the instability defects has been detected. Finally, the induced crystalline structure and the thermo-mechanical properties of blown films were shown to be improved.

KEYWORDS

biopolymers; shear and elongation rheology; structural modification; blown film extrusion; crystallization; thermo-mechanical behavior

1. INTRODUCTION

In packaging applications, the blown film extrusion represents an efficient way to make films with thin thickness and the polymer needs to possess relatively high melt strength [1,2]. Unfortunately, compared to polyolefins, poly(lactic-acid) (PLA) is typically a linear polymer with a rather low molecular weight distribution [3]. PLA rheology is not well suited for the blowing extrusion because it has relatively low melt strength, resulted from its thermal, oxidative and hydrolytic degradations [4,5]. These degenerative effects lead, during processing, to the cleavage of the polymer manifested by a loss of the molecular weight and deterioration of rheological properties when higher shear and extensional deformations are required in such process [6]. Another shortcoming of PLA is its brittle properties. PLA should be therefore melt strengthened under processing. Many strategies have motivated considerable research efforts. In particular, two strategies are presented here: (i) modification of PLA through branched architecture with chain-extension reactions to compensate for the molecular weight decrease caused by the processing degradation [7,8] and (ii) blending PLA with other natural or synthetic materials such as Polycaprolacton (PCL), starch, poly (butylene adipate-*co*-terephtalate) (PBAT), poly(butylene succinate-*co*-adipate) (PBSA), *etc*. [9].

A better understanding of the blowing extrusion process is still of high actuality. Rheological measurements, in shear and in elongation flows, should be well established to characterize polymers with respect to processing. Extensional flow of polymer melts is a more challenging task in comparison to the shearing deformations. The subtle changes in molecular architecture are readily detected in the processing performance of PLA and more readily visible in extensional deformations [10].

On the one hand, the chemical chain extension of PLA using a chain extender is highly discussed. Some experiments have been conducted by our team researchers in previous works [5,6,7,8,9,11,12,13]. We showed that the structural modification of PLA highly increased the shear rheological properties (melt strength, zero-shear viscosity, elasticity, shear-thinning behavior) and contributed to very long relaxation process. A complex structure *i.e.*, mixture of linear and randomly branched chains, was deduced according to Van–Gurp–Palmen plots, molar measurements and viscometric properties. During elongation, linear PLA does not exhibit any strain-hardening. This was due to linear chains being free from any branch points and was not prevented from slipping over each other. Several modification routes have been discussed in the literature in order to create a strain-hardening. Liu *et al*. [14] found that the strain hardening was not observed in PLA without long chain branches (LCB) or even in the PLA with star branched structure. Palade *et al*. [15] reported that the presence of a low fraction of high molecular of PLA makes the strain-hardening occurred. Yamane *et al*., showed that the addition of small amounts of poly(D,L-lactic acid) (PDLA) into PLLA could lead to strain-hardening in poly(L-lactic acid) (PLLA) melts due to the stereocomplex formation acting as branching points on the PLA chains. The effects of chain branching of PLA on foaming ability have been investigated by extensional flow experiment [14,16,17]. The authors found that the chemical chain-extension enhanced the extensional flow of PLA where a relevant strain hardening appeared due to the presence of long chain branches when the macromolecular chain disentanglement rate is too low in relation with the deformation rate. This resistance to stretching became stronger as the chain extender content was increased. Additionally, the presence of LCB in PLA has shown the relationship between the extension rate and the extent of strain hardening, namely that strain hardening occurred faster at higher extension rates. Moreover, the extensional rheological experimental results of branched PLAs elucidated that the LCB–PLA has shown the foaming process-ability with fine cell structures. However, they reported that some processing limitations arose at a very high extension ratio (a high content of chain extender) because of the high melt elasticity. Meanwhile, Mallet *et al.* [13] discussed the improved elongational properties and the stability blowing extrusion properties of modified PLA using a ternary system (chain-extenders, plasticizers and nucleating agents).

On the other hand, blending PLA with PBAT presents a suitable approach to improve its mechanical properties. In our previous work [5], we have reported an improvement of the elongation at break and toughness of PLA–PBAT blends even at low concentrations of PBAT. The effect of the chain extension reaction on the interfacial properties of PLA–PBAT blends has been also studied [9]. However, few works have been dedicated to study the elongational behavior of blends based on PLA in the molten state. Eslami *et al.* [18] evaluated the elongational rheology of PLA/PBSA blends with different blend compositions. They found that PLA/ PBSA blends containing less than 50 wt% PBSA showed no clear strain-hardening behavior. However, a well-defined strain hardening behavior was observed for blends containing PBSA at concentrations of 50 wt% or higher.

Compared to polyethylene [19,20], literature works covering molecular characterization, elongational rheology and film blowing are very rare. That is why new insight can be expected from the results reported in this paper.

The ultimate objective of the present work is thus dedicated to the film blowing process for PLA and its blends. It deals with the better understanding of the structure modification and compatibilization on the physical and morphological properties of PLA–PBAT blends, with special emphasis on their elongational rheology. A clear correlation between the induced shear and elongation properties on the blown extrusion stability process has been also discussed. The micro-structural and thermo-mechanical properties of the blown biopolymer films have been finally investigated.

2. EXPERIMENTAL SECTION

2.1. Polymers Used

PLA (Grade: 4032D) with a weight- average molar weight of 100.000 g/mol (Gel Permeation Chromatography GPC analysis), was purchased from

Natureworks (Minnetonka, MN, USA), and consisted of 2% D-Lactide units. PBAT (Grade: Ecoflex FBX 7011) was obtained from BASF (Ludwigshafen, Germany). It exhibits a weight-average molar weight of 40.000 g/mol (GPC analysis), a glass transition temperature and melting point of −30 °C and 110–120 °C (Differential Scanning calorimetry DSC analysis), respectively. Commercial multi-functional styrene-acrylic oligomers (BASF, Joncryl ADR®-4368) accepted by the Food & Drug Administration (FDA) for food packaging, were used. It is an epoxy functional oligomeric acrylic with the following physical characteristics: Transition temperature T_g = 54 °C, EEW (epoxy equivalent weight) = 285 g/mol, M_w = 6800 g/mol, obtained in flake form. Table 1 summarizes the compositions of the various studied samples.

Table 1. Composition of the different studied samples.

Notations	Composition
PLA	100 wt% PLA
PBAT	100 wt% PBAT
PLA-0.7	99.3 wt% PLA + 0.7 wt% Joncryl
PBAT-0.7	99.3 wt% PBAT + 0.7 wt% Joncryl
80–20	80 wt% PLA + 20 wt% PBAT
80–20–0.25	79.8 wt% PLA+ 19.95 wt% PBAT + 0.25 wt% Joncryl
80–20–0.5	79.6 wt% PLA + 19.9 wt% PBAT + 0.5 wt% Joncryl
80–20–0.7	79.44 wt% PLA + 19.86 wt% PBAT + 0.7 wt% Joncryl
80–20–1	79.2 wt% PLA + 19.8 wt% PBAT + 1 wt% Joncryl

2.2. Sample Preparation and Preliminary Work

2.2.1. Blends Preparation

Before compounding, both PBAT and PLA pellets were dried under vacuum at 80 °C for 12 h to remove moisture. Joncryl ADR® was also dried under vacuum at 60 °C. Numerous papers have been published by our research group on the reactivity of Joncryl on both PLA and PBAT polymers [5,11,12,13]. Since its higher functionality, Joncryl reacts via its epoxy rings with carboxyl and hydroxyl groups of both polyesters.

The preparation of samples was carried out under nitrogen flow in a corotating twin-screw extruder having a clam shell barrel design with a length:diameter ratio of 25:1 (Thermo Electron Polylab System Rhecord

RC400P, Courtaboeuf, France). The screw rotation speed was set at 40 rpm during 3 min. The temperature profile was set at 140 °C (feeding zone), 190 °C, 180 °C (melting zones) and 180 °C (die). After melt blending, each extrudate was quenched in a cold water bath and granulated. A special sealed and heated hopper was used. The twin screw profile was optimized to obtain good dispersion properties. For the sake of clarity, only the results of the modified polymers and blends at the end of the reaction were studied here, for which the viscoelastic parameters (complex viscosity and storage modulus) remain unchanged during time.

2.2.2. Validation of Reaction Completion with Multifunctional Epoxide

The optimal reaction times with Joncryl were evaluated based on the monitoring of the experimental *in situ* stabilization of the torque *versus* time to be sure that the samples are not changing in time and no thermal degradation can occur. Moreover, the samples stability with time was studied during a dynamic time sweep test (DTS) by plotting the normalized complex viscosity modulus $\eta^*(t)/\eta$ ($t = 0$) *versus* time at various temperatures ranging from 180 to 200 °C and the rheo-kinetic properties as well the apparent conversion and the end reaction time are determined. In a previous work [5], the stabilization region was thus determined when achieving a viscosity plateau *i.e.*, when the reaction between PLA (or PBAT) with chain extender has finished. We have previously concluded that for the modified PLA with less than 0.7% of multi-functionalized epoxy, 3 min are quite sufficient with 180 °C. With higher Joncryl amount, the reaction need longer time (higher than 10 min) to be stabilized. Therefore, the compounds with 1% of Joncryl were re-granulated and passed once again in the twin screw extruder until achieving a stable torque.

2.2.3. Blown Film Extrusion

Blowing extrusion was performed using and annular 25 mm diameter blow die coupled to a melt pump and twin—extruder to process the PLA, PBAT and their blends into a tubular shape. The molten polymer, in the form of a tube, exiting from the die (T^0_{die} = 180 °C) is drawn upward by a take-up device. When the process starts up, the air is introduced at the bottom of the die to inflate the tube and form a bubble. The size of the bubble is maintained at a certain internal pressure. The tube is then flattened in the nip rolls and taken up by the winder. An air ring is also used to rapidly cool the hot bubble and solidify it at some distance above the die exit. This distance is usually called Frost Line Height (FLH). The film dimensions are determined by the blow-up ratio (BUR) and the take-up ratio (TUR) (*cf.* Figure 1). BUR is defined as the ratio of the final bubble diameter (D_f) to die diameter (D_0) and is controlled by varying the air pressure; TUR is defined as the ratio of the speed of the take-up device (V_f) to the extruded material velocity (V_0) at the die exit. These two key parameters will be useful to plot the stability map (BUR *vs.* TUR).

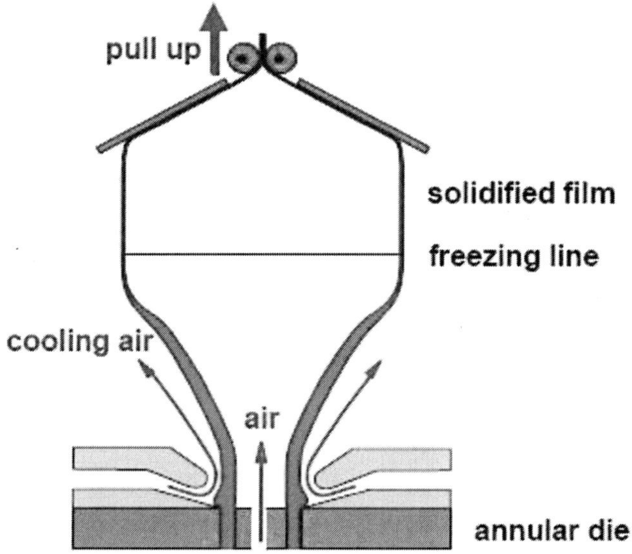

Figure 1. Schematic diagram of the film blowing experimental set-up.

2.3. Interfacial Tension Measurements

The effect of the multifunctional epoxide on the interfacial tension of the PLA–PBAT blends was determined with the deformed drop retraction method (DDRM).

DDRM-sandwich preparation: Molten polymer droplets are obtained from finely grinding powder and dispersed between two circular matrix films to form a sandwich, as presented in Figure 2. The thickness of these films is about 500 μm to 1 mm. The prepared sandwich is then heated up to the molten state of the both dispersed phase and matrix. The preparation of our sample for the measurement of interfacial tension is described schematically:

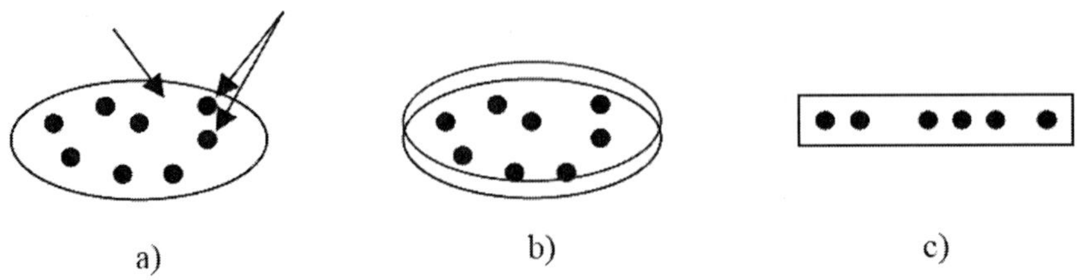

Figure 2. Three steps of sandwich preparation for the DDRM.

Step (a) consists in depositing homogeneously the dispersed phase in the form of powder over the entire surface of a matrix disk. In step (b), a second disk covers the inclusions. The sample sandwich is then introduced in the shear plate and heated (Step c); so that the inclusion and the matrix are melted. Step strain with and relaxation experiments were then performed by the Linkam device® (Surrey, UK). In order to obtain well-defined spherical droplets, the dispersed particles were subjected to a shear treatment for a few seconds (30 s; steady shear ranging from 0.1 to 5 s^{-1} for various model systems in the Newtonian rheological zone) at 180 °C. According to the higher capillary numbers, the drop first deforms into a long fibril and then breaks up into a large number of droplets through Rayleigh capillary instabilities. The sheared droplets were observed during their relaxation process through two holes made in the heating plates using a camera.

Here, we present the various experimental sandwiches (droplet/matrix):

-Model A: it is used as a reference where neat PBAT is dispersed into PLA matrix;

-Model B: it represents the more complicated systems where Joncryl reacts with both PLA and PBAT at the same time.

More details about the calculation of the interfacial tension according to DDRM carried out with the other model sandwiches were given in our previous work [9].

2.4. Morphological Study

To investigate the morphology of the blends, scanning electron microscopy (SEM) and transmission electron microscopy (TEM) were carried out. For the SEM micrographs, the cryo-fractured specimens were coated with a thin gold layer to improve the sample conductivity for observation. They were then scanned under an accelerating voltage of 8 kV. Image analysis was successfully performed on the obtained micrographs. For the TEM investigations, the microtomed specimens were stained by ruthenium tetraoxide (RuO_4) vapor for 30 min before observations.

2.5. Shear and Elongation Rheological Properties

2.5.1. Small-Amplitude Oscillatory Shear (SAOS) Measurements

The measurements were investigated with a stress-controlled rotational rheometer TA Instruments (Guyancourt, France) DHR-2 with a parallel plate geometry. The linear domain, performed at 0.1 rad/s, extends from 0.2 to 1000 Pa. During the rheological experiments, the heated chamber was continuously purged with nitrogen to avoid the potential degradation and oxidation of the polymers. Dynamic frequency sweep tests were performed at 20 Pa over an angular frequency range of 10^{-3}–628 rad/s.

2.5.2. Capillary Flow Rheology

Experimental runs with a non-isothermal flow were performed by using a pressure—controlled CEAST (Compagnia Europea Apparecchi Scientifici, Torino) capillary rheometer at 180 °C through a die with a 180° entry angle and various Length/Diameter (L/D) atios (30–20 and 5) to obtain Bagley corrections. The capillary measurements were conducted in a closed environment under nitrogen atmosphere. For this kind of experiments, the elongation viscosity was determined using Cogswell model [21].

2.5.3. Uniaxial Elongational Measurements

Experiments in uniaxial extension flow were performed using a second generation SER-2 (Sentmanat Extensional Rheometer Universal Testing

Platform from Xpansion Instruments, Tallmadge, OH, USA) mounted on the DHR-2 stress-controlled rotational rheometer (Guyancourt, France). All experiments in extension were performed at 180 °C. For elongational viscosity measurements, the samples were prepared by compression molding at 180 °C into the sheet shape of 1 mm. The test specimens were cut in a width of 10 and the length of 25 mm. Rectangular plates were clamped between the two counter-rotating wind-up drums. The sample width *w*(t) was recorded *versus* strains to confirm the linearity and thus the uniform elongation. Different samples were elongated in the melt up to very large strains. The uniaxial extensional measurements were carried out at constant strain rates from 10^{-3} to 20 s^{-1}. Therefore, the duration of tests varies from 0.19 to 38 s. The test chamber was maintained by dry nitrogen to avoid the degradation of samples. The samples were allowed to relax before each experiment to ensure that the samples were stress-free. The maximum Hencky strain was set at 3.8 by applying a given angular velocity to turn the dual wind-up drums. Most importantly, the reproducibility of the results was examined by repeating the measurements on three different specimens for each sample.

The following equation was used to calculate the elongational viscosity of molten samples in steady extension:

$$\eta_E^+(t) = \frac{T}{2R\dot{\varepsilon}_H A_0 \left(\frac{\rho_s}{\rho_M}\right)^{2/3} \exp(-\dot{\varepsilon}_H t)} \tag{1}$$

where T is the torque, R is the drum radius, A_0 is the initial area of the sample measured in the solid state, and t is the time data. The density of PLA in solid state, ρ_s, was considered 1.25 g/cm^3 and the value for PLA melt at 180 °C, ρ_M, was 1.05 g/cm^3. Finally, the elongational viscosity of the neat and branched PBAT was hardly measured due to sagging effects prior or during the measurement. All the results shown in this paper were obtained in reproducible experiments.

Because the sample's cross-section becomes reduced during stretching, a video recorder was used to monitor this variation. The sample's section should decrease exponentially with the time when the strain rate is kept constant. For all the rate strain deformations with the Sentmanat Extensional Rheometer Generation 2, SER-G2 geometry, we experimentally validated that the strain rate was constant, thanks to a camera installed in the oven enabling to measure the real dimensions of the sample.

Monitoring of the section rendered it possible to check for a constant strain rate and to determine the real strain rate applied to the polymer.

2.5.4. Start-up Shear Experiments (Transient Step Shear Rate Experiments)

The experiments were carried out in the same Discovery Hybrid Rheometer DHR-2 stress controlled rheometer using parallel-disk geometry. In this case, strain rate is commanded to increase instantaneously at time zero and the transient shear stress is recorded over time. The envelope $3\eta_0^+(t)$ in uniaxial elongation measurements has been plotted according to the experiments with 0.01 s^{-1} as a start up shear rate.

2.6. Non-Isothermal Characterization

The thermal properties of neat and processed PLA, PBAT and their blends were investigated with the help of a differential scanning calorimeter Q10 DSC (TA Instruments). The DSC cell was constantly purged with nitrogen at a flow rate of 50 mL/min. A set of heating/cooling ramps was carried out following a three step process; the samples were heated to 200 °C at 10 °C/min. The crystallinity was calculated according to the Equation (2):

$$X_c(\%) = \frac{\Delta H_m - \Delta H_{cc}}{\Delta H_{m\infty}} \times (1/w) \times 100 \qquad (2)$$

where ΔH_m is the melt enthalpy; ΔH_{cc} is the cold crystallization enthalpy; $\Delta H_{m\infty}$ is theoretical specific enthalpy of fusion of the perfect crystal. It was taken as 93 J/g for PLA. It represents the. *W* is weight percentage of PLA in the blend.

2.7. Thermo-Mechanical Properties

The investigation of the thermo-mechanical properties was conducted through dynamic mechanical thermal analysis (DMTA) performed on a Rheometer Solids Analyser RSA II (TA instruments). Oscillating tensile-compressions tests under a strain of 0.01% at a fixed oscillatory frequency of 1 Hz were performed during temperature sweeps from −80 up to 110 °C at a rate of 3 °C/min.

3. RESULTS AND DISCUSSION

3.1. Interfacial Tension Measurements and Morphological Properties

Before addressing the elongation properties, it would be wise to note that the interfacial properties of PLA–PBAT blends are modified with the incorporation of Joncryl.

On the one hand, the interfacial tension "α" has been measured from the deformed drop retraction method (DDRM) according to Taylor model given in Equation (3):

$$D = D^{\circ}\exp-\left\{\frac{40(p+1)}{(2p+3)(19p+16)}\frac{\alpha}{\eta_m R^{\circ}}t\right\} = D^{\circ}\exp\left\{-\frac{t}{\tau_d}\right\} = \frac{L-B}{L+B} \tag{3}$$

Here, D is the drop deformation parameter defined as $D = (L-B)/L+B$, and L and B are the major and minor axes of the ellipsoid drop, respectively. $D°$ is the initial deformation parameter, p is the viscosity ratio of the dispersed to the matrix phase, α is the interfacial tension, η_m is the viscosity of the matrix phase, and R is the radius of the drop at the equilibrium. Plotting ln(D) *versus* (t) renders possible an estimation of the interfacial tension α from the plot's slope ($-1/\tau_d$). τ_d is the relaxation time for deformed droplet retraction. More details about the calculation method are presented in our previous paper [9].

A decrease of "α" has been observed with the incorporation of Joncryl into blends, as plotted in Figure 3, highlighting once again the role of Joncryl as a compatibilizer. Therefore, the coexistence of chain extension/branching chains coupled to the PLA–Joncryl–PBAT copolymer formation had to be taken into account when studying such blends.

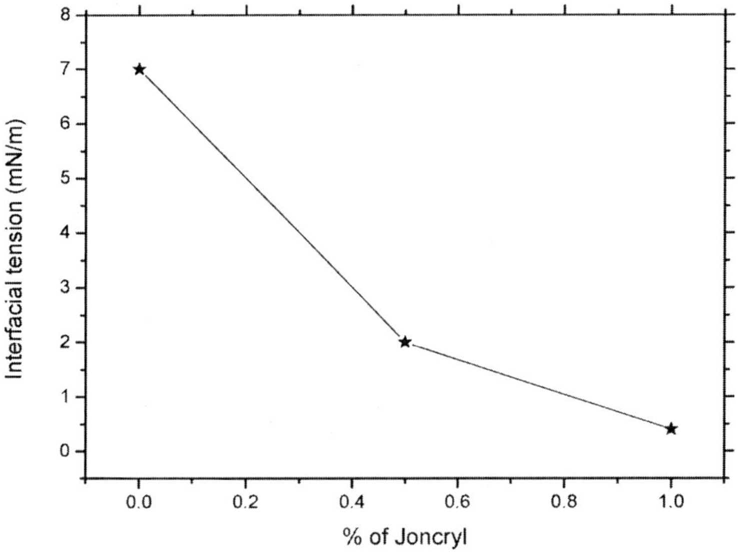

Figure 3. Evolution of the interfacial tension given by DDRM as a function of the Joncryl amount into PLA–PBAT blends.

Besides, a significant change on both deformation mechanism and break-up conditions has been elucidated, as shown inFigure 4.

On the other hand, according to SEM and TEM micrographs (see Figure 5a,b), the addition of Joncryl leads to a reduction of the average size of the dispersed PBAT phase. The ultimate volume average size was reduced from 0.25 μm for the uncompatibilized blends to 0.07 μm for the modified blend with 0.7 wt% of Joncryl. This drastic decrease could be generated from a decrease of the mobility of the interface which slowed down the coalescence rate, related to the formation of the copolymer at the interface.

3.2. Linear Viscoelastic and Shear Capillary Flow Properties of Neat and Modified PLA, PBAT Polymers and Their Respective Blends

Shear rheological properties of the neat PLA, PBAT polymers and their stable blends (up to reaction completion) were measured and compared in terms of dynamic viscosity modulus *versus* angular frequency as shown in Figure 6.

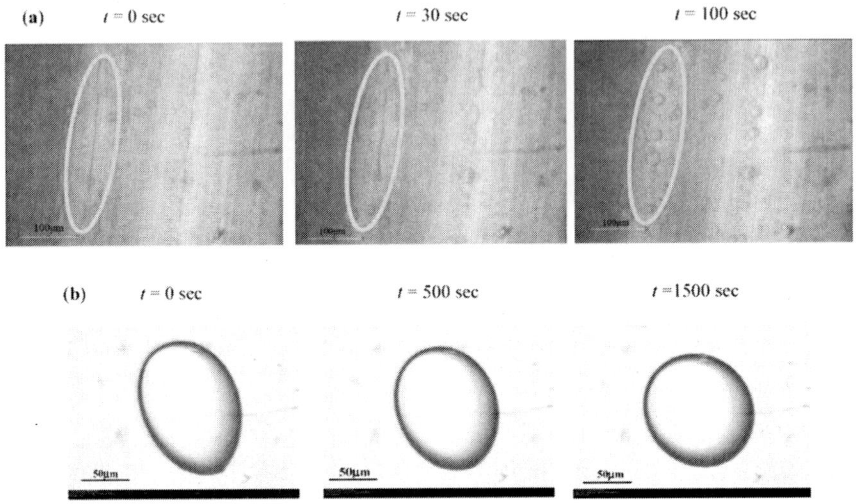

Figure 4. Examples of illustrations of the ellipsoid drop retraction, immersed in a fluid matrix at 180 °C of (**a**) uncompatibilized and (**b**) compatibilized PLA–PBAT blends. The measurement time (in second) is noted on each micrograph.

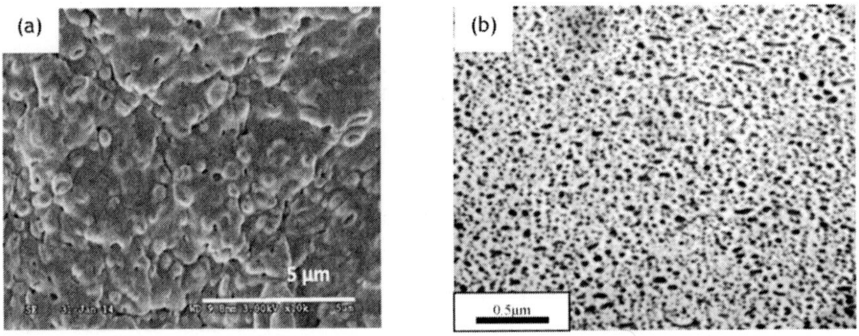

Figure 5. Examples of (**a**) SEM micrograph of PLA–PBAT (80–20) blends; Volume average diameter (D_v) = 0.25 µm and (**b**) TEM micrograph of PLA–PBAT–Joncryl (80–20–0.7); D_v = 0.07 µm.

From Figure 6a,b, a significant increase of the complex viscosity, η*, in the low frequency region, is firstly observed with the addition of Joncryl into both PLA and PBAT. This evolution can be analyzed by the chemical reaction which can happen between each polymer with the chain extender. In fact, since Joncryl has multiple reactive end groups, several polymer chains could be chemically bounded with one Joncryl molecule. In this way, a chain extended and branched structure could be developed [5]. Secondly, a pronounced shear thinning behavior was detected. An increase of the long chain branching amount, a broadening of the molar mass distribution and a shift of the chains relaxation time for both PLA and PBAT polymers to higher values were resulted, as already demonstrated in our previous paper [5]. The modified PLA and PBAT exhibited thus a typical feature of a mixture of linear and randomly branched polymers.

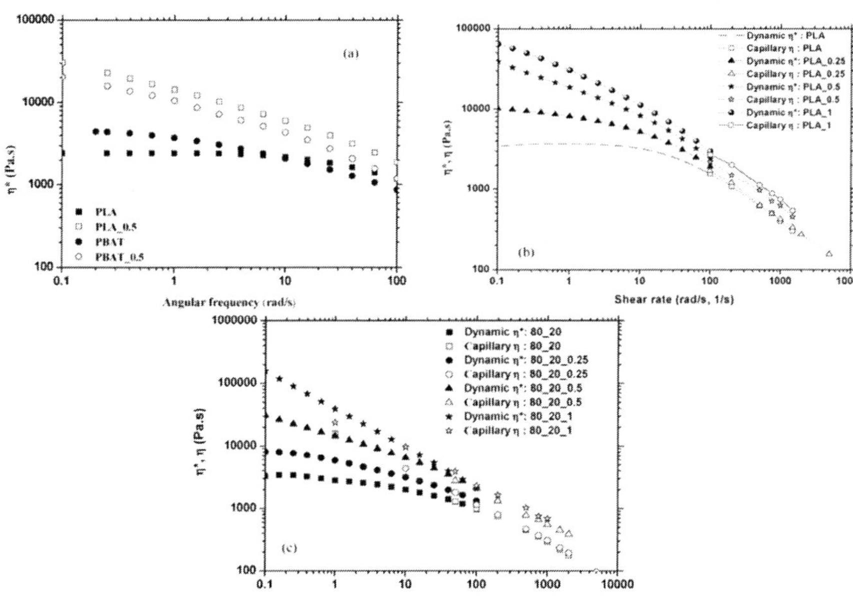

Figure 6. (**a**) Complex viscosity modulus *versus* angular frequency for neat and modified PLA and PBAT at 180 °C; (**b**) Validity of Cox–Merz rule: Dynamic viscosity modulus and shear viscosity *vs.* angular frequency (rad/s), shear rate (1/s) for the unmodified and modified PLA, PBAT polymers; (**c**) and for the uncompatibilized and compatibilized PLA–PBAT blends (Full symbol: Dynamic rheology/Open symbol: capillary rheology).

From Figure 6c, the results show that the shear viscosity and the shear-thinning tendency become stronger with increasing Joncryl content into blends. According to Figure 6b,c, the Cox–Merz rule works fairly well with neat polymers into the shear thinning region but still shows slightly deviation for the modified polymers and their blends. The polydispersity and the heterogeneity explain this deviation from the rule [12,13,14,15,16,17,18,19,20,21,22]. By this way, the elongation properties during processing can ben also influenced.

3.3. Elongation Properties of Neat and Modified PLA, PBAT Polymers and Their Respective Stable Blends

Our challenge is thus to make this process stable in order to obtain stable films at higher BUR/TUR couples. To achieve this objective; a proper correlation between melt shear and extensional rheological properties should be a suitable tool to get a better handle of this process. It is well accepted that the simple shear data cannot provide direct relations with blowing extrusion flows because they only reflect the linear viscoelastic part of the material behavior. For this reason; uniaxial elongation rheology is complementary to the shear data as they capture the non-linear aspects of the melt flow and are not very far from the actual material deformation.

3.3.1. Indirect Method of Determining the Elongation Viscosity—Cogswell Elongation Viscosity

Using the capillary rheological data of our different samples, the elongational viscosity η_E, was evaluated by the indirect Cogswell method. It is a simple rheological technique based on the entrance pressure drop method and based on the relationship between the elongational viscosity and the pressure at the die governed by the melt elasticity [21]. Figure 7depicts an example of the evolution of the η_E after Bagley correction for the modified PLA with three different amounts of Joncryl (up to 1% by weight) at a fixed elongation rate (30 s^{-1}). It can be seen that the η_e is increasing with the incorporation of Joncryl. The enhancement of melt strengthening of the modified PLA, as already discussed in detail [5], was thus confirmed by the drastic increase of the elongational viscosity.

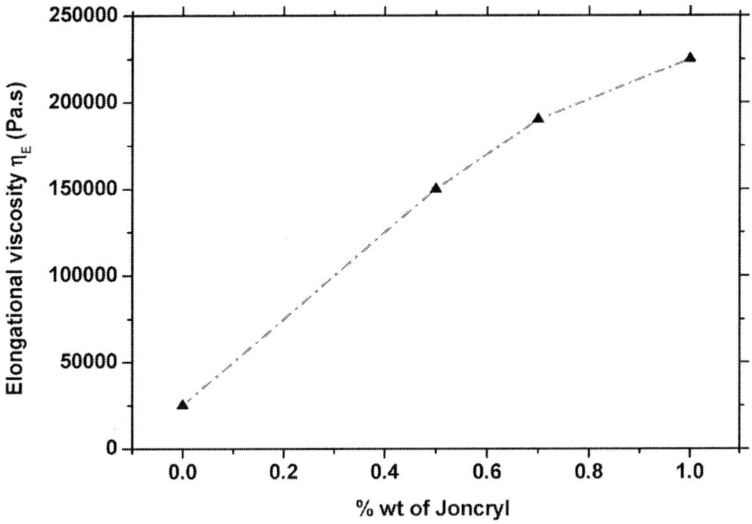

Figure 7. Example of the evolution of the elongational viscosity at 180 °C given by Cogswell method *versus* the Joncryl® amount of the modified PLA at 30 s^{-1}.

For PLA–PBAT blends, the elongational viscosity determined by Cogswell is increasing with the amount of Joncryl. This increase is related to the reaction between PLA, PBAT polymers and the multifunctional epoxide as presented in Figure 8. Therefore, an improvement of the blown film extrusion ability can be expected.

In the next section, the transient elongational viscosity of the samples will be quantified more accurately.

3.3.2. Uniaxial Extensional Experiments

Neat PLA and PBAT Polymers

Figure 9 shows double logarithmic plots of the transient elongation viscosity η_E *versus* time *t*, observed for both PLA and PBAT polymers at 180 °C. Uniaxial elongation experiments were conducted at various shear rates ranging from 0.1 to 20 s^{-1}. The extensional data are superposed well with the linear viscoelastic envelope (LVE) ($3\eta_0^+(t)$) at lower times. This indicates that the Trouton ratio ($T_R = \eta_e/\eta_0$) is equal to 3, the limiting case of

Newtonian polymer. For PLA, the absence of a deviation could be explained by the Brownian motion where macromolecules tend to return to their equilibrium configuration after being subjected to deformations. The PLA and PBAT polymers did not exhibit any strain-hardening up to 1 s^{-1}. This is the ordinary behavior of the linear polymers which is called strain-softening phenomena, as discussed by Liu *et al.* [14]. The presence of free volume within polymer chains could hinder the formation of entanglements through reduced contact surface and ineffective overlapping. Consequently, the linear chains are being free from any branch points and are not prevented from slipping over each other. The absence of strain hardening behavior of both PLA and PBAT could confirm that they are not favorable for blown film application. However, at high values of the Hencky strain rate (>5 s^{-1}), it was speculated that an apparent strain hardening was observed. The Trouton rule does not hold; at longer times, the η_e^+deviates from the linear viscoelastic behavior. The observed increase could be due to shorter overlapping time of macromolecules during elongational flow. Thus, the occurrence of the strain hardening behavior of linear PLA has been reported in few studies and might be due to its stereo-regularity [15,16,17,18,19,20,21,22,23,24,25,26,27]. In fact, the stereo-regularity of high L-content PLA (98% of L-content in our case) may allow them to form a range of structures when stretched, where helical conformations may form in the melt as a precursor to strain-induced crystallization. This behavior was not observed in our previous work with other grade of PLA.

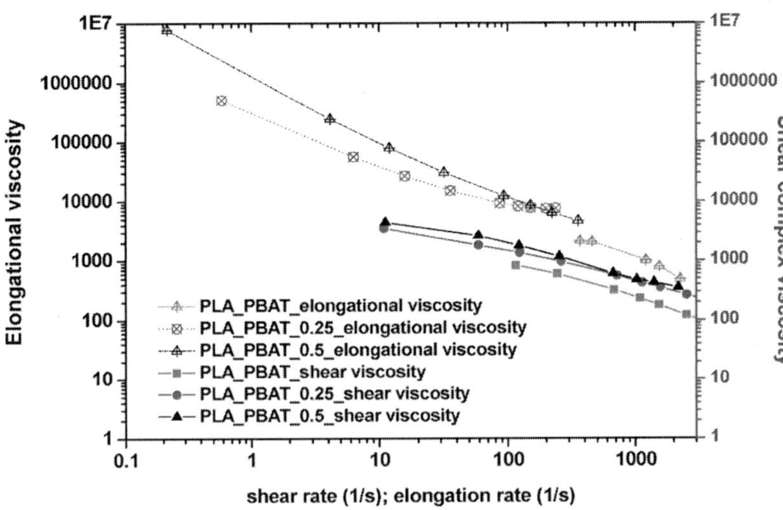

Figure 8. Elongational and shear viscosities *versus* elongation (and shear) rate for the non compatibilized and compatibilized PLA–PBAT blend at 180 °C.

For PBAT, care should be taken when analyzing the results since (i) it is not strong enough to be tested at 180 °C and (ii) Its higher sagging during experiments. The higher molar masses and the presence of rigid chains with phenyl component could explain the apparent strain hardening observed at high strain rate.

Modified PLA System

Figure 10 displays the transient elongational viscosities for the branched PLA. For the clarity purpose, only the results with 0.7 wt% of Joncryl will be shown thereafter. In all cases, the extensional viscosities raise above the linear viscoelastic envelope. A distinguished strain hardening is observed even at low strain rates whereas for linear PLA no strain hardening has been reported. It increases with increasing strain rate; in the early stage, η_e gradually increases with time but is almost independent of the elongation rates. After a certain time *t*, elongation viscosity shows a rapid upward deviation (which correspond to higher Trouton ratio) due to the resistance caused by the branches intermingled with the main chains of the polymer. Thus, the modified polymer resisted up to the elongational deformation without failure for the complete test durations. Considering the wide polydispersity for the modified PLA, this gives rise to the assumption that the effect of the branching structure on the strain hardening dominates [28,29]. Depending on the number of branches and molecular weight, entanglement density may be increased or decreased by branching. In this study, based on the data presented in the paper and the observed positive deviation, entanglement density was increased. Moreover, the strain hardening is not only accorded to the presence of branch which they reduce the rate of chain disentanglement. Indeed, some gripping forces can happen depending of branch density and length. The literature suggests that the strain-hardening is exhibited when the rate of the deformation considerably exceeds the rate of molecular relaxation [30]. The longest relaxation time was obtained by analyzing the linear viscoelastic data in terms of its discrete relaxation spectrum, as presented in our previous paper [11]. As the longest relaxation time of a modified PLA is larger than the solicitation time, strain hardening effects are expected.

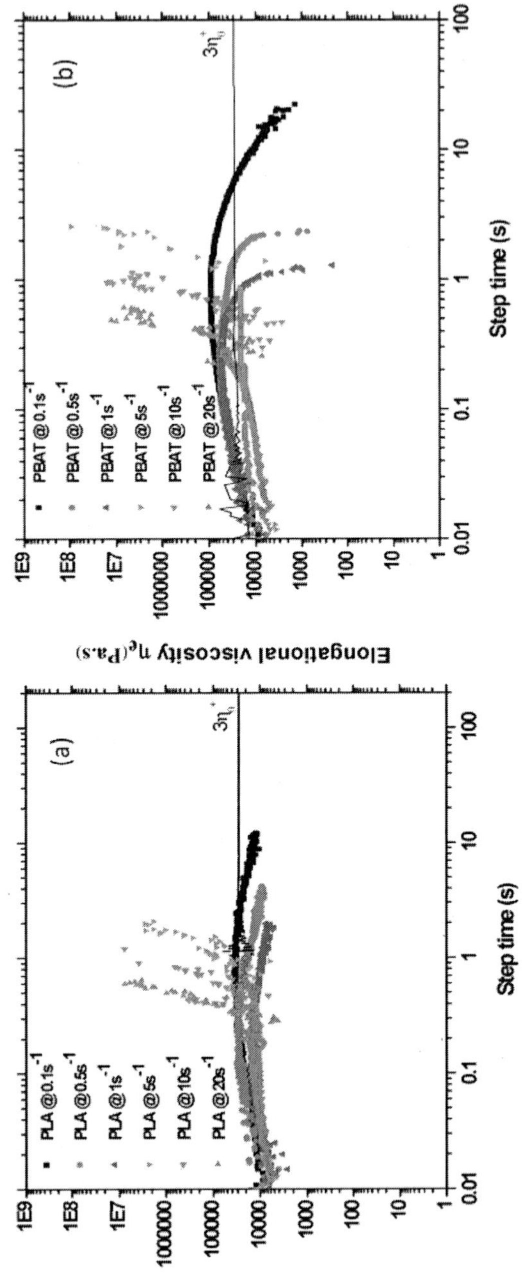

Figure 9. Start-up elongation viscosity *versus* step time at 180°C for the neat (**a**) PLA polymers and (**b**) PBAT polymers at different elongation rates.

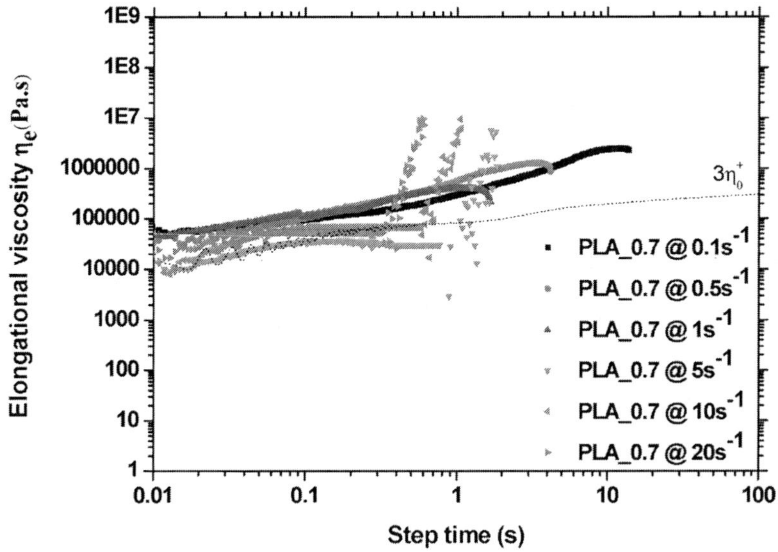

Figure 10. Start-up elongation viscosity *versus* step time for the modified PLA with 0.7 wt% of Joncryl at different elongation rates at 180 °C.

PLA-PBAT Blends

Elongational flow properties of PLA/PBAT 80–20 blends are given in Figure 11. The results illustrate that their behavior is more dominated by the PLA matrix. At a strain rate less than 1 s^{-1}, the extensional viscosity of the uncompatibilized blends tends to hold the Trouton rule. The envelope of compatibilized PLA–PBAT blends is over 10,000 Pa·s in comparison of non-compatibilized systems. Therefore, an increase of the elongational viscosity was observed for the whole range of elongation rates for the compatibilized blends. The indications are that the stronger the interface, the highest the elongational deformation. At high elongation rates, the strain hardening was expected as is the case of neat polymers. It could be attributed to the occurrence of trans-esterification reactions in the uncompatibilized PLA–PBAT blend and to the strong interfacial adhesion in the compatibilized PLA–PBAT–Joncryl blend [9]. Like Cogswell indirect results (Figure 8), the addition of Joncryl improve the elongational behavior of the blends.

According to the transient elongational behavior of the uncompatibilized and compatibilized PLA–PBAT blends presented in the previous section, the measured engineering stress for each PLA and its blends is firstly plotted in Figure 12according to Equation (4):

$$\sigma_{engineering} = \frac{F_{elongation}}{S} = \eta_{elongation} \times \dot{\varepsilon} \quad (4)$$

where $F_{elongation}$ is the elongational force and S is the section of the specimen.

It was clearly observed that the neat PLA and its uncompatibilized blend break at complete sample failure at 0.5 Hencky. Hence, a pronounced decrease of the engineering stress is observed.

However, the compatibilized blend breaks at around 2 of strain Hencky deformation amplitude. Note that it takes more stress, or more elongational force, to extend them at a constant strain rate, for which the interfacial adhesion between matrix and dispersed phase is strong. For instance, at 0.5 s^{-1}, the force is 10 times higher than for the uncompatibilized counterpart.

3.4. Application to Blown Film Extrusion: Investigation of the Stability Processing of Neat/Modified Polymers and Their Blends

3.4.1. Stability Processing of Neat PLA and PBAT Polymers

The instability defects in blown film were firstly reported by Han and co-workers [23,24]. They observed that depending on the processing parameters and the rheological properties of materials, different instabilities can appeared, *i.e.*, lowering the extrusion temperature stabilized the blown bubble for high density polyethylene (HDPE) and low density polyethylene (LDPE). Some other authors have more extensively discussed the instabilities during blown film processing [25,26].

In our case, the study of the stability of the process was conducted by establishing the stability map (BUR *vs.* TUR). Thereby, these two main parameters were changed and bubble stability or instability was observed. The BUR(s) values have changed according to air flow inflated into the die. The TUR(s) have changed based on the variation of the chill roll velocity and the melt flow was kept constant. The process was considered as stable since no defect appeared for a given BUR–TUR couple during at least 10 min.

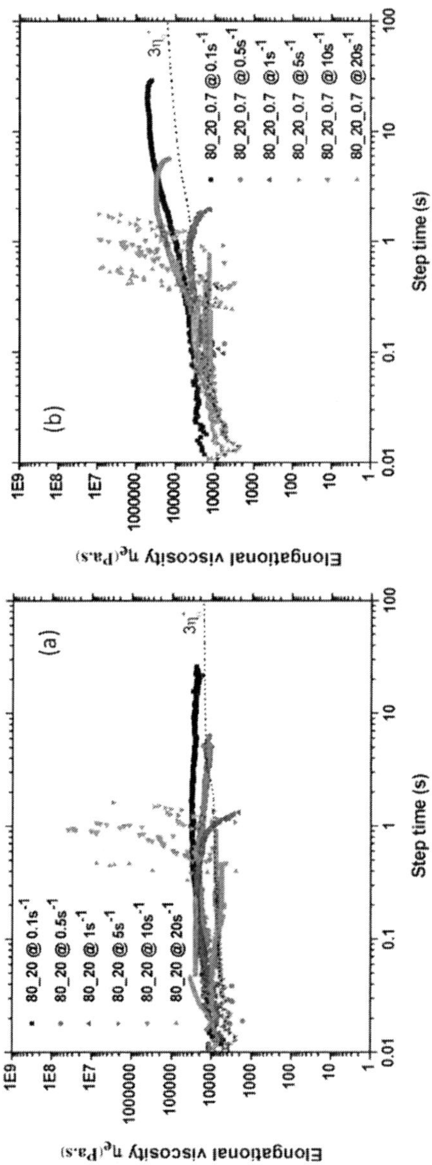

Figure 11. Elongation viscosity *versus* step time (*t*) 180 °C for the (**a**) uncompatibilized and (**b**) compatibilized PLA–PBAT blends at different elongation rates.

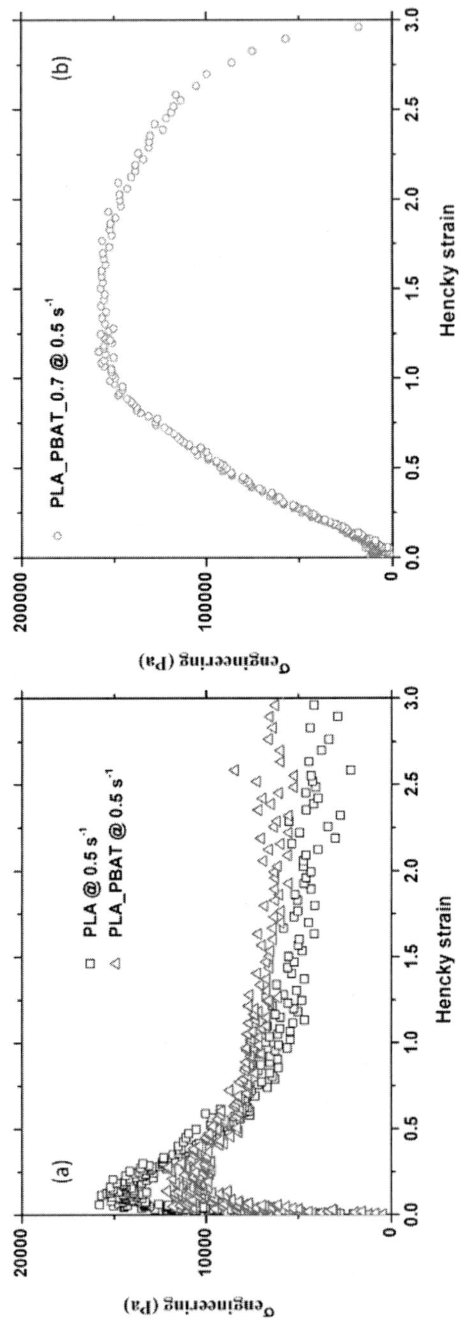

Figure 12. Example of engineering stress evolution at 0.5 s^{-1} at 180 °C for (**a**) PLA and its uncompatibilized blend and (**b**) its compatibilized blend.

PLA shows a limited blowing ability according to its stability map plotted in Figure 13. Any stable blown film has been obtained neither at lower BUR–TUR couple nor at higher ones. The map presents unstable points related to different instabilities including (i) draw resonance; (ii) helical instability; (iii) frost line oscillation; (iv) breathing; (v) dancing and; (vi) bubble sag. Moreover, PBAT has no ability to be blown. Its stability map cannot be plotted due to sagging problem. The poor shear and elongation properties of PLA and PBAT could explain their inadaptability to the blown film process.

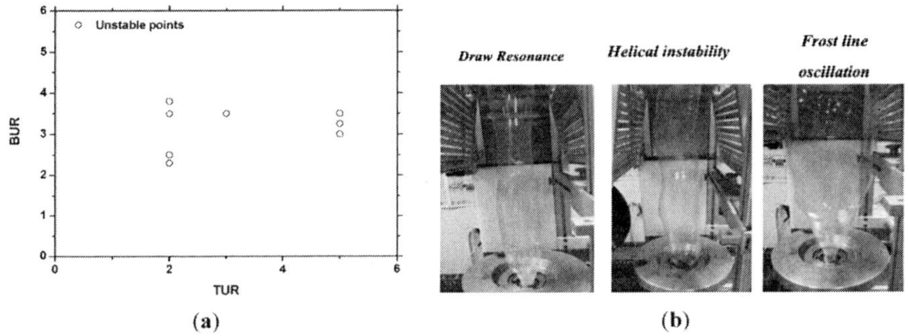

Figure 13. Stability map (BUR *versus* TUR) of neat PLA at 180°C (**a**). Various instabilities observed (**b**).

3.4.2. Stability Processing of Chain Extended/Branched PLA Polymers

We have shown that the structural modification of PLA (PLA with Joncryl) leads to an increase in elongational viscosity by 6 orders of magnitude at 0.1 s. This is the property that mostly determines the bubble stability and ease of processing as presented in Figure 14. The Blow-Up Ratio (BUR on *y* axis) values rose from 2.5 to 7. Large bubbles were formed. Stable bubbles were obtained at high Take-Up Ratio (TUR on *x* axis) of 6. The occurred instabilities are less chaotic compared to pure PLA.

3.4.3. Stability Processing of PLA–PBAT Blends

The stability map of the uncompatibilized blend is established and presented in Figure 15. The results show that the incorporation of PBAT into PLA matrix leads to an enlargement of the stability domain. BUR raises to 3.5. Three types of defects have been detected which are breathing, draw resonance and dancing. But, breathing instability remains the main occurred defect. It is important to note that the PLA–PBAT blends present a more stable region for higher TUR compared to the modified PLA polymers. Self-compatibilization behavior between PLA and PBAT according to some trans-esterification reaction can occur. Indeed, a bit increase of its elongation properties was noted in the previous section and could explain a stable blown behavior. However, it retracts when the biaxial stress is suppressed. Therefore, the obtained bubbles are not very large and BUR is lower compared to the modified PLA (PLA–Joncryl) film which reach 7 but at not at a higher TUR.

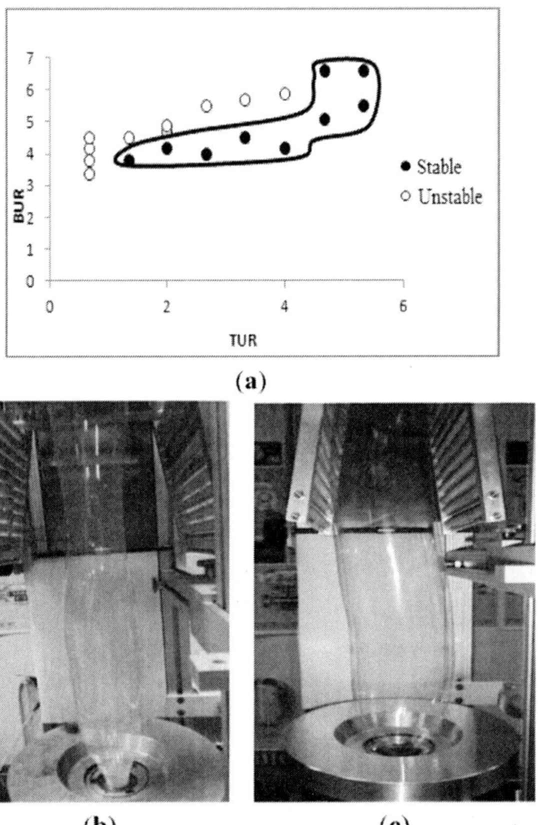

Figure 14. (**a**) Stability maps (BUR *versus* TUR) of modified PLA with 0.7 wt% of Joncryl. Comparison of the bubble shape between (**b**) PLA and (**c**) PLA–0.7; Unlike PLA, large and stable bubbles were obtained with the incorporation of Joncryl.

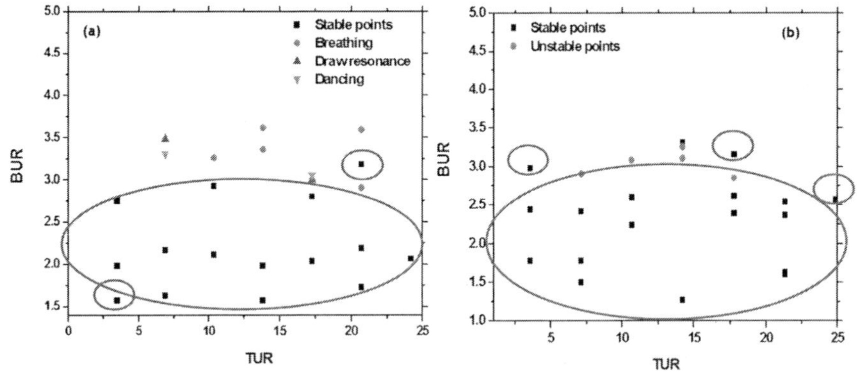

Figure 15. Stability maps of the (**a**) uncompatibilized PLA-PBAT blends and (**b**) compatibilized PLA–PBAT blends with 0.7 wt% of Joncryl.

With the incorporation of Joncryl ADR®, the breathing instability zone is reduced and the draw resonance instability is eliminated (Figure 16).

(a) (b) (c)

Figure 16. Comparison of the bubble shape between (**a**) PLA; (**b**) PLA–PBAT and (**c**) PLA–PBAT–0.7.

Once can conclude that Joncryl helps process stabilization at lower and higher TUR which ensure an achievement of high BUR values. Meanwhile, the addition of Joncryl leads to an increase in engineering stress and elongational viscosity according to the stronger interfacial cohesion between matrix and dispersed phase. This is the main difference with the uncompatibilized systems. Moreover, a new zone of stable bubble can be obtained with higher TUR values (25 of amplitude) and BUR reaching 3.5. The compatibilization effects of the Joncryl play a real role to eliminating dancing and draw resonance defects according to the induced higher elongation properties at great strain Hencky deformation amplitude.

3.5. Thermo-Mechanical and Crystalline Properties of the Blown Films

Table 2 includes the thermal transitions and crystalline properties of our studied systems. It is worthy concluded that the blowing extrusion altered the crystalline properties of neat and modified PLA due to the orientation of polymer chains upon biaxial strain. The crystallinity level increases from 4% to 6.5% and from 5.5% to 13% after blowing for neat and modified PLA, respectively. A depression of the cold crystallization temperature, T_{cc} is also observed. It should be pointed out that the glass transition temperature, occurring within the 60–65 °C range, is nearly unaffected by the processing.

The incorporation of PBAT leads to a significant increase of the crystallinity of the PLA–PBAT blown film, explaining its ability to be blown and to be crystallizing. However, with the addition of Joncryl, the crystallinity level increases slightly. The development of crystalline structure in modified PLA–PBAT blends is effectively hindered. This can be related to the higher branching density, as demonstrated for PET [31], despite it exhibits a slightly increase of the cold crystallization compared to the unmodified one.

Representative curves of DMTA recording tanδ and log*E*' for the neat and modified blown PLA films are shown in Figure 17. The solid-state rheological properties of PLA typically show in its glassy state, a storage modulus (*E*') of about 3×10^9 Pa. As the sample is heated above T_α, a sharp drop in modulus occurs between 55 and 75 °C and *E*' decreases to 5×10^6 Pa by softening the film. This region is accompanied by a maximum peak of tanδ of about 65 °C, which corresponds to the α-relaxation of PLA. The obtained data regarding the relaxation α of PLA and modified PLA corroborate those in the literature [32,33].

Table 2. Thermal parameters and crystallinity values of unblown and blown systems.

Films	T_g (°C)	T_{cc} (°C)	ΔH_{cc} (J/g)	T_m (°C)	ΔH_m (J/g)	X_c (%)
PLA before blowing	60	115	24	170	25	4
PLA–0.7 before blowing	59	111	33	169	38	5.5
80–20 before blowing	−31/62	98	20.5	169	28	11
80–20–0.7 before blowing	−30/62	108	20	167.5	25	7
PLA after blowing	61.5	110	27	170	33	6.5
PLA–0.7 after blowing	63	98	28	170	40	13
80–20 after blowing	−29/64	103	11	171	29	24
80–20–0.7 after blowing	−29/63	95	13	169	19	9

The storage modulus then increases at a temperature around 83–85 °C. This recovery is attributed to the cold crystallization of the PLA chains, being higher with increasing the crystallinity of PLA when comparing with DSC thermograms.

From Figure 17a, we can conclude that the curve of PLA with Joncyl almost has the same shape as the neat PLA below the melting temperature. An increase of the storage modulus (*E*') below the α-relaxation is observed. A strain-induced crystallization may had occurred upon biaxial blowing where Joncryl acts as precursor of new nucleation sites, enhancing the growth of the crystalline phase. Moreover, the tanδ curves is nearly unaffected by the structural changes induced upon reaction with the chain extender indicating that the Joncryl ADR® did not act as a plasticizer agent.

The thermo-mechanical behavior of the modified and unmodified PLA–PBAT blends showed that the high storage modulus is observed for the PLA–PBAT blend compared to its modified counterpart. This is probably to the high obtained crystallinity upon blowing as shown in Table 2. In both cases, an increase of the storage modulus is observed at around 80 °C, highlighting the cold crystallization behavior of PLA. Such evolution indicates that the prepared blown films are semi-crystalline, as also confirmed by DSC data.

At low temperature relaxation (−27 °C) and for low contents of PBAT, the PBAT α-relaxation is not clearly visible. Moreover, the reactive agent (Joncryl) did not affect the tanδ of PLA–PBAT blends. The loss factor peak of blends (tanδ = 2.2) showed a decreased magnitude of tanδ in comparison to virgin PLA (tanδ = 3.7). This behavior is mainly due to the rubbery phase of PBAT which is well dispersed in PLA matrix.

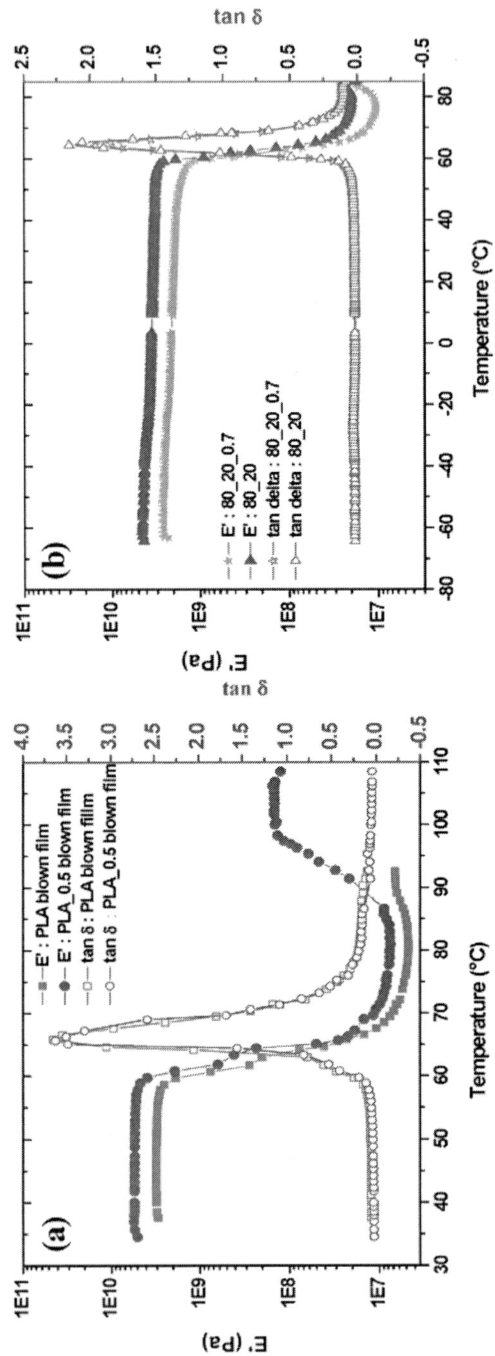

Figure 17. Storage modulus and loss factor evolution *versus* temperature for (**a**) PLA and PLA–0.7 and (**b**) PLA–PBAT blends.

4. CONCLUSIONS

Two main routes has been investigated in the present work: (i) a structural modification through chain extension and branching mechanisms by adding a reactive multifunctional epoxide (Joncryl) and (ii) blending with PBAT in presence (or nor) of this reactive agent. Through this paper, structure–processing–properties relationships have been addressed. Several conclusions can be derived.

On the one hand, both PLA and PBAT are not able to be blown at lower Hencky strain rate (>1 s^{-1}) since their respective elongational properties are weak. For PBAT, care should be taken when analyzing the results since it is not strong enough to be tested at 180 °C.

On the other hand, the change of the molecular structure of PLA by incorporating 0.7 wt% of Joncryl resulted in significant shear and elongational behavior (higher shear and elongation viscosity, presence of strain hardening behavior even at lower strain rates). Consequently, its ability to blowing was enhanced. Stables bubbles with higher BUR and TUR were obtained.

For PLA–PBAT blends, the effect of Joncryl on the shear and elongation rheology, morphological, and interfacial properties of the blends was investigated. The role of Joncryl as a compatibilizer was highlighted thanks to scanning and transmission electron microscopy (SEM and TEM) observations and the DDRM.

For the first time, the elongational measurements of the non-compatibilized and compatibilized PLA–PBAT blends were assessed. The results showed that the elongational behavior has been governed by the matrix. A strain-hardening was thus expected. In all, the incorporation of PBAT into PLA matrix leads to an enlargement of the stability domain.

The incorporation of PBAT into PLA matrix leads to an enlargement of the stability domain according to some self-compatibilization behavior (trans-esterification reaction can occur).

Furthermore, the addition of Joncryl leads to an increase in engineering stress and elongational viscosity according to the stronger interfacial cohesion between matrix and dispersed phase. A reduction of the instability zone and a new zone of stability was obtained. The draw resonance and the dancing defects are totally eliminated. Moreover, higher values of BUR have been reached although TUR is quite higher.

Finally, it has been demonstrated that the blown film extrusion process leads to semi-crystalline films with improved thermo-mechanical properties.

ACKNOWLEDGMENTS

The authors gratefully acknowledge the 'DGCIS' (Direction Générale de la Compétitivité de l'Industrie et des Services) for the Financial support. The authors also express their appreciation to the reviewers for their meticulous review and assessment of this work.

AUTHOR CONTRIBUTIONS

The authors contributed equally to this work. They all participated to the writing of the present manuscript. Racha Al-Itry performed the overall experimental work as a PhD Student in INSA de Lyon, France. The settings up of the experimental protocols as well as the interpretation of the obtained results were performed under the supervision of khalid Lamnawar and Abderrahim Maazouz.

REFERENCES

1. Kanai, T.; Campbell, G.A. *Film Processing*; Carl Hanser Verlag: Munich, Germany, 1999.
2. Lim, L.T.; Auras, R.; Rubino, M. Processing technologies for poly (lactic acid). *Progr. Polym. Sci.* **2008**, *33*, 820–852.
3. Jung, H.W.; Hyun, J.C. Instabilities in extensional deformation polymer processing. *Rheol. Rev.* **2006**, *2006*, 131–164.
4. Arraiza, A.L.; Sarasua, J.R.; Verdu, J.; Colin, X. Rheological behavior and modeling of thermal degradation of poly(ε-caprolactone) and poly(L-lactide). *Int. Polym. Process. XXII* **2007**, *22*, 389–394.
5. Al-Itry, R.; Lamnawar, K.; Maazouz, A. Improvement of thermal stability, rheological and mechanical properties of PLA, PBAT and their

blends by reactive extrusion with functionalized epoxy. *Polym. Degrad. Stab.* **2012**, *97*, 1898–1914.

6. Baird, D.J. The role of extensional rheology in polymer processing. *Korea-Aust. Rheol. J.* **1999**, *11*, 305–311.

7. Mihai, M.; Huneault, M.; Favis, B.D. Rheology and extrusion foaming of chain-branched poly(lactic acid). *Polym. Eng. Sci.* **2010**, *50*, 629–642.

8. Japon, S.; Luciani, A.; Nguyen, Q.T.; Leterrier, Y.; Manson, J.A.E. Molecular characterization and rheological properties of modified poly(ethylene terephthalate) obtained by reactive extrusion. *Polym. Eng. Sci.* **2001**, *41*, 1299–1309.

9. Al-Itry, R.; Lamnawar, K.; Maazouz, A. Rheological, morphological, and interfacial properties of compatibilized PLA/PBAT blends. *Rheol. Acta* **2014**, *53*, 501–517.

10. La Mantia, F.P.; Valenza, A.; Scargiali, F. Non isothermal elongational behavior of blends with liquid crystalline polymers. *Polym. Eng. Sci.* **1994**, *34*, 799–803.

11. Al-Itry, R.; Lamnawar, K.; Maazouz, A. Reactive extrusion of PLA, PBAT with a multi-functional epoxide: Physico-chemical and rheological properties. *Eur. Polym. J.* **2014**, *58*, 90–102.

12. Corre, Y.M.; Duchet, J.; Maazouz, A.; Reignier, J. Melt strengthening of poly (lactic acid) through reactive extrusion with epoxy-functionalized chains. *Rheol. Acta* **2011**, *50*, 612–629.

13. Mallet, B.; Lamnawar, K.; Maazouz, A. Improvement of blown film extrusion, of poly (lactic acid): Structure-processing properties relationships. *Polym. Eng. Sci.* **2014**, *54*, 840–857.

14. Liu, J.; Lou, L.; Yu, W.; Liao, R.; Li, R.; Zhou, C. Long chain branching polylactide: Structures and properties. *Polymer***2010**, *51*, 5186–5197.

15. Palade, L.L.; Lehermeier, H.J.; Dorgan, J.R. Melt rheology of high L-content poly(lactic sacid). *Macromolecules* **2001**, *34*, 1384–1390.

16. Corre, Y.M.; Maazouz, A.; Duchet, J.; Reignier, J. Batch foaming of chain extended PLA with supercritical CO_2: Influence of the rheological properties and the process parameters on the cellular structure. *J. Supercrit. Fluids* **2001**,*58*, 177–188.

17. Zhu, W.L.; Wang, J.; Park, C.B.; Pop-Iliev, R.; Randall, J. Effects of chain branching on the foamability of polylactide. In Proceedings of 67th Annual Technical Conference of the Society of Plastics Engineers (SPE-ANTEC), Chicago, IL, USA, 22–24 June 2009; Volume 1, pp. 41–45.

18. Eslami, H.; Kamal, M.R. Effect of a chain extender on the rheological and mechanical properties of biodegradable poly(lactic acid)/ poly(butylene succinate-*co*-adipate) blends. *J. Appl. Polym. Sci.* **2013**, *129*, 2418–2428.

19. Kim, Y.M.; Park, J.K. Effect of short chain branching on the blown film properties of linear low density polyethylene. *J. Appl. Polym. Sci.* **1996**, *61*, 2315–2324.

20. Majumder, K.K.; Hobbs, G.; Bhattacharya, S.N. Molecular, rheological, and crystalline properties of low-density polyethylene in blown film extrusion. *Polym. Eng. Sci.* **2007**, *47*, 1983–1991.

21. Cogswell, F.N. Converging flow of polymer melts in extrusion dies. *Polym. Eng. Sci.* **1972**, *12*, 64–73.

22. Lehermeier, H.J.; Dorgan, J.R. Melt rheology of poly (lactic acid): Consequences of blending chain architectures.*Polym. Eng. Sci.* **2001**, *41*, 2172–2184.

23. Han, C.D.; Park, J.Y. Studies on blown film extrusion. III. Bubble instability. *J. Appl. Polym. Sci.* **1975**, *19*, 3291–3297.

24. Han, C.D.; Shetty, R. Flow instability in tubular film blowing. 1. Experimental study. *Ind. Eng. Chem. Fundam.* **1977**, *16*, 49–56.

25. Kanai, T.; White, J.L. Kinematics, dynamics and stability of tubular film extrusion of various polyethylenes. *Polym. Eng. Sci.* **1984**, *24*, 1185–1201.

26. Minoshima, W.; White, J.L. Instability phenomena in tubular film and melt spinning of rheological characterized high density, low density and linear low density polyethylenes. *J. Non-Newton. Fluid Mech.* **1986**, *19*, 275–302.

27. Yamane, H.; Sasai, K.; Takano, M. Poly(D-lactic acid) as a rheological modifier of poly(L-lactic acid): Shear and biaxial extensional flow behavior. *J. Rheol.* **2004**, *48*, 599–609.

28. Mekhilef, N.; Hedhli, L.; Moyses, S. Effect of rheological strain hardening on extrusion blown film of polyvinylidene fluoride. *J. Plast. Film Sheeting* **2007**, *23*, 203–219.

29. Stange, J.; Uhl, C.; Münstedt, H. Rheological behavior of blends from a linear and a long-chain branched polypropylene. *J. Rheol.* **2005**, *49*, 1059–1080

30. Kasehagen, L.J.; Macosko, C.W. Nonlinear shear and extensional rheology of long-chain randomly branched polybutadiene. *J. Rheol.* **1998**, *42*, 1303–1327.

31. Raffa, P.; Coltelli, M.B.; Savi, S.; Bianch, S.; Castelvetro, V. Chain extension and branching of poly(ethylene terephthalate) (PET) with di- and multifunctional epoxy or isocyanate additives: An experimental and modelling study. *React. Funct. Polym.* **2012**, *72*, 50–60.

32. Henton, D.E.; Gruber, P.; Lunt, J.; Randall, J. Polylactic acid technology. In *Natural Fibers, Biopolymers, and Biocomposites*; Mohanty, A.K., Misra, M., Drzal, L.T., Eds.; Taylor & Francis: Boca Raton, FL, USA, 2005; pp. 527–577.

33. Huda, M.S.; Yasui, M.; Mohri, N.; Fujimura, T.; Kimura, Y. Dynamic mechanical properties of solution-cast poly(L-lactide) films. *Mater. Sci. Eng. A* **2002**, *333*, 98–105.

Chapter 6

Engineering Behavior and Characteristics of Water-Soluble Polymers: Implication on Soil Remediation and Enhanced Oil Recovery

Shuang Cindy Cao [1,†], Bate Bate [2,†], Jong Wan Hu [3,4,†] and Jongwon Jung [5,*]

[1]Department of Civil and Environmental Engineering, Louisiana State University, Baton Rouge, LA 70803, USA

[2]Department of Civil, Architectural and Environmental Engineering, Missouri University of Science and Technology, Rolla, MO 65409, USA

[3]Department of Civil and Environmental Engineering, Incheon National University, Incheon 22012, Korea

[4]Incheon Disaster Prevention Research Center, Incheon National University, Incheon 22012, Korea

[5]Department of Civil and Environmental Engineering, Louisiana State University, Baton Rouge, LA 70803, USA

ABSTRACT

Biopolymers have shown a great effect in enhanced oil recovery because of the improvement of water-flood performance by mobility control, as well as having been considered for oil contaminated-soil remediation thanks to their mobility control and water-flood performance. This study focused on the wettability analysis of biopolymers such as chitosan (85% deacetylated power), PEO (polyethylene oxide), Xanthan (xanthan gum), SA (Alginic Acid Sodium Salt), and PAA (polyacrylic acid), including the measurements of contact angles, interfacial tension, and viscosity. Furthermore, a micromodel study was conducted to explore pore-scale displacement phenomena during biopolymer injection into the pores. The contact angles of biopolymer solutions are higher on silica surfaces submerged in decane than at atmospheric conditions. While interfacial tensions of the biopolymer solutions have a relatively small range of 25 to 39 mN/m, the viscosities of biopolymer solutions have a wide range, 0.002 to 0.4 Pa·s, that dramatically affect both the capillary number and viscosity number. Both contact angles and interfacial tension have effects on the capillary entry pressure that increases along with an applied effective stress by overburden pressure in sediments. Additionally, a high injection rate of biopolymer solutions into the pores illustrates a high level of displacement ratio. Thus, oil-contaminated soil remediation and enhanced oil recovery should be operated in cost-efficient ways considering the injection rates and capillary entry pressure.

KEYWORDS

biopolymer; enhanced oil recovery; contact angle; interfacial tension; micromodel; capillary pressure

1. INTRODUCTION

Recently, organic agents such as polymers, biopolymers, and surfactants have been developed for soil improvement and have demonstrated their abilities to improve the shear strength, stiffness, soil remediation, and erosion resistance of geomaterials [1,2,3,4,5,6,7,8,9]. Furthermore, biopolymers, such as polyacrylamide (PAM) and xanthan gum, have shown a great effect in enhanced oil recovery (EOR) thanks to its mobility control in EOR that improves water-flood performance [10,11,12]. Thus, water flooding by biopolymers and polymers has become a superior method in enhancing the oil recovery according to laboratory and field tests [13,14,15]. Additionally, biopolymers have been considered for oil-contaminated soil remediation thanks to their mobility control and water flood performance [16,17,18,19]. However, engineering behavior and characteristic of biopolymers have yet to be well recognized. Thus, in order to improve the capacity of biopolymers for oil-contaminated soil remediation and enhanced oil recovery (EOR), we explored the wettability of biopolymers, including contact angle, surface tension, interfacial tension, and viscosity that influences on capillary pressure, biopolymer solution flow, and biopolymer solutions–oil displacement in porous media. Additionally, the micromodel study was intended to understand the flow behavior of biopolymer solutions and biopolymer solutions–oil displacement in porous media.

2. LITERATURE REVIEW

2.1. Utilization of Biopolymer in Soil Remediation and Enhanced Oil Recovery

Soil contamination in urban and rural environments is usually caused by industrial activities, such as mining wastes, dumping and landfill settlement, or quarries [20]. Important characteristics and remediation techniques have been developed over the years to deal with soil contamination [21]. Various technologies are under development for remediation of soils and sediments [22]. For instance, soil remediation technologies, such as soil excavation, soil vapor extraction, bioremediation, and steam injection are basic methods used for soil remediation for years [23,24,25,26]. Due to the limitation in soil excavation, including high cost and lack of available landfill sites [22,23,24,25,26,27], flushing and bioremediation methods that do not require the soil excavation have become more popular recently [22]. For example, the surfactant flush has been used to improve the rate of soil remediation [22]. Recently, more suitable eco-friendly soil remediation methods have been required. Thus, application of bio-surfactant and biopolymers could provide the sustainability because they are mainly obtained from plants containing eco-friendly properties [28,29,30,31]. Biopolymer flushing was originally developed for petroleum recovery areas which, afterwards, had been used for the remediation of petroleum waste-contaminated sites [32,33,34,35,36,37,38]. Biopolymers are biologically-available compounds that could be obtained using the basic principles of lowering the mobility ratio and increasing the capillary number of fluids [39,40]. They aim to increase the viscosity and the mobility and to reduce the interfacial tension to enhance contamination recovery and accelerate contamination mobilization [41].

A large portion of the world's oil reserves still have low permeability. Thus, enhanced oil recovery (EOR) technology has been used for exploitation [42]. Polymer solutions have been used to improve the water-flood performance in enhanced oil recovery (EOR) for 70 years [13]. In general, two types of polymers could be applied to reservoir application, which are synthetic polymers and biopolymers [43,44,45]. While synthetic polymers are not stable at high temperatures and high salinity, biopolymers are more stable, relatively [43]. The basic principles of biopolymer flooding are to increase the viscous forces, control mobility, as well as to reduce the interfacial tension in the reservoir for enhanced oil recovery [39,40,41]. Considering their lower cost and higher viscosity, the most widely-used biopolymers for enhanced oil recovery (EOR) application are xanthan gum [10,12], pullulan [46], levan [47], curdlan [48], dextran [49], and scleroglucan [50]. However, better understanding of wettability of biopolymers, including contact angle, surface tension, interfacial tension, and viscosity, are required for both oil-contaminated soil remediation and enhanced oil recovery (EOR).

2.2. Capillary Pressure

Wettability of mineral and interfacial tension between two fluids are key factors controlling the mobility of biopolymer solution, biopolymer solution–oil displacement, and capillary pressure in porous media. Due to the differences in mineral surface wettability by biopolymer solution and oil, a capillary pressure difference (P_c) between these two phases must exceed in order for biopolymer solution to enter the pores in the porous media. As described by the Young–Laplace equation (Equation (1)), capillary entry pressure of biopolymer solution into the original oil-filled pores is determined by the pore radius (R), the biopolymer solution-oil interfacial tension (σ), and the contact angle (ϑ) among the biopolymer solution–oil–mineral surface [51]:

$$P_c = P_{biopolymer} - P_{oil} = \frac{2\sigma \cos\theta}{R} \quad (1)$$

where, the values of pore radius (R) are quite fixed depending on in-site geological conditions. However, both the biopolymer solution–oil interfacial tension (σ) and the contact angle (ϑ) among the biopolymer solution–oil–mineral surface have not been well identified. Therefore, the biopolymer solution–oil interfacial tension (σ) and contact angle (ϑ) cause the greatest uncertainty to the predictions of biopolymer solution mobility, saturation, and capillary entry pressure.

2.3. Multiphase Fluids Flow in Porous Media

In multiphase fluids flow, displacement ratios in porous media is determined by two dimensionless numbers [52,53]:

$$N_m = \frac{\mu_{inv}}{\mu_{def}} \quad (2)$$

$$N_c = \frac{\mu_{inv} v}{\sigma \cos\theta} \quad (3)$$

where, viscosity number (N_m) is defined as the ratio of the injected fluid viscosity (μ_{inv}) and the defensed fluid viscosity (μ_{def}); in addition, capillary number (N_c) represents the ratio of viscous force over capillary force and is associated with the injected fluid velocity (v_{inv}), injected fluid viscosity (μ_{inv}), the contact angle on mineral surface (ϑ), and the interfacial tension between injected- and defensed- fluids (σ). These two dimensionless numbers govern three dominant regions with distinct displacement patterns and efficiencies (Figure 1): capillary fingering, viscous fingering, and stable displacement [54,55,56].

3. EXPERIMENTAL METHODS AND RESULTS

3.1. Material

Five different biopolymers were selected in this study that are popular in soil remediation and enhanced oil recovery: chitosan (85% deacetylated power, Alfa Aesar), PEO (polyethylene oxide, Acros), xanthan (xanthan gum, Pfaltz and Bauer), SA (Alginic Acid Sodium Salt, MP Blomedicals), and PAA (polyacrylic acid, Polysciences) (Figure 2). Chitosan is obtained by deacetylating chitin which is the structural element in the exoskeleton of crustaceans and cell walls of fungi [57,58]. PEO is produced through the interaction of ethylene oxide with water, ethylene glycol, or ethylene glycol oligomers [59]. Xanthan gum is obtained from common allergens such as corn, wheat, dairy, and soy [10,12]. SA is refined from brown seaweeds.

PAA is the generic name for synthetic high molecular weight (~1,000,000 g/mol) polymers of acrylic acid which is produced from propene, which is the byproduct of ethylene and gasoline production [60]. The biopolymers were mixed with deionized (DI) water and biopolymer solutions were applied in this study (Table 1).

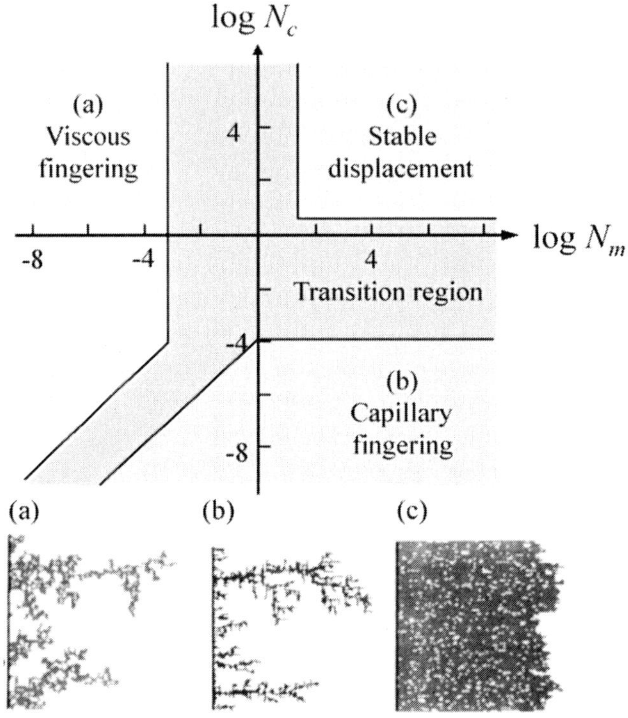

Figure 1. Invading patterns of immiscible fluids in porous media. The displacement pattern of non-reactive flow in porous media is determined by the viscosity number (N_m) and capillary number (N_c) [52,53].

3.2. Contact Angle of Biopolymer Solution

3.2.1. Material and Test Procedure

The sessile drop technique was used to measure the contact angle of biopolymer solution on a silica surface. Smooth-fused pure silica plates (VWR Vista Vision—Cover Glasses, amorphous SiO_2) were used as the substrates that represent Ottawa sand. Figure 3 shows the apparatus used in this research. Contact angles of different biopolymer solutions were measured both on silica plates submerged in decane (alkane hydrocarbon, anhydrous, ≥99%, $C_{10}H_{22}$), which are the main component consisting of natural oil, and on the silica plates at atmospheric conditions. The temperature was maintained constantly (24 °C) at room temperature. To prepare the experiment, a silica plate was cleaned using ethanol (Malickrodt Baker, ACS reagent grad) and handled with gloved-hand. Each silica plate was not reused in order to avoid uncertainty from reaction history with decane and biopolymer solution. The droplet of biopolymer solution was foamed on the silica surface at atmospheric conditions and the contact angle was measured. In other tests, biopolymer was introduced into the decane-filled chamber and released on the silica plate which was placed in the chamber (Figure 3b). The change in the biopolymer droplet was monitored using high-resolution time-lapse photography (12.3 megapixel, Nikon D90). The tests were repeated 3–4 times under same conditions. Images of the droplets captured were used for measuring the contact angle and analyzed with ImageJ (Figure 3b).

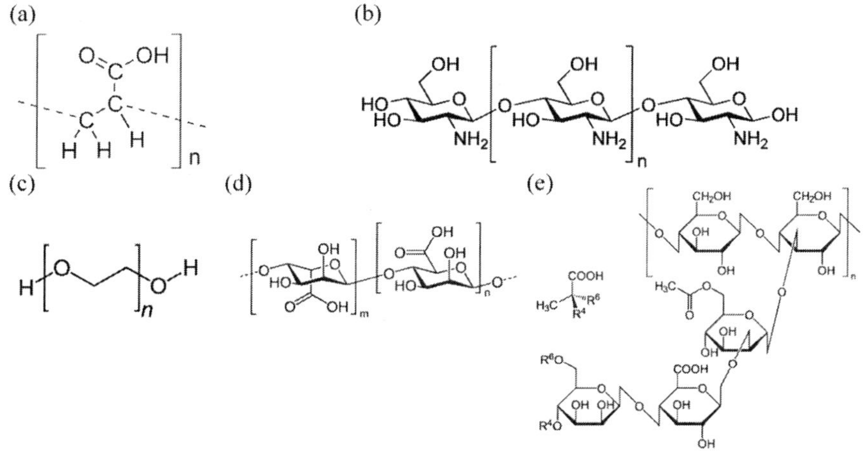

Figure 2. Chemical composition structure (**a**) PAA (polyacrylic acid) (~1,000,000 g/mol); (**b**) chitosan (chitosan, 85% deacetylated power); (**c**) PEO (polyethylene oxide) (600,000 g/mol); (**d**) SA (alginic acid sodium salt); and (**e**) xanthan (xanthan gum).

Table 1. Biopolymer Solutions.

	PEO-1	PEO-10	SA-2	SA-20	Xanthan-2	PAA-2	Chitosan-2
Concentration (g/L)	1	10	2	20	2	2	2

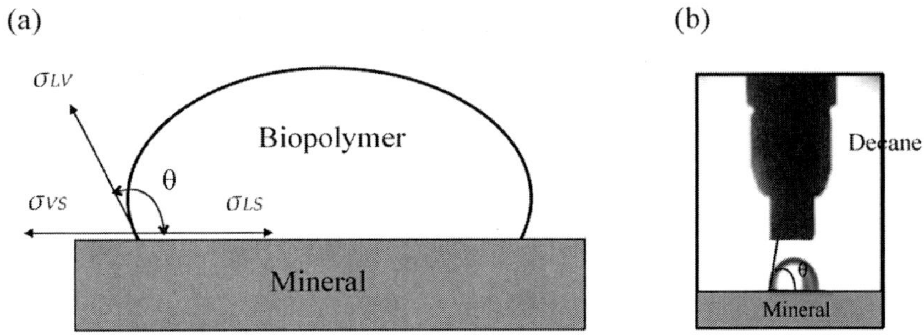

Figure 3. Contact angle of biopolymer solution; (**a**) schematic diagram of contact angle of biopolymer solution on silica surface, and (**b**) biopolymer droplet on silica surface submerged in decane.

3.2.2. Result

Contact angles obtained using ImageJ are presented in Figure 4. The result shows that: (1) all biopolymer solutions used in this study are hydrophilic on the silica surface at atmospheric conditions and overall contact angles are within the range of 31.7° to 41.2°, which is similar to deionized (DI) water (~39.1°); (2) contact angles on a silica surface submerged in decane are higher than that at atmospheric conditions. Even biopolymer solutions such as PEO-10, SA-20, Xanthan-2, including deionized (DI) water, are hydrophobic on a silica surface in decane (>90°), (3) while contact angles of PEO at atmospheric conditions decrease in line with the increase in PEO concentration, but increase in decane with its concentration; and (4) contact angles of SA increased both at atmospheric conditions and in decane with its concentration.

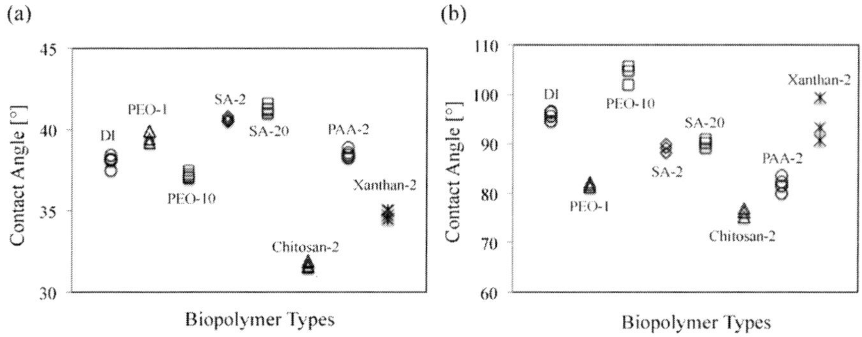

Figure 4. Contact angles of biopolymer solutions. (**a**) Contact angles on silica surfaces at atmospheric conditions, and (**b**) contact angle on silica surfaces submerged in decane (DI: deionized water).

3.3. Interfacial Tension and Surface Tension of Biopolymer Solution

3.3.1. Material and Test Procedure

SensaDyne QC 6000 Surface Tensiometer (SensaDyne Instrument Division, Flagstaff, AZ, USA) was used to measure the surface tension of biopolymer solutions. This device can blow a nitrogen bubble using two probes with different radii into the liquid, which generates the differential pressure. The dynamic surface tension can be calculated using Equation (5) with consideration of both differential pressure measured and the calibration factor *k* supported by the device [61].

$$\Delta p = p_{max} - p_{min} \tag{4}$$

$$\sigma = k \times \Delta p \tag{5}$$

where, σ is surface tension, *k* is calibration factor, Δp is a differential pressure between the maximum pressure p_{max} and the minimum pressure p_{min}.

Additionally, interfacial tension was measured using a force Tensiometer (Sigma 703D) based on the Du Nouy ring method. In the Du Nouy ring method, the container including two fluids (decane and biopolymer solution) is prepared (note: the Du Nouy ring method may not give an accurate value compared to the Wilhelmy plate method for viscoelastic water-soluble polymers). Decane is placed on the biopolymer solution due to its lower density. The platinum ring (with a diameter of 6 cm) is initially submerged into biopolymer solution and then raised to the decane phase forming a fluid meniscus. The meniscus tears from the ring and the maximum force required to support the ring is measured. The force measured is the interfacial tension between decane and the biopolymer solution.

3.3.2. Result

Figure 5 shows the surface tensions and interfacial tensions measured. Figure 5a shows that (1) PEO-1 and PEO-10 have lower surface tensions than deionized (DI) water; and (2) surface tensions of other biopolymer solutions used in this study (*i.e.*, SA-2, SA-20, Chitosan-2, and Xanthan-2) are in the range of 74–81 (mN/m) and are slightly higher than deionized (DI) water (73 mN/m). Figure 5b shows that (1) the interfacial tensions among all fluids and decane are lower than its surface tension, and (2) the interfacial tensions of all biopolymer solutions (25–39 mN/m) are lower than deionized (DI) water (51 mN/m).

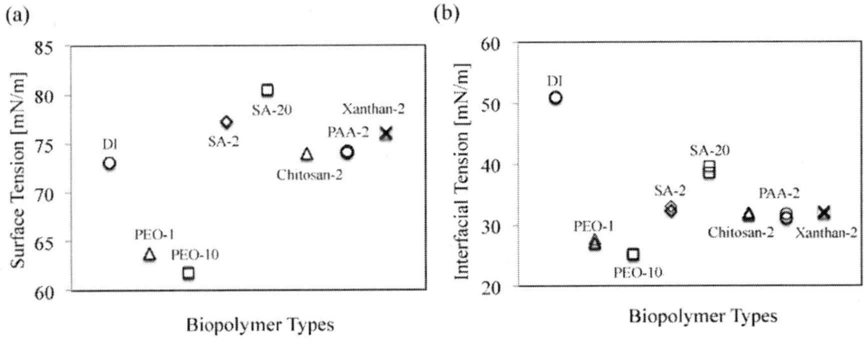

Figure 5. Surface and interfacial tensions. (**a**) Surface tensions of biopolymer solutions, and (**b**) interfacial tensions between biopolymer solution–decane.

3.4. Viscosity of Biopolymer Solution

3.4.1. Material and Test Procedure

The values of the biopolymer solutions' viscosities were measured through the spindle rotation method using the Anton Paar MCR302 rheometer at room temperature and atmospheric pressure. Each solution was placed in Anton Paar MCR302 rheometer that controls the shear rate at the range of 0.1 to 10 (*1/s*) and measured shear stress (*Pa*) at the same time. The values of viscosity were estimated using both shear rate and shear stress.

3.4.2. Result

Figure 6a shows that (1) while biopolymer solutions such as PEO-1, SA-2, and Chitosan-2 have similar viscosity to deionized (DI) water, the rest of them (PEO-10, SA-20, PAA-2, Xanthan-2) have higher viscosities than deionized (DI) water; (2) while viscosities of PAA-2 and Xanthan-2 decrease in line with the increase of shear rate that represent non-Newtonian fluids, the rest of them have relatively constant viscosities; and (3) the viscosities of both SA and PEO increase in line with their concentration.

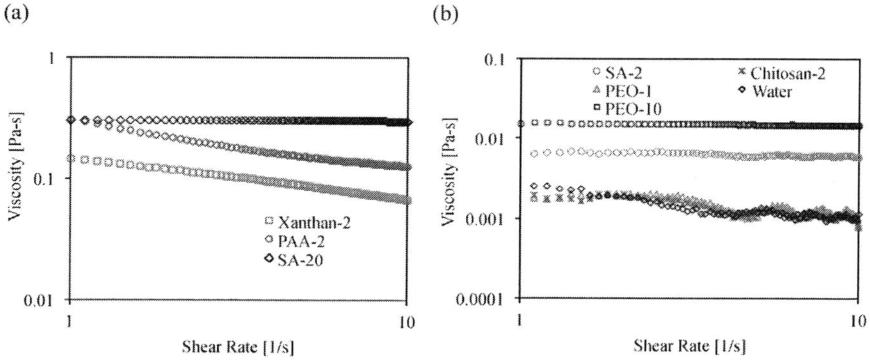

Figure 6. Shear rate effect on viscosity of biopolymer solution. (**a**) Viscosity of Xanthan-2, PAA-2, and SA-20. (**b**) Viscosity of SA-2, PEO-1, PEO-10, Chitosan-2, and deionized (DI) water.

3.5. Micromodel Test

3.5.1. Apparatus and Materials

The experimental configuration of this study is schematically shown in Figure 7. The silica micromodel is placed horizontally on a customized jack stage. A syringe (2.5 mL with 1/16 inch fitting), controlled by a precision syringe pump (Kats Scientific, NE-1010), is connected with the micromodel for biopolymer solution injection. To monitor the flow processes in the micromodel, a high-resolution camera (Nikon D7000, 16.2 megapixels), with image and video- capturing function controlled by the computer, is used. SA-20 and PAA-2 were not used in the micromodel test due to excessively high viscosity.

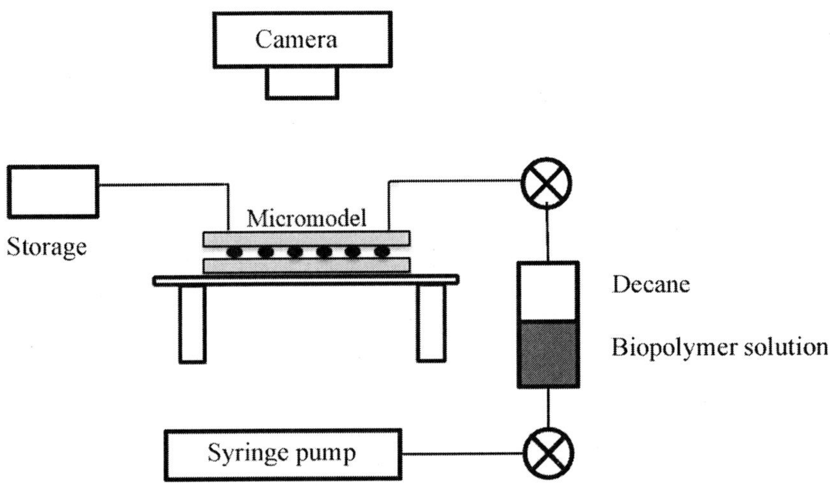

Figure 7. Schematic design for micromodel setup (Note: Figure not drawn to scale).

3.5.2. Micromodel

The micromodel fabricated by Micronit Microfluidics BV is made of fused silica. Customized pore networks are etched on two symmetrically-patterned silica plates, which are then fused together to form a two-dimensional porous network. The patterned area is 20 mm × 10 mm, containing 576 discoid silica grains (590 μm in diameter and 1186 pore bodies). The pore throat is 20 μm deep and around 30 μm wide. Figure 8a shows the top view of the micromodel within a chip holder for protection and tubing connection.

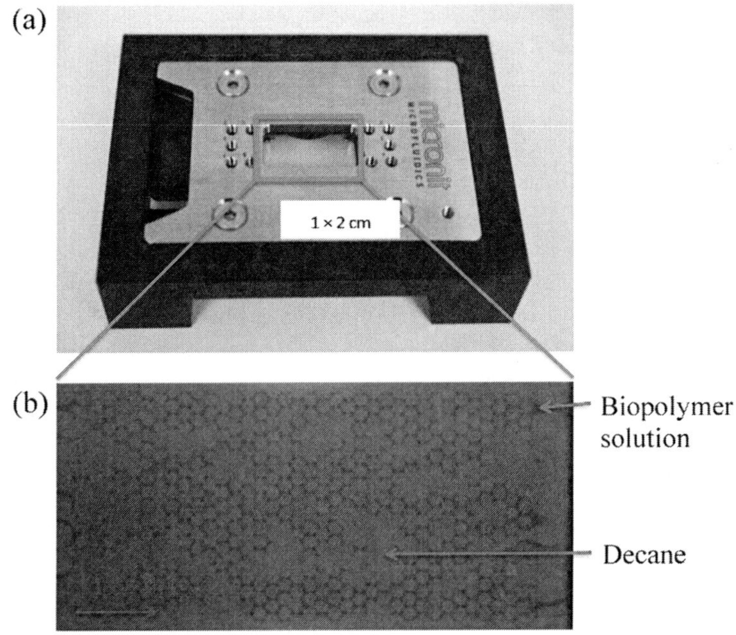

Figure 8. Biopolymer solution–decane (or air) displacement using micromodel, (**a**) micromodel, and (**b**) the injection of biopolymer solution into the decane-saturated micromodel.

3.5.3. Experimental Procedure

The micromodel was firstly cleaned by way of injecting 10 mL of absolute ethanol (Mallinckodt Baker, ACS reagent grad) which was followed by 30mL of deionized (DI) water and then oven-dried at 120 °C for 48 h. After cleaning, the experimental system was assembled accordingly as illustrated in Figure 7. Two types of tests were conducted using the micromodel. The first test was the biopolymer injection into the air-filled micromodel at atmospheric conditions. Biopolymer injection into the micromodel was initiated at a constant flow rate in the range of 0.7 to 70 μL/min. The injection continued until biopolymer percolated through the micromodel. Additional one hundred pore volume (PV) of biopolymer was eventually injected into the micromodel after the biopolymer had percolated through the micromodel. The second test was the biopolymer injection into the decane-saturated micromodel. A small tubing chamber was placed between the micromodel and the syringe pump that was filled with biopolymer solution and decane (Figure 7). Decane was always placed on the upper layer in this chamber considering its lower density than the biopolymer solutions. Thus, the micromodel was initially saturated with decane (Sigma-Aldrich, anhydrous, >99%) first. After that, biopolymer injection into the micromodel was initiated at a constant flow rate in the range of 0.5 to 50 μL/min. Injection continued until no more decane was drained from the micromodel. Furthermore, an additional one hundred pore volume (PV) of biopolymer was finally injected into the micromodel after the biopolymer had percolated through the micromodel. Additional injection of 100 PV is intended to clarify the displacement change after the percolation of fluid in the micromodel.

Captured images and videos during such injection processes were used to compute the time-lapse biopolymer saturations and biopolymer–decane displacement ratios in the micromodel. Biopolymer saturation is defined as the ratio between the volume of biopolymer solution and the pore volume in micromodel obtained from biopolymer injection into the air-filled micromodel test. Additionally, the biopolymer–decane displacement ratio is defined as the ratio between the volume of biopolymer solution and the pore volume in the micromodel obtained from the biopolymer injection into the decane–saturated micromodel test.

3.5.4. Result

Figure 8b shows the snapshot of biopolymer injection into the initially decane-saturated micromodel. With this image, the biopolymer–decane displacement ratio was obtained. In this case, the injected biopolymer solution percolates through the micromodel within 2.0 s. While additional 100 pore volume (PV) biopolymer solution is injected into the micromodel after percolation (>2.0 s), distribution of residual decane remains unchanged. Biopolymer solution injection was controlled at constant flow rates (from 0.5 μL/minto ~70 μL/min) during the tests. Figure 9 shows that (1) biopolymer saturation (biopolymer-air displacement) was obviously increased in line with the increase in injection rate, which agrees with previous study by Kuo *et al.* (2011); (2) the pore saturations of SA-2 and PEO-1, according to injection rate, have a similar pattern with deionized (DI) water; (3) the pore saturation of Xanthan-2 and PEO-10 are higher than others, including SA-2, PEO-1, and deionized (DI) water at the same injection rate; and (4) at higher injection rate (>0.1 mL/s), the pore saturations of all fluids are relatively high (>90%).

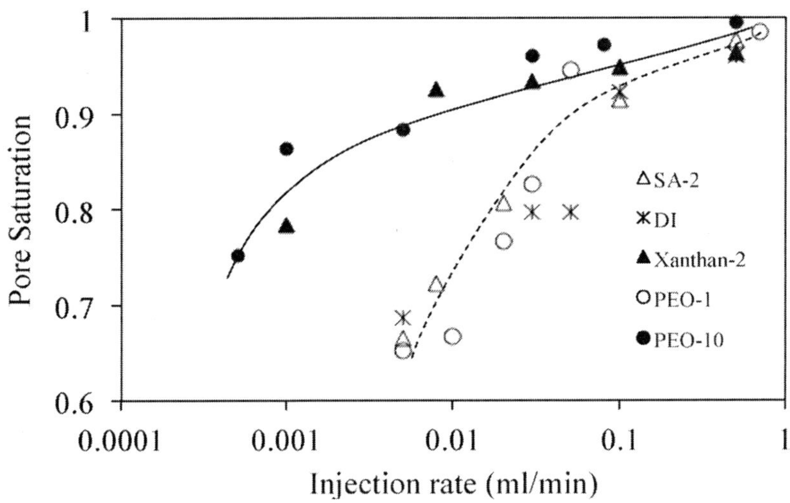

Figure 9. Effects of biopolymer injection rate on pore saturation of biopolymer solutions on a logarithmic scale. The pore saturation generally increases with the injection rate of biopolymer solution.

Figure 10 shows that: (1) biopolymer–decane displacement was obviously increased in line with the increase in the injection rate, which also agrees with previous study; (2) at a higher injection rate (>0.1 mL/s), the biopolymer–decane displacements of all fluids are relatively high (>90%); and (3) the biopolymer–decane displacement ratios with SA-2, PEO-10 and chistosan are immensely high even at a lower injection rate (~0.00002 mL/min).

4. ANALYSES AND DISCUSSION

4.1. Viscous Number (N_m) and Capillary Number (N_c)

N_m and N_c values of each test in this study are calculated based on Equations (2) and (3), respectively. Injection velocity of biopolymer solution v_{inv} is computed using the injection rate divided by the cross-sectional pore area (20 μm × 30 μm) of the micromodel. For example, an injection rate of 10 μL/min in this study corresponds to the injection velocity of 104.16 cm/min.Figure 11 and Figure 12 show the N_c and N_m values in our experiments, which are also relevant to field conditions, located at the transition region among the invading patterns represented by Lenormand (1990) [52]. Within this region, while the changes of N_m do not show any clear tendency in this study (Figure 11a and Figure 12a), biopolymer solution–decane displacement ratio (or biopolymer saturation) was clearly increased with N_c (Figure 11b and Figure 12b). This finding also agrees with previous pore network simulation results [62]. Thus, increasing the N_c value is recommended to increase the biopolymer solution–decane displacement ratio (or biopolymer saturation). N_c is proportional to injection velocity v_{inv}, injected biopolymer solution viscosity μ_{inv}, and contact angle on the mineral surface ϑ, but inversely proportional to the interfacial tension σ (Equation (3)). The following techniques which increase N_c values are recommended for biopolymer solution injection into the porous media for oil-contaminated soil remediation and enhanced oil recovery: (1) increasing the biopolymer solution injection velocity (note: while excessively increasing velocity may cause potential plugging problems or others in the field, however, this study focuses on relations between injection velocity and displacement efficiency in the laboratory); (2) increasing the viscosity of biopolymer solutions by adding more biopolymer; and (3) increasing the contact angle and decreasing interfacial tension (or surface tension) controlling the concentration of biopolymer solution (note: preliminary results show the proportional changes of viscosity, contact angle, and interfacial tension with the concentration of biopolymer solutions). Conversely, any event that decreases the N_c value will impede the biopolymer solution injection efficiency for oil-contaminated soil remediation and enhanced oil recovery.

4.2. Effects of Effective Stress on Capillary Pressure

The capillary pressure curve is inherently dependent on pore-scale characteristics, including pore connectivity, size distribution, and gas–fluid–mineral contact properties in the manifestations of contact angle, interfacial tension (or surface tension), and hysteresis. To prove the effect of pore size distribution on the capillarity development during biopolymer solution–decane displacement, the following analysis is conducted to identify the influential factors and their interrelation. Capillary pressure as a function of effective stress p', sediment compressibility (e_{1kps}, C_c), pore structure (S_s, $\alpha\sigma_x$), pressure-dependent surface tension σ, and contact angle θ are expressed (Equations (6) and (7)) [63]:

$$P_c^* = \psi \frac{S_s \, \rho\sigma \, cos\theta}{e_{1kpa} - C_c log \frac{p'}{1kPa}} \tag{6}$$

$$\psi = \frac{4}{k10^{\alpha\sigma_x}} \tag{7}$$

where, *p'* is the effective stress of sediments that are defined with a depth and unit weight of soil, C_c is the compressive index of sediments, e_{1kpa} is the void ration of sediments at 1kPa effective stress, S_s is the specific surface, *ρ* is the density of mineral, *k* is the geometric factor between 6 and 12, σ_x is the standard deviation of pore size distribution, *α* is the factor of standard deviation, *σ* is the interfacial tension, and *ϑ* is the contact angle.

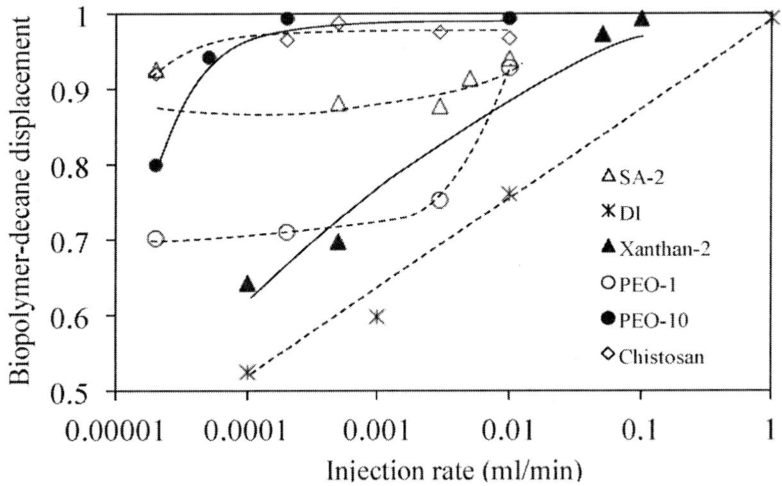

Figure 10. Effects of biopolymer injection rate on biopolymer solution–decane displacement ratios on a logarithmic scale. The displacement ratio was generally increased with the injection rate of the biopolymer solution.

In general, the factor ψ = 4/ k10ασx is in the range of 0.04 to 0.08 [63]. Table 2 shows soil properties of sand, silt, kaolinite, illite, and montmorillonite. For example, considering various effective stress conditions, the capillary pressure for each biopolymer solution to inject into silt sediments can be interpreted using both surface tension and contact angle measured in this study and the soil properties in Table 2. The results show that capillary pressure was increased with applied effective stress (Figure 13). It also shows that PEO-10 causes smallest capillary pressure, while Chitosan-2 causes the largest capillary pressure. Furthermore, higher capillary pressures are expected with (1) higher density minerals, (2) increase in the effective stress of sediments p' that can be increased with the depth of sediments, (3) higher specific surface of mineral S_s, like clay materials, (4) higher compressive index C_c material, like clay, and (5) lower void ratio at 1kPa effective stress e_{1kpa}.

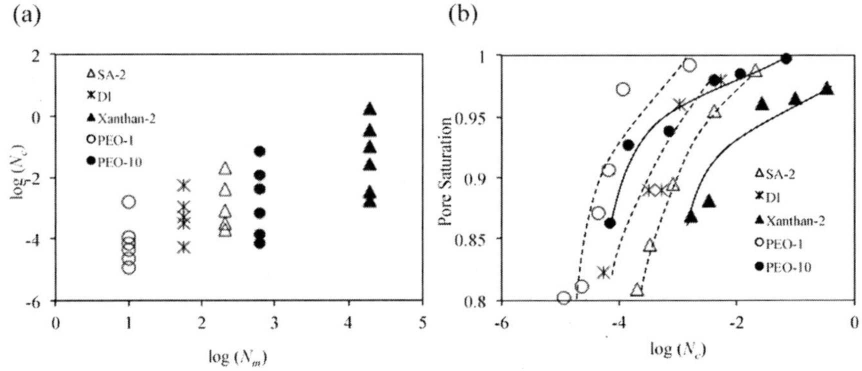

Figure 11. Pore saturation of biopolymers at various N_c and N_m conditions. (**a**) The experimental conditions are within the transition region of the log(N_c)-log(N_m) plot. (**b**) Pore saturation of biopolymers generally increases along with the log(N_c).

5. CONCLUSIONS

Biopolymer solutions, such as chitosan, PEO, xanthan, SA, and PAA, are tested to identify the contact angle, surface tension, interfacial tension, and viscosity. Flow characteristics have been studied using a micromodel. The implications from the test results with regard to soil remediation and enhanced oil recovery are discussed below.

In the air, overall contact angles of all biopolymer solutions on silica surfaces are within the range of 31.7° to 41.2°, which is similar to deionized (DI) water (39.1°). Contact angles on silica surfaces submerged in decane are higher than those at atmospheric conditions and biopolymer solutions such as PEO-10, SA-20 and Xanthan-2, including deionized (DI) water (94.5°), even present as hydrophobic in decane. Overall contact angles of all biopolymer solutions on silica surfaces submerged in decane are within the range of 76.1° to 102.6°. Within the given range of contact angles, the effect of contact angles on flow behavior in porous media would be minor compared to the fluid injection rate.

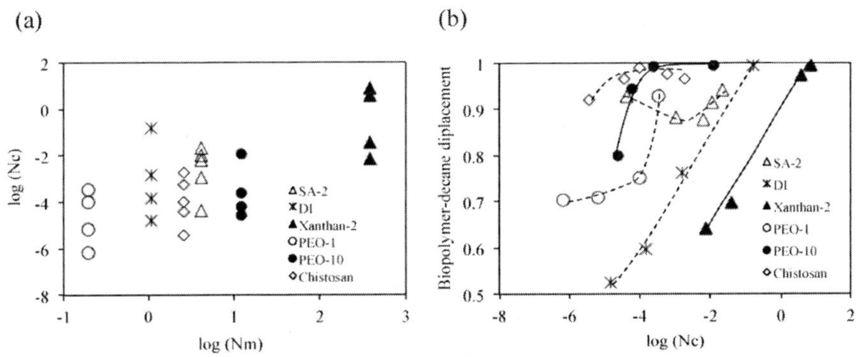

Figure 12. Biopolymer–decane displacement ratios at various N_c and N_m conditions. (**a**) The experimental conditions are within the transition region of the log(N_c)-log(N_m) plot. (**b**) The biopolymer–decane displacement generally increases along with the log(N_c).

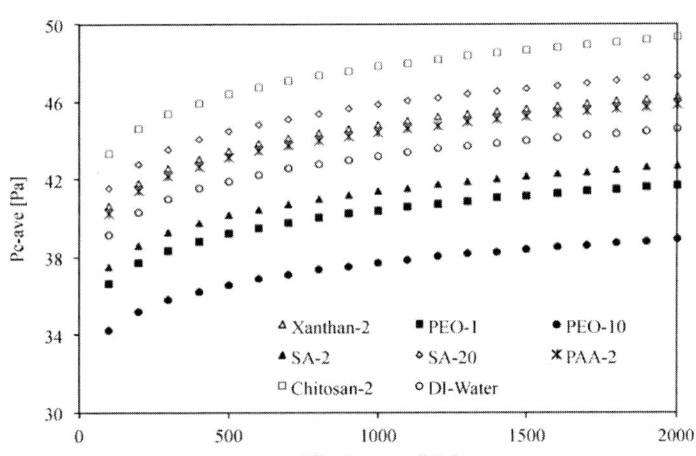

Figure 13. Capillary pressure caused by different biopolymers into silt sediments as function of effective stress.

Table 2. Soil properties in relationship between capillary pressure and effective stress [63,64,65,66,67,68].

Soil Properties	Sand	Silt	Kaolinite	Illite	Montmorillonite
S_s (m_2/g)	0.044	0.045–1.0	10.0–20.0	65–100	300–780
$\rho_{particle}$ (t/m^3)	2.655–2.659	2.798	2.65	2.6–2.9	2.35
ρ_{bulk} (t/m^3)	1.586–2.083	1.3	7.69	-	-
e_{100} (at 100 Kpa)	0.35–0.85	0.6–0.8	0.9–1.1	2.0–3.0	2.5–4.0
C_c	<0.02	0.02–0.09	0.2–0.4	0.5–1.1	1.0–2.0

While PEO-1 and PEO-10 solutions have lower surface tensions than deionized (DI) water, the rest of them (*i.e.*, SA-2, SA-20, Chitosan-2, PAA-2, and Xanthan-2) have higher surface tensions than deionized (DI) water. Overall surface tensions are within the small range of 62 mN/m to 81 mN/m. Thus, the influence of surface tension for capillary pressure (P_c) and capillary number (N_c) is significant and more obvious than the contact angle (ϑ).

The viscosities of PAA-2 and Xanthan-2 have higher values than deionized (DI) water, which causes biopolymer solution–decane (or air) displacements to increase. Biopolymers have lower viscosities effect on less biopolymer solution–decane (or air) displacement at the same flow rates. Additionally, the viscosities of PAA-2 and Xanthan-2 solutions are highly dependent on shear rate. Therefore, it is important to properly determine the shear rate (that is, in turn, injection rate) in order to design soil remediation process and calculate the efficiency.

The biopolymer solution–decane (or air) displacement ratio was increased with the capillary number N_c, while its dependency on the viscosity number N_m remains unclear. The capillary number N_c in microscopic multiphase flow is different from that of the macroscopic flow, which is also associated with length scale, saturation history, and relative permeability. However, the biopolymer solution–decane (or air) displacement ratio was clearly increased with the increase in flow rate. Low flow rate results in low levels of displacement ratio, but the displacement ratio is higher for biopolymer solutions than deionized (DI) water. Thus, biopolymer injection would possibly lead to cost-efficient high flow rates.

In high-pressure reservoirs, the capillarity at the early stage is relatively lower, which is attributable to both increased contact angle and decreased interfacial tension. Within the function of effective stress of capillary pressure, estimation of capillary pressure can be achieved with measured interfacial tension and contact angle of fluids at different effective stress levels. In this research, the capillary pressure for each biopolymer has been estimated in silt sediments at different effective stress levels. Compared with all biopolymers, PEO-10 could result in the smallest capillary pressure at the same effective stress level compared with other biopolymers. On the contrary, Chitosan-2 leads to the largest capillary pressure. Furthermore, higher capillary pressure can be expected within (1) a decrease in contact angle θ, such as PEO-1 and PEO-10; (2) an increase in surface tension σ, like SA-2 and SA-20; (3) a higher density mineral; (4) an increase in effective stress of sediments p' that can be increased with the depth of sediments; (5) a higher specific surface of mineral S_s, like clay materials; (6) a higher compressive index C_c material, like clay; and (7) lower void ratio at 1kPa effective stress.

ACKNOWLEDGMENTS

This work was supported by FIER (The Fund for Innovative Engineering Research - Round V) grand award in Louisiana State University. This research was also supported by Basic Science Research Program through the National Research Foundation of Korea (NRF) funded by the Ministry of Science, ICT, and Future Planning (Grant No. 2015R1A2A1A15054704). Xiaoyi Zhao, a former master student at Missouri University of Science and Technology performed the viscosity, surface tension and contact angle measurements at water-air interface in low pressure.

AUTHOR CONTRIBUTIONS

Bate Bate and Jongwon Jung conceived and designed the experiments; Bate Bate, Shuang Cindy Cao and Jong Wan Hu performed the experiments. Shuang Cindy Cao, Jong Wan Hu and Jongwon Jung analyzed the data and wrote the paper. All authors read and approved the final manuscript.

REFERENCES

1. Mitchell, R.; Nevo, Z. Effect of bacterial polysaccharide accumulation on infiltration of water through sand. *Appl. Microbiol. Biotechnol.* **1964**, *12*, 219–223.

2. Bate, B.; Zhao, Q.; Burns, S. Impact of organic coatings on frictional strength of organically modified clay. *J. Geotech. Geoenviron. Eng.* **2013**, *140*, 228–236.

3. Briscoe, W.H.; Klein, J. Friction and adhesion hysteresis between surfactant monolayers in water. *J. Adhes.* **2007**, *83*, 705–722.

4. Cabalar, A.F.; Canakci, H. *Ground Improvement by Bacteria*; Taylor and Francis Group: Norman, OK, USA, 2005; pp. 707–712.

5. Kang, X.; Kang, G.C.; Bate, B. Measurement of Stiffness Anisotropy of Kaolinite Using Bender Element Tests in a Floating Wall Consolidometer. *Geotechn. Test. J.* **2014**, *37*, 869–883.

6. Kavazanjian, E., Jr.; Iglesias, E.; Karatas, I. Biopolymer soil stabilization for wind erosion control. In Proceedings of the 17th International Conference Soil Mechnics Geotechnical Engineering, Alexandria, Egypt, 5–9 October 2009; pp. 881–884.

7. Martin, G.; Yen, T.; Karimi, S. Application of biopolymer technology in silty soil matrices to form impervious barriers. In Proceedings of the 7th Australia New Zealand Conference on Geomechanics: Geomechanics in a Changing World, Melbourne, Australia, 1–5 July 1996; p. 814.

8. Nugent, R.; Zhang, G.; Gambrell, R. *The Effect of Exopolymers and Void Ratio on the Erosional Resistance of Cohesive Sediments*; In Proceedings of Geo-Frontiers 2011@ Advances in Geotechnical Engineering, Dallas, TX, USA, 13–16 March 2011; ASCE: Reston, VA, USA, 2011; pp. 1493–1502.

9. Yoshikawa, T.; Naito, Y.; Kishi, A.; Tomii, T.; Kaneko, T.; Iinuma, S.; Ichikawa, H.; Yasuda, M.; Takahashi, S.; Kondo, M. Role of active oxygen, lipid peroxidation, and antioxidants in the pathogenesis of gastric mucosal injury induced by indomethacin in rats. *Gut* **1993**, *34*, 732–737.

10. Hove, A.O.; Nilson, V.; Loknes, J. Visualization of xanthan flood behavior in core samples by means of x-ray tomography. *SPE Reserv. Eng.* **1990**, *5*, 475–480.

11. Philips, J.C.; Miller, J.W.; Wernau, W.C.; Tate, B.E.; Auerbach, M.H. A high-pyruvate xanthan for eor. *Soc. Pet. Eng. J.***1985**, *25*, 594–602.

12. Pollock, T.J.; Thorne, L. Xanthomonas campestris strain for production of xanthan gum. Google Patents US5279961 A, 1994.

13. Sandiford, B.B. Laboratory and field studies of water floods using polymer solutions to increase oil recoveries. *J. Pet. Technol.* **1964**, *16*, 917–922.

14. Strom, T.E.; Paul, J.M.; Phelps, C.H.; Sampath, K. A New Biopolymer for High-Temperature Profile Control: Part 1-Laboratory Testing. *Soc. Pet. Eng. J.* **1991**, *6*, 360–364.

15. Blokker, N.C.M. Analysis of Alginate-like Exopolysaccharides for the Application in Enhanced Oil Recovery. Master's Thesis, Delft University of Technology, Delft, The Netherlands, 2014.

16. Bailey, S.A.; Bryant, R.S.; Duncan, K.E. *Use of Biocatalysts for Triggering Biopolymer Gelants*; Society of Petroleum Engineers: Tulsa, OK, USA, 2000

17. Kosaric, N. Biosurfactants and their application for soil bioremediation. *Food Technol. Biotechno.* **2001**, *39*, 295–304.

18. Leveratto, M.A.; Lauri, J.; Sanz, C.; Sigal, J.; Farouq Ali, S.M. EOR Polymer Screening For An Oil Field With High Salinity Brines. *Soc. Pet. Eng. J.* **1996**, *4*, 73–81.

19. Stepp, A.K.; Bailey, S.A.; Bryant, R.S.; Evans, D.B. *Alternative Methods for Permeability Modification Using Biotechnology*; Society of Petroleum Engineers: Denver, CA, USA, 1996.

20. Meuser, H. *Soil Remediation and Rehabilitation*; Springer: Dordrecht, Germany, 2010; Volume 23, pp. 1–2.

21. Swartjes, F.A. *Dealing with Contaminated Sites*; Springer: Dordrecht, Germany, 2011.

22. Mulligan, C.; Eftekhari, F. Remediation with surfactant foam of pcp-contaminated soil. *Eng. Geol.* **2003**, *70*, 269–279.

23. Chai, J.C.; Miura, N. Field vapor extraction and long-term monitoring at a pce contaminated site. *J. Hazard. Mat.* **2004**,*110*, 85–92.

24. Cho, J.S.; DiGiulio, D.C.; Wilson, J.T. *In-situ* air injection, soil vacuum extraction aqnd enhanced biodegradation: A case study in a jp-4 jet fuel contaminated site. *Environ. Prog.* **1997**, *16*, 570–576.

25. Hall, B.L.; Lachmar, T.E.; Dupont, R.R. Field monitoring and performance evaluation of a field-scale in-well aeration system at a gasoline-contaminated site. *J. Hazard. Mater* **2001**, *82*, 197–212.

26. Tse, K.K.C.; Lo, S.L.; Wang, J.W.H. Pilot study of in-situ thermal treatment for the remediation of pentachlorophenol-contaminated aquifers. *Environ. Sci. Technol.* **2001**, *35*, 4910–4915.

27. Mann, M.; Dahlstrom, J.D.; Esposito, P.; Everett, G.; Peterson, G.; Traver, R.P. Soil washing/soil flushing. In *Innovative Site Remediation Technology*; Anderson, W.C., Ed.; WASTECH (American Academy of Engineers): Annapolis, MD, USA, 1993.

28. Khatami, H.R.; O'Kelly, B.C. Improving mechanical properties of sand using biopolymers. *J. Geotech. Geoenviron. Eng.***2013**, *139*, 1402–1406.

29. Lee, M.; Chung, S.; Kang, D.; Choi, S.; Kim, M. Surfactant enhanced in-site soil flushing pilot test for the soil and groundwater remediation in an oil contaminated site. *J. Korean Soc. Soil Groundw. Environ.* **2002**, *7*, 77–86

30. Rouse, J.D.; Sabatini, D.A.; Brown, R.E.; Harwel, L.J.H. Evaluation of ethoxylated alkylsulfate surfactants for use in subsurface remediation. *Water Environ. Res.* **1996**, *68*, 162–168.

31. Sabatini, D.A.; Knox, R.C. *Transport and Remediation of Subsurface Contaminants: Colloidal, Interfacial and Surfactant Phenomena*; ACS: Washington, DC, USA, 1992; p. 834

32. Jafvert, C.T. *Surfactants/Cosolvents, Technology Evaluation Report TE-96-02*; Groundwater Remediation Technologies Analysis Center: Pittsburgh, PA, USA, 1996; p. 45.

33. Jawitz, J.W.; Annable, M.D.; Rao, P.S.C.; Rhue, R.D. Field implementation of a winsor typesurfactant/alcohol mixture for *in situ* solubilization of a complex lnapl as a single-phase microemulsion. *Environ. Sci. Technol.* **1998**, *32*, 523–530.

34. Lee, M.; Kang, H.; Do, W. Application of nonionic surfactant-enhanced *in situ* flushing to a diesel contaminated site.*Water Res.* **2005**, *39*, 139–146.

35. Mulligan, C.N.; Yong, B.R.; Gibbs, B.F. Surfactant-enhanced remediation of contaminated soil: A review. *Eng. Geol.***2001**, *60*, 371–380.

36. Park, S.K.; Bielefeldt, A.R. Non-ionic surfactant flushing of pentachlorophenol from napl-contaminated soil. *Water Res.***2005**, *39*, 1388–1396.

37. Tsai, T.T.; Kao, C.M.; Yeh, T.Y.; Liang, S.H.; Chien, H.Y. Application of surfactant enhanced permanganate oxidation and biodegradation of trichloroethylene in groundwater. *J. Hazard. Mater.* **2009**, *161*, 111–119.

38. Zhou, W.; Zhu, L. Enhanced soil flushing of phenanthrene by anionic-nonionic mixed surfactant. *Water Res.* **2008**, *42*, 101–108.

39. Lake, L.W.; Johns, R.; Rossen, B.; Pope, G. *Fundamentals of enhanced oil recovery*; Society of Petroleum Engineers: Richardson, TX, USA, 2014.

40. Hong, C.H.; Jang, C.H.; Choi, H.J. Turbulent drag reduction with polymers in rotating disk flow. *Polymers* **2015**, *7*, 1279–1298.

41. Sorbie, K.S. *Polymer-Improved Oil Recovery*; Blackie & Son: Glasgow, UK, 1991; p. 359

42. Delamaide, E.; Zaitoun, A.; Gérard, R.; Tabary, T. Pelican lake field: First successful application of polymer flooding in a heavy-oil reservoir. *SPE Reserv. Elval. Eng.* **2013**, *17*, 340–354.

43. Moawad, T.; Elhomadhi, E.; Gawish, A. Novel promising, high viscosifier, cheap, available and environmental friendley biopolymer (polymtea) for different applications at reservoir conditions under investigation. Part a: Polymer properties. In Processing of The Seventh

Egyptian-Syrian Conference in Chemical and Petroleum Engineering, Suez, Egypt, 29–31 October 2007.

44. Jung, J.C.; Zhang, K.; Chon, B.H.; Choi, H.J. Rheology and polymer characteristics of partially hydrolyzed polyacrylamide for enhanced heavy oil recovery. *J. Appl. Polymer Sci.* **2013**, *127*, 4833.

45. Jang, H.Y.; Zhang, K.; Chon, B.H.; Choi, H.J. Enhanced oil recovery performance and viscosity characteristics of polysaccharide xanthan gum solution. *J. Ind. Eng. Chem.* **2015**, *21*, 741–745.

46. Cho, D.H.; Chae, H.J.; Kim, E.Y. Synthesis and characterization of a novel extracellular polysaccharide by rhodotorula glutinis. *Appl. Biochem. Biotechnol.* **2001**, *98*, 183–193.

47. Akit, R.J.; Cooper, D.G.; Neufield, R.J. Effects of oil reservoir conditions on the production of water-insoluble levan by bacillus licheniformis. *Geomicrobiol. J.* **1989**, *7*, 115–165.

48. Canella, W.J.; Huh, C.; Seright, R.S. Prediction of xanthan rheology in porous media. In Processing of SPE Annual Technical Conference and Exhibition, Houston, TX, USA, 2–5 October, 1988; Society of Petroleum Engineers: Richardson, TX, USA, 1988.

49. Kim, D.S.; Fogler, H.S. The effects of exopolymers on cell morphology and culturability of l. Mesenteroides during starvation. *Appl. Microbiol. Biotechnol.* **1999**, *52*, 839–844.

50. Sandford, P.A. Exocellular microbial polysaccharides. *Adv. Carbohydr. Chem. Biochem.* **1979**, *36*, 265–312.

51. Bear, J. *Dynamics of Fluids in Porous Media*; Dover Publications Inc.: New York, NY, USA, 1972

52. Lenormand, R. Liquids in porous media. *J. Phys. Condens. Matter* **1990**, *2*, 79.

53. Lenormand, R.; Touboul, E.; Zarcone, C. Numerical models and experiments on immiscible displacements in porous media. *J. Fluid Mech.* **1988**, *189*, 165–187.

54. Buchgraber, M. An enhanced oil recovery micromodel study with associative and conventional polymers. Master's Thesis, University of Leoben, Austria, 2008; pp. 16–30.

55. Daripa, P.; Pasa, G. On capillary slowdown of viscous fingering in immiscible displacement in porous media. *Transp. Porous Media* **2008**, *75*, 1–16.

56. Lenormand, R.; Zarcone, C.; Sarr, A. Mechanisms of the displacement of one fluid by another in a network of capillary ducts. *J. Fluid Mech.* **1983**, *135*, 337–353.

57. Goosen, M.F.A. *Applications of Chitin and Chitosan*; CRC: Boca Raton, FL, USA, 1996; pp. 132–139.

58. Woodmansey, A. *Chitosan Treatment of Sediment Laden Wate-Washington State I-90 Issaquah Project*; Federal Highway Administration: Washington, DC, USA, 2012.

59. Kahovec, J.; Fox, R.B.; Hatada, K. Nomenclature of regular single-strand organic polymers. *Pure Appl. Chem.* **2002**, *74*, 1921–1956.

60. Orwoll, R.A.; Yong, C.S. *Poly (Acrylic Acid)*; Oxford University Press, Inc: Oxford, UK, 1999; pp. 252–253

61. Adamson, A.W.; Gast, A.P. *Physical Chemistry of Surfaces*, 6th ed.; Wiley: New York, NY, USA, 1997.

62. Ferer, M.; Bromhal, G.S.; Smith, D.H. Pore-level modeling of carbon dioxide sequestration in brine fields. *J. Energy Environ. Res.* **2002**, *2*, 120–132.

63. Espinoza, D.N.; Santamarina, J.C. Water-CO_2-mineral systems: Interfacial tension, contact angle, and diffusion-implications to CO_2 geological storage. *Water Resour. Res.* **2010**, *46*, W07537.

64. Burland, J.B. On the compressibility and shear strength of natural clays. *Geotechnique* **1990**, *40*, 329–378.

65. Lambe, T.W.; Whitman, R.V. *Soil Mechanics*; Wiley: New York, NY, USA, 1969. Mitchell, J.K.; Soga, K. *Fundamentals of Soil Behavior*; Wiley: New York, NY, USA, 2005.

66. Santamarina, J.C.; Klein, K.A.; Fam, M.A. *Soils and Waves: Particulate Materials Behavior, Characterization and Process Monitoring*; Wiley: New York, NY, USA, 2001; pp. 250–280.

67. Skempton, A.W. The consolidation of clays by gravitational compaction. *Q. J. Geol. Soc.* **1970**, *125*, 373–411.

Chapter 7

Application of Spectroscopic Methods for Structural Analysis of Chitin and Chitosan

Jolanta Kumirska [1,*], Małgorzata Czerwicka [1], Zbigniew Kaczyński [1], Anna Bychowska [1], Krzysztof Brzozowski [1], Jorg Thöming [2] and Piotr Stepnowski [1]

[1]Faculty of Chemistry, University of Gdansk, Sobieskiego 18/19, PL-80-952 Gdansk, Poland

[2]UFT-Centre for Environmental Research and Sustainable Technology, University of Bremen, Leobener Straße UFT, D-28359 Bremen, Germany

ABSTRACT

Chitin, the second most important natural polymer in the world, and its *N*-deacetylated derivative chitosan, have been identified as versatile biopolymers for a broad range of applications in medicine, agriculture and the food industry. Two of the main reasons for this are firstly the unique chemical, physicochemical and biological properties of chitin and chitosan, and secondly the unlimited supply of raw materials for their production. These polymers exhibit widely differing physicochemical properties depending on the chitin source and the conditions of chitosan production. The presence of reactive functional groups as well as the polysaccharide nature of these biopolymers enables them to undergo diverse chemical modifications. A complete chemical and physicochemical characterization of chitin, chitosan and their derivatives is not possible without using spectroscopic techniques. This review focuses on the application of spectroscopic methods for the structural analysis of these compounds.

KEYWORDS

chitin; chitosan; chemical modification; physicochemical parameters; structural analysis using spectroscopic techniques

1. INTRODUCTION

Chitin, the second most abundant natural polymer in the world, functions as a natural structural polysaccharide [1]. A major component of the carapaces, crusts and shells of crustaceans such as shrimps, crabs and lobsters, it is also an ingredient of cell walls in fungi and yeast [2]. Its estimated production is 10^{10}–10^{12} tonnes per year [3]. Chitin is a linear polymer consisting mainly of β-(1→4)-linked 2-acetamido-2-deoxy-β-D-glucopyranose units and partially of β-(1→4)-linked 2-amino-2-deoxy-β-D-glucopyranose. In this form, chitin is insoluble in water and common organic solvents, dissolving only in specific solvents such as *N*,*N*-dimethylacetamide (DMAc)-LiCl [4], hexafluoroacetone or hexafluoro-2-propanol [5]. When the degree of *N*-acetylation (defined as the average number of *N*-acetyl-D-glucosamine units per 100 monomers expressed as a percentage) is less than 50%, chitin becomes soluble in aqueous acidic solutions (pH < 6.0) and is then called chitosan [6,7]. Thus, chitosan is a collective name for a group of fully and partially deacetylated chitins, but a rigid nomenclature with respect to the degree of *N*-deacetylation between chitin and chitosan has not been defined [8]. The structures of "ideal" chitin and "ideal" chitosan, and the "real" structures of these compounds are shown in Figure 1.

According to the nomenclature proposed by the European Chitin Society (EUCHIS) [9], chitin and chitosan should be classified on the basis of their solubility and insolubility in 0.1 M acetic acid; the soluble material is named chitosan, whereas chitin is insoluble. Chitin and chitosan have a molecular mass of up to several million g/mol. Commercially available chitosan has an average molecular weight ranging between 3,800 and 500,000 g/mol and its degree of *N*-acetylation is 2% to 40% [8,10].

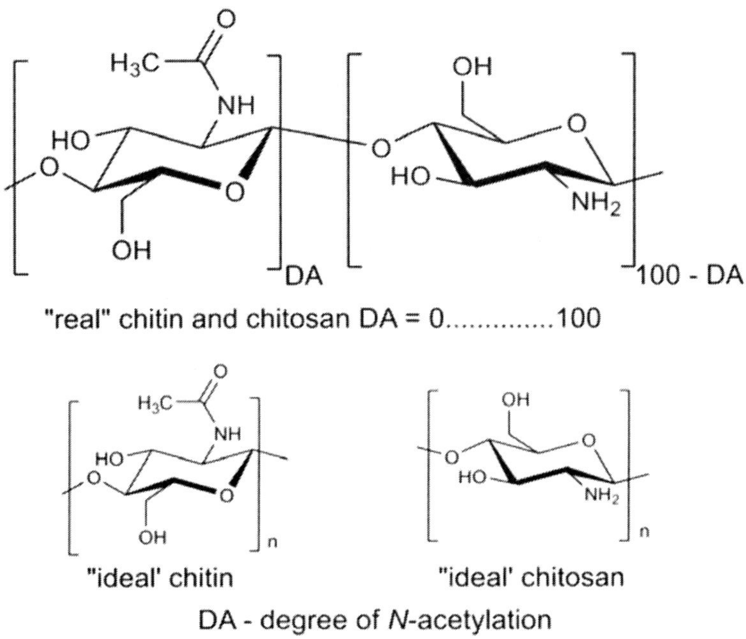

Figure 1. The chemical structures of chitin and chitosan.

Chitin and chitosan are of commercial interest because of their high nitrogen content (6.89%) and their excellent properties such as biocompatibility, biodegradability, non-toxicity and adsorptive abilities [2,8,11]. These compounds are of very low toxicity: LD_{50} of chitosan in laboratory mice is 16 g/kg body weight, which is similar to that of sugar and salt. In rats, chitosan is safe up to 19% in the diet [12]. As a material that is highly insoluble and has a low chemical reactivity, chitin has limited applicability. Recently, however, chitosan has come back into the spotlight because of its numerous and extensive applications–in biomedicine, waste water treatment, food, cosmetics and the fiber industry [7,8,11,13–17]. Most of the characteristic properties of chitosan are due to the high content of primary amino groups with a pKa of 6.3. At low pH, the positive charge on the-NH_3^+ groups converts chitosan to a water-soluble cationic polyelectrolyte; when pH increases to above 6.0 the positive charge on the amino groups is lost and chitosan becomes insoluble [7]. The soluble-insoluble transition of chitosan occurs around pH 6.0–6.5 at the pKa of its primary amino groups. The pKa depends closely on the degree of *N*-acetylation (DA), hence the solubility of chitosan is dependent on DA and the method of *N*-deacetylation [18]. The amino groups are also responsible for several straightforward chemical modifications of chitosan, which predisposes its ongoing development for many applications [19–22]. The improved solubility in water and organic solvents of chemically modified chitin and chitosan has been reported by many scientists [23–25].

Chitin is isolated from the exoskeletons of crustaceans, molluscs, insects and certain fungi, but the main commercial sources of chitin are crab and shrimp shells [2,26]. Depending on the source, chitin occurs in two allomorphs, the α and β-forms, and additionally as γ-chitin, which appears to be a combination of the α and β structures rather than a different allomorph [3]. α-Chitin is by far the most abundant and is usually isolated from the exoskeleton of crustaceans, particularly from shrimps and crabs. β-Chitin can be extracted from squid pens, and γ-chitin from fungi and yeast [27]. β-Chitin is easily converted to α-chitin by alkaline treatment followed by flushing in water [28].

Several techniques to extract chitin from different sources have been published [26,29–31]. Crustacean shells consist of proteins (30–40%), calcium carbonate (30–50%), chitin (20–30%) and pigments (astaxanthin, canthaxanthin, lutein or β-carotene). These proportions vary from species to species and from season to season [32]. The most common method for chemically isolating chitin from crustacean shells involves a number of major steps: the washing, grinding and sieving of raw shells, followed by their demineralization (elimination of calcium carbonate in dilute acidic acid) and deproteinization in aq NaOH or KOH. The use of enzymatic hydrolysis for deproteinization [29] and microorganisms for both demineralization and deproteinization has been also reported [29,33].

Industrially chitin is converted into the more readily applicable chitosan by partial or complete deacetylation of chitin in both the solid (heterogeneous process) and dissolved (homogeneous process) states under alkaline conditions or by enzymatic hydrolysis (using a chitin deacetylase). The source of natural chitin used to produce chitosan affects the production parameters and chitosan preparations. It has been shown that β-chitin is more reactive in *N*-deacetylation than α-chitin [34]. The difference in the semicrystalline morphology of chitin means, for example, that chitosans obtained in a solid-state reaction have a heterogeneous distribution of *N*-acetyl groups along the molecular chains [35]. Differences in the chitosan production process (e.g., temperature, alkali concentration, ratio of alkali solutions to the shells) also mean that chitosan preparations consist of a mixture of chitosans varying in molecular weight and degree of *N*-acetylation. Those produced from chitin may also contain impurities such as heavy metals, protein residues and acid/alkaline residues.

For most of the commercially available chitosans, parameters like weight-average molecular weight (M_W), polydispersity (M_W/M_N), degree of *N*-acetylation (DA), pattern of acetylation (PA) and impurity content (protein, heavy metal) are usually unknown [32,36]. Knowledge of the microstructure of chitosan samples is thus essential for an understanding of the structure–property–activity relationships in them, and special emphasis in this respect is placed on the chitosans used in biomedical applications [32,36,37]. The complete structural characterization of synthesized, biologically active and/or water-soluble derivatives of chitin and chitosan is also very important.

The present paper reviews the applications of spectroscopic methods for the structural analysis and physicochemical characterization of chitin, chitosan and their derivatives. The most important techniques are X-ray spectroscopy, infrared (IR) and UV-Vis-spectroscopy, mass spectrometry (MS), and nuclear magnetic resonance spectroscopy (NMR). The usefulness of these methods for determining and confirming molecular structures, for monitoring reactions and controlling the purity of these compounds is discussed.

2. APPLICATION OF SPECTROSCOPIC METHODS FOR ANALYZING THE STRUCTURE AND DETERMINING THE PHYSICOCHEMICAL PROPERTIES OF CHITIN, CHITOSAN AND THEIR DERIVATIVES

2.1. X-ray spectroscopy

X-ray spectroscopy is unarguably the most versatile and widely used means of characterizing materials of all forms [38]. There are two general types of structural information that can be studied by X-ray spectroscopy: electronic structure (focused on valence and core electrons, which control the chemical and physical properties, among others) and geometric structure (which gives information about the locations of all or a set of atoms in a molecule at an atomic resolution). This method encompasses several spectroscopic techniques for determining the electronic and geometric structures of materials using X-ray excitation: X-ray absorption spectroscopy (XAS), X-ray emission spectroscopy (XES), X-ray photoelectron spectroscopy (XPS) and X-ray Auger spectroscopy. Which type of X-ray spectroscopy is employed depends on whether the target information is electronic, geometric or refers to oxidation states: for instance, XAS (first developed by de Broglie) to probe empty states and the shapes of molecules or local structures [39], and XPS (first developed by Siegbahn) to investigate occupied electronic states [40]. X-ray spectroscopy is thus a powerful and flexible tool and an excellent complement to many structural analysis techniques such as UV-Vis, IR, NMR or Raman.

2.1.1. Typical conditions of X-ray measurements

In an X-ray diffraction measurement, a crystal is mounted on a goniometer and gradually rotated while being bombarded with X-rays, producing a diffraction pattern of regularly spaced spots known as reflections [41]. The two-dimensional images taken at different rotations are converted into a three-dimensional model of the density of electrons within the crystal using the mathematical method of Fourier transforms, combined with the chemical data obtained for the sample. If single crystals of sufficient size cannot be obtained, various other X-ray methods, like fiber and powder diffraction, can be applied to record less detailed information.

Clark and Smith in 1937 were the first to make crystal studies of chitin and chitosan using X-ray diffraction (XRD) [42]. They carried out those investigations using a commercial copper-target diffraction tube operated at 30 kV and 25 mA as the X-ray sources, which generated principally Cu-K_α lines. The diffraction patterns were recorded on a flat film perpendicular to the beam with the sample 5.0 cm from the film. In later research, the conditions for X-ray measurements of chitin and chitosan were mostly modifications of the ones used by Clark and Smith [42]. For instance, X-ray diffraction measurements were done at 100% relative humidity in a helium atmosphere [43] to avoid the X-ray scattering that led to a "dirty" background on the X-ray film. Present-day X-ray analyses of chitin, chitosan and their derivatives are carried out on advanced X-ray diffractometers.

Apart from traditional X-ray diffraction (XRD), other X-ray techniques for determining chitin/chitosan and their derivatives have been applied: X-ray photoelectron spectroscopy (XPS), the second most popular X-ray spectroscopy technique–for determining the bonding energies of C, O and N atoms on the surface of chitosan and its metal chelate, and for other chitin and chitosan investigations [44–49]; X-ray emission spectroscopy (XES)–perfect for studying the chemical bonding in chitosan and cross-linked chitosan derivatives [50]; X-ray absorption spectroscopy (XAS)–for determining the coordination number of Fe atoms in chitosan-metal complexes [51].

2.1.2. X-ray spectra of chitin and chitosan

Numerous X-ray spectroscopic studies of chitin and chitosan have yielded the diffractive patterns of these compounds [52–59]. However, different sources have characterised these patterns with differently indexed crystalline peaks: these can be labeled either using d, the centre-to-centre spacing of the crystallites, or Miller indices, e.g., (020). It is also very common to describe the diffraction pattern using the values of angles.

Typical spectra of chitin fiber and chitosan fiber are shown in Figure 2 [56]. The spectrum of the former exhibits broad peaks at d = 0.34, 0.45, 0.50 and 1.09 nm with a shoulder at 0.71 nm; in the latter the spectral peaks are at d = 0.45, 0.88 and 2.93 nm.

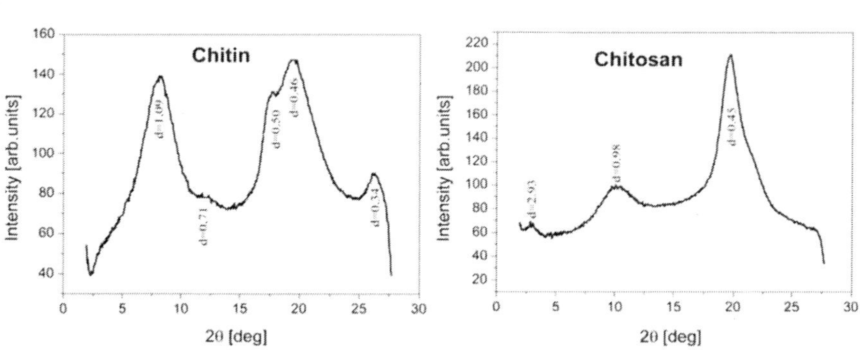

Figure 2. X-ray diffraction spectra of chitin and chitosan fibers. Reprinted from Carbohydrate Polymers 56, 2004, Muzzarelli, C., Francescangeli, O., Tosi, G., Muzzarelli, R.A.A., Susceptibility of dibutyryl chitin and regenerated chitin fibers to deacetylation and depolymerization by lipases, 137–146, Copyright (2010), with permission from Elsevier [56].

The Miller indices of the diffraction peaks characteristic of chitin were (020), (110), (120), (101) and (130) [57,58], although Muzzarelli *et al.* reported the lack of a strong (020) peak for chitin. [59]. In turn, the typical chitin diffraction pattern, given in angle form, showed strong reflections at 2ϑ around 9–10° and 2ϑ of 20–21° and minor reflections at higher 2ϑ values, e.g., at 26.4° and higher [60].

2.1.3. X-ray analysis of chitin and chitosan polymorphs

As already mentioned, the crystal structures of chitosan have been examined since the work of Clark and Smith in 1937 [42]. A great number of diffraction experiments have been undertaken in an attempt to elucidate the molecular geometry of chitosan [42,61–71]. The first X-ray studies showed that the chitosan molecule can adopt at least two different conformations in crystals–a 2-fold [42,61–63] and an eight-fold right-handed helical structure [64,65]. Apart from these two helical conformations, other polymorphs of chitosan have been characterized [66–71]. In 1994 Yui *et al.* determined the detailed crystal structure of the anhydrous form of chitosan by combining X-ray diffraction analysis with a stereochemical model refinement [67]. Chitosan chains crystallized in an orthorhombic unit cell of the following dimensions: a = 0.828 nm, b = 0.862 nm and c (fiber axis) = 1.043 nm. The X-ray diffraction pattern was recorded on the imaging plate, and the intensity of each reflection was estimated by two-dimensional measurement and subsequent background removal. The results showed that the chain conformation of this anhydrous form of chitosan was an extended two-fold helix stabilized by intramolecular O3···O5 hydrogen bonds and that the adjacent parallel chains were connected by O6···N2 hydrogen bonds.

At the same time Mazeau *et al.* examined the conformation and packing of a chitosan polymorph crystallized at a high temperature on the basis of diffraction data [68]. In this kind of polymorph, the chitosan chains form orthorhombic crystals with a = 0.807 nm, b = 0.844 nm and c (chain axis) = 1.034 nm. This chitosan molecule adopted a two-fold helical conformation stabilized by two intramolecular hydrogen bonds, a strong one between O5′ and O3, and a weaker one between O5′ and O6. These two anhydrous forms of chitosan have relatively similar structures.

In 1997 Yui's group published another paper on the X-ray fiber diffraction method used to determine the crystal structure of a hydrated form of chitosan [69]. The results indicated that the hydrated form of chitosan molecules had a two-fold helical symmetry reinforced by a O3····O5 hydrogen bond with a repeating period of 10.34 Å; this is a structure typical of β (1→4) linked polysaccharides such as cellulose, mannan and chitin.

The following crystalline polymorphs of chitosan have so far been found using X-ray diffraction measurements: the most abundant "tendon-chitosan" [42,69], as well as the "annealed" [63], "1–2", "L-2" [66], "form I" and "form II" [61] and "8-fold right-handed" forms [64,65]. Apart from the last one, all the polymorphs of chitosan molecules have the extended 2-fold helix configuration; the 8-fold polymorph is unstable and is easily converted into the 2-fold helix [72].

Depending on the source, chitin can occur in the α-, β- and γ-forms. The differences among them depend on the arrangement of chains in the crystalline regions [52]. The most abundant and stable form is α-chitin, which Minke and Blackwell studied in detail using XRD in 1978 [45]. These authors determined the α-chitin structure on the basis of intensity data from deproteinized lobster tendon. They discovered that α-chitin chains form orthorhombic crystals with a = 0.474 nm, b = 1.886 nm and c (fiber axis) = 1.032 nm. Additionally, the chains form hydrogen-bonded sheets linked by C=O...H–N bonds approximately parallel to the a axis, and each chain is stabilized by an C(3')O–H····OC(5) intramolecular hydrogen bond, as in cellulose. These data also indicated that a statistical mixture of CH_2OH orientations was present, equivalent to half an oxygen on each residue, each forming inter- and intramolecular hydrogen bonds. As a result, Minke and Blackwell found that the α-chitin structure contained two types of amide groups, differing in their hydrogen bonding. In addition, the inability of this polymorph to swell in the presence of water was explained by the extensive intermolecular hydrogen bonding. The modes of hydrogen bonding in α- and β-chitin are illustrated in Figures 3 and 4 [18].

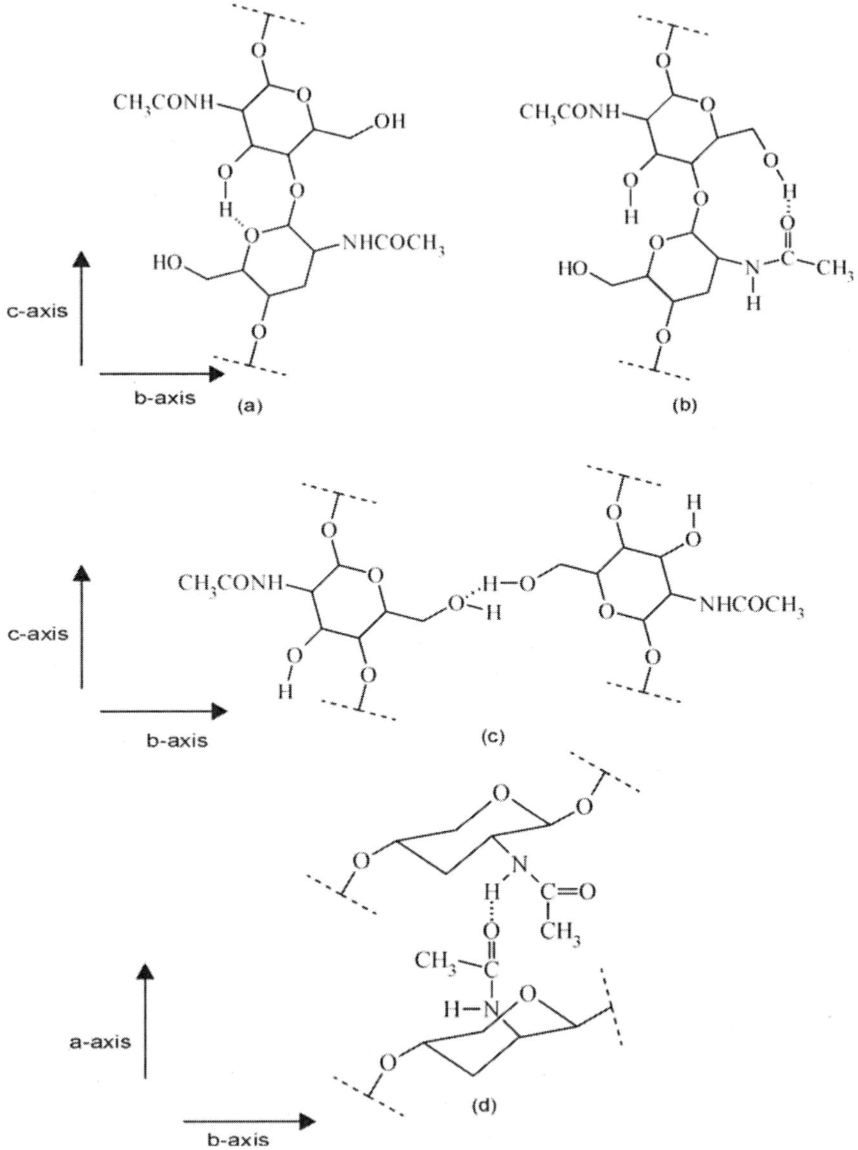

Figure 3. Modes of hydrogen bonding in α-chitin: (a) intrachain C(3′)OH···OC(5) bond; (b) intrachain $C(6'_1)OH \cdots O=C(7_1)$ bond; (c) interchain $C(6'_1)O \cdots HOC(6_2)$ bond; (d) interchain $C(2_1)NH \cdots O=C(7_3)$ bond. Adapted with permission from *Biomacromolecules*.**2000**, *1*, 609–614. Copyright 2010, American Chemical Society [18].

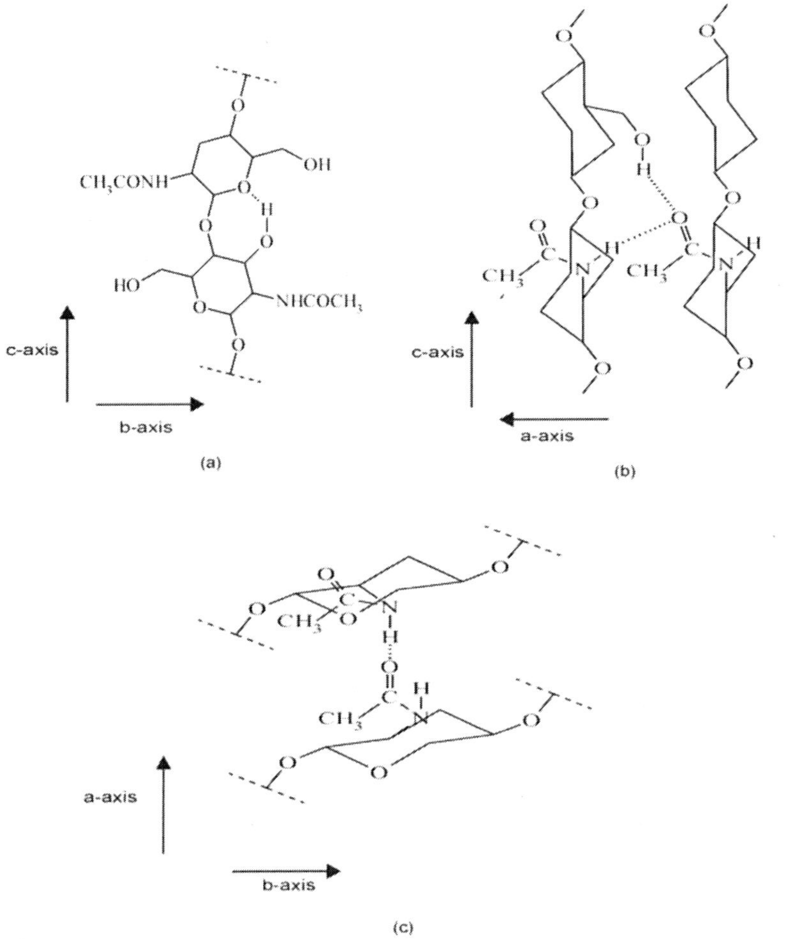

Figure 4. Modes of hydrogen bonding in β-chitin: (a) intrachain C(3′)OH···OC(5) bond; (b) interchain C(2_1)NH···O=C(7_3) bond and C($6'_1$)OH···O=C(7_3) bond (*ac* plane projection); (c) interchain C(2_1)NH···O=C(7_3) bond (*ab* plane projection). Adapted with permission from *Biomacromolecules*. **2000**, *1*, 609–614. Copyright 2010, American Chemical Society [18].

As mentioned above, the differences among chitin polymorphs are due to the arrangement of the chains in the crystalline regions: α-chitin has a structure of antiparallel chains [45], β-chitin has intrasheet hydrogen-bonding by parallel chains [52], and γ-chitin, being a combination of α- and β-chitin [52], has both parallel and antiparallel structures. Because of these differences each chitin polymorph has different properties specific to it. For example, β-chitin is more soluble in and more reactive towards solvents and has a greater affinity towards them; it is also more susceptible to swelling than α-chitin. β-Chitin is also more amenable to *N*-deacetylation than α-chitin. In this case, comparing the diffraction profiles of α- and β-chitin before and after *N*-deacetylation using X-ray spectroscopy seems to be a relatively good solution for distinguishing these forms of chitin. Abdou *et al.* used X-ray diffraction to study two chitin polymorphs, the α- and β-forms, obtained from different sources [60]. The chitin samples were converted into the more soluble chitosan by steeping them in solutions of NaOH of various concentrations and for extended periods of time. The X-ray diffraction patterns of the α-chitin samples and the corresponding hydrolyzed chitosans showed strong reflections at 2ϑ around 9–10° and 2ϑ of 20–21° and minor reflections at higher 2ϑ values at e.g., 26.4° and higher. The chitin bands were sharper than the chitosan bands, even though there was only a slight decrease in the crystallinity percentage. In turn, the X-ray diffraction patterns of the β-chitin samples and their corresponding hydrolyzed chitosans showed that the band at $2\vartheta = 9.9°$ decreased significantly after deacetylation, and that this was followed by a dramatic decrease in the crystallinity percentage. It was therefore concluded that β-chitin is much more amenable to *N*-deacetylation than the α-form. The X-ray diffraction patterns of α- and β-chitin and the corresponding hydrolysed chitosans make them easily distinguishable. The crystallinity index (*CI*) can also be calculated on the basis of X-ray diffractograms. This takes different values for different forms of chitin; for example, Lima showed that *CI* for α-chitin is 28.3% but that for β-chitin it is 20.8% [73].

Many other studies of chitin polymorphs have revealed differences in crystallinity peaks between α-, β- and γ-chitins obtained from various sources [52–55]. For example, Jang *et al.* found crystalline peaks at 9.6, 19.6, 21.1 and 23.7° for α-chitin, at 9.1 and 20.3° for β-chitin, and at 9.6 and 19.8° for γ-chitin [52]. Similarly, Cárdenas *et al.* [53] reported that WAXD patterns of α-chitin (chitins from shrimp, lobster, prawn and king crab) and β-chitin (chitin from squid) exhibited their major characteristic peak at 19.2–19.3° and 18.8° respectively. Kim [54] found that β-chitin from squid pen exhibited crystalline peaks at 9.8° and 19.3°. Yen and Mau [55] found that fungal chitin (γ-chitin) showed two crystalline reflections at 5.4–5.6° and 19.3–19.6°. Irrespective of their origin, the three types of chitin consistently display a major peak at ~19° in their crystallinity structure.

2.1.4. Physicochemical characterization of chitin and chitosan using X-ray diffraction

In 1937, Clark and Smith, in their pioneer X-ray diffraction studies of chitin/chitosan and their derivatives [42], were the first to investigate the physicochemical properties of chitin. They reported the action of hydrochloric acid, lithium thiocyanate and nitric acid on chitin. Their data showed that even at room temperature the ether linkages of chitin were hydrolysed in hydrochloric acid; concurrently, but more slowly, the amide groups were also hydrolysed. In addition, those authors investigated the dispersion of chitin in lithium thiocyanate; at a temperature of 200 °C chitin formed a definite addition compound with lithium thiocyanate, but at lower temperatures only intramicellar swelling was observed. In turn, chitin nitrate was roughly as soluble in hydrochloric acid as the original chitin–no substantial hydrolysis of the acetyl groups had occurred.

Subsequent studies revealed that the properties of chitin and chitosan depended mostly on the degree of *N*-acetylation, molecular weight, polydispersity and crystallinity [31]. Commonly used to measure crystallinity, XRD is also applied to determine the degree of *N*-acetylation of chitin and chitosan [74].

In 1990, Focher *et al.* [75] used XRD to study chitin and postulated the following equation for determining the crystallinity index (*CI*):

$$CI\,(\%) = [(I_{110} - I_{am})/I_{110}] \times 100 \qquad (1)$$

where I_{110} (arbitrary units) is the maximum intensity of the (110) peak at around $2\vartheta = 19°$, and I_{am} (arbitrary units) is the amorphous diffraction at $2\vartheta = 12.6°$. This expression had in fact been employed three years earlier by Struszczyk to determine the *CI* of chitosan [76]. Currently this equation is routinely applied during investigations of chitin, chitosan and their derivatives [73,77,78]. In most cases, *CI* provides information about the crystal state, but it is also very useful for distinguishing α-chitin from β-chitin [73].

On the basis of X-ray powder diffractograms of chitin and chitosan with different degrees of *N*-acetylation, Zhang noted two maximum peaks of the following intensities: one at the (020) reflection and the other at the (110) reflection [74]. He therefore postulated a crystallinity index (*CI*) expressed by two equations:

$$CI_{020} = [(I_{020} - I_{am})/I_{110}] \times 100 \quad (2)$$

$$CI_{110} = [(I_{110} - I_{am})/I_{110}] \times 100 \quad (3)$$

Further chitin and chitosan studies indicated that crystallinity could also be assigned from an X-ray diffractogram by dividing the area of the crystalline peaks by the total area under the curve (background area) [60,79,80]. In these calculations, the crystallinity percentage supplied information on relative crystallinity.

A lot of studies have been carried out in which X-ray measurements were applied to determine which parameters affect the crystalline structure of chitin and chitosan and how they do so. In 1991 Ogawa reported an increase in crystallinity with a decrease in the M_W of chitosan [71]. One year later,he also determined how chitosan polymorphism and its crystallinity in the membrane depended on the membrane preparation procedure and the molecular weight of the chitosan [43]. XRD measurements demonstrated that when an acetic acid solution of chitosan was dried in air and then soaked in an alkaline solution, both hydrated and anhydrous polymorphs of chitin were present in the resulting membrane. On the other hand, when a highly concentrated solution of chitosan in aqueous acetic acid was neutralized with an alkaline solution, no anhydrous polymorphs were detected in the membrane because drying was incomplete. In another paper from 1993 [81], Ogawa compared the crystallinities of partially *N*-deacetylated chitin (PDC) and partially *N*-acetylated chitosan (PAC) samples with a similar degree of *N*-acetylation and their behavior by heating them in water. He discovered that the *N*-acetylation of pure chitosan is far superior to the solid-state *N*-deacetylation of chitin for producing a less crystalline sample, and in particular, for obtaining a less anhydrous crystal.

The effect of DA on solubility in relation to the crystal structure of deacetylated chitin was also discussed by Cho *et al.* in 2000 [18]. Wide-angle X-ray diffractometry (WAXD) revealed that chitin with ca 72% DA retained the crystal structure of α-chitin with significantly reduced crystallinity and crystallite perfection. The water-soluble chitin with *circa* 51% DA had a new crystal structure resembling that of β-chitin rather than that of either α-chitin or chitosan, suggesting that homogeneous deacetylation converted the crystal structure of chitin from the α- to the β-form.

Using X-ray powder diffraction Zhang *et al.* [74] also tried to look for a relationship between the crystalline state and DA of chitin. Figure 5 presents the XRD patterns of chitin and chitosan with different degrees of *N*-acetylation [74]. These authors noted that the peak of maximum intensity at the (020) reflection diminished together with the decrease in DA and moved to a higher angle. The second intensive peak at the (110) reflection also diminished with the decrease in DA. Consequently, CI_{020} decreased linearly with the decrease in DA. This linear relationship between CI_{020} and the degree of *N*-acetylation suggested the possible use of XRD for determining DA of macromolecular chitin and chitosan.

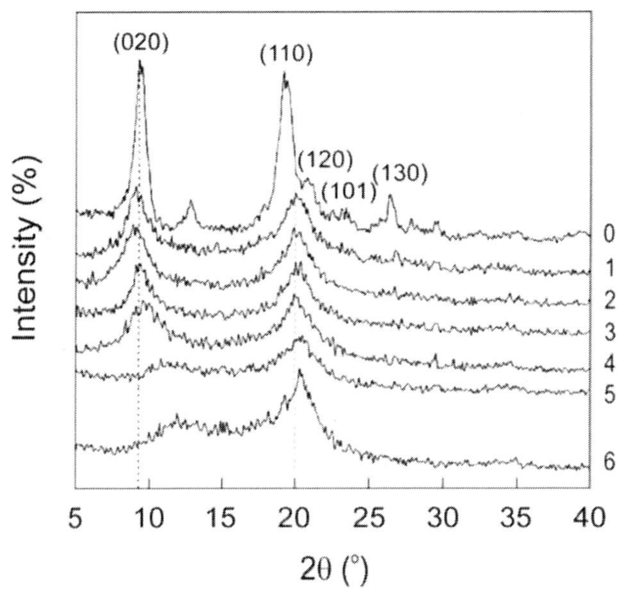

Figure 5. Comparison of X-ray powder diffractograms of chitin and chitosan with different degrees of *N*-acetylation. Figures 0–6 imply different DA (%): 0-83.1, 1-40.6, 2-36.5, 3-41.3, 4-28.6, 5-13.0, 6-7.2. Reprinted from Carbohydrate Research 340, 2005, Zhang, Y., Xue, C., Xue, Y., Gao, R., Zhang, X., Determination of the degree of deacetylation of chitin and chitosan by X-ray powder diffraction, 1914–1917, Copyright (2010), with permission from Elsevier [74].

As demonstrated, molecular weight and DA are the major parameters significantly influencing the crystal structure of chitin and chitosan, although it has also been reported that the crystallinity index depends on other factors. For example, Seoudi reported that *CI* decreased after chitin was treated with HCl and NaOH [82], and Wada and Saito [57] found that when the biopolymer was heated from room temperature to 250 °C the α-chitin remained structurally stable. Moreover, the influence of alkali-freezing treatment on the solid state structure of chitin was examined by Feng [58]. XRD revealed that during this treatment, the crystal space parameters of chitin changed, and the order of the hydrogen bonds in chitin was modified.

X-ray diffraction was used to measure the elastic moduli E_l of the crystalline regions of α-chitin and chitosan [83]. This parameter provided important information on the molecular conformation in the crystal lattice and the mechanism of deformation in the crystalline regions. The data enabled the elastic moduli E_l of the crystalline regions in the direction parallel to the chain axis at 20 °C to be assigned as 41 GPa for α-chitin and 65 GPa for chitosan. These E_l values, which are low compared to those for cellulose I (138 GPa), were due to the contracted skeletons of α-chitin and chitosan in the crystal lattice. In addition, the elastic moduli calculated from X-ray data showed that the molecular chain of α-chitin in the crystal lattice was mechanically stable from − 190 °C to 150 °C.

2.1.5. X-ray analysis of chitosan salts

Chitosan has a regular distribution of aliphatic primary amino groups along its chain. These produce salts when the molecule reacts with inorganic or organic acids. Ogawa and Inukai used XRD to study several crystallized inorganic acid salts of chitosan [64]. On the basis of X-ray fiber diffraction, these authors suggested that chitosan acid salts took up two different conformations. One, which they called “type I salt”, retained the extended two-fold helix of the unreacted chitosan molecule, although they were different crystals. The second one, the “type II salt”, had an eight-fold helical conformation in the crystal. The salts forming with HNO_3, HBr and HI took up the former structure, and those with HF, HCl, and H_2SO_4 the latter one. Crystals of type I salts were anhydrous, whereas those of type II were hydrated. Despite the different anion sizes, all the type II salts gave fiber patterns that were very similar to each other. They crystallized in a monoclinic unit cell with a helical repetition of 4.073 nm. The chirality of the eight-fold helix was right-handed, since their fiber patterns were very similar to those obtained by Cairns *et al.* [65].

Apart from inorganic salts, chitosan can easily form organic salts, for example, when it reacts with L-ascorbic acid. In 1996 Ogawa carried out an X-ray study of the chain conformation of the ascorbic acid salt of chitosan [84]. He determined that both L- and D-ascorbate chitosan salts retained the extended 2-fold helical conformation of the unreacted chitosan chain and that both crystals were anhydrous. In view of these results, he classified them as type I salts of the chitosan acid salts [64], although during the preparation of the L- and D-ascorbates of chitosan, that author discovered that D-ascorbic acid displayed a higher reactivity towards chitosan than the L-isomer. Those differences in reactivity between L- and D-ascorbic acids could be connected with the optical resolution of ascorbic acid.

Chitosan salts were also examined in 1999 by Kawada *et al.* [72]. They studied the spontaneous water-removing action of acid by preparing chitosan salts of many different (monocarboxylic, inorganic and organic) acids, and examined their structures using X-ray diffraction. The results indicated that the temperature required for salt formation depended on the hydrophobicity of the acid; for instance, the chitosan formic acid salt could be prepared at room temperature, whereas the formation of the propionic acid salt had to be carried out at 4 °C. Moreover, type II salts of monocarboxylic acids, the hydrated crystals of chitosan, could be dehydrated even at room temperature without any loss of orientation or decomposition of the chitosan specimen.

2.1.6. X-ray analysis of chitosan derivatives

Chitin and chitosan are a family of polymers with highly variable chemical and physical properties. These compounds and their derivatives have at least 200 potential and current applications [32] in the biomedical, food, biotechnological, agricultural and cosmetics industries.

Muzzarelli and co-workers found that chitosan exhibited good adsorption selectivity towards some transition and post-transition metal ions from aqueous solution [85]. A chitosan-metal complex dissociates easily when the pH is lowered; therefore, chitosan is very often used in the recovery of useful transition metals from waste. Ogawa and Oka reported on the behavior of chitosan-metal complex formations examined by X-ray diffraction [86]. The unit cells of all the salt complexes studied were orthorhombic, although the cupric salt complexes showed some unindexed reflections. Their lattice parameters and the number of water molecules in the cell depended on the counteranions of the metal salt and not on the metal ion. The ratio of glucosamine residues to metal salts was 2:1. On the basis of the fiber diffraction patterns of various chitosan-transition metal salt complexes, these authors postulated a coordination mode which they named the "pendant model". This model had already been put forward by Ogawa *et al.* back in 1984, but unequivocal experimental evidence to support it was lacking [70]. In this "pendant model" metal anions were coordinated to the amino groups of the chitosan chain like a pendant. Additionaly there was another contrasting model called "the bridge model", in which metal ions were coordinated with four nitrogen atoms of the intra- and inter-chitosan chains [87]. In spite of these differences, chitosan exhibits a high affinity for metal ions, a property that has been used to recover transition metals from waste water. At present, the adsorption properties of metal ions on chitin and chitosan derivatives are still routinely examined by X-ray diffraction [49,88–90].

The majority of current studies on chitin and chitosan are seeking to discover new derivatives with unusual properties and different potential applications [56,91–97]. X-ray measurements are still very often applied to characterize most of these new derivatives. For example, chitosan-based nanocomposite films, containing chitin nanocrystals as functional components, were successfully prepared and cross-linked using glutaraldehyde [97]. XRD showed that chitin nanocrystals retained their crystalline morphology in the nanocomposites before and after cross-linking, and that chitosan also retained its amorphous characteristics in the nanocomposites. Another example, novel chitosan/gelatin membranes were prepared using a suspension of chitosan hydrogel mixed with gelatin [95]. XRD studies showed that the chitosan and gelatin in these membranes are compatible and interact well with each other. In addition, the incorporation of gelatin reduced the crystallinity of chitosan. Finally, natural rubber/chitosan blends were studied by XRD analysis [91], the measurements indicating that vulcanization enhanced the crystallinity.

2.1.7. Other X-ray techniques used in chitin and chitosan analysis

As mentioned at the beginning of this section, X-ray spectroscopic techniques other than the traditional X-ray diffraction measurements have been successfully used to analyze chitin, chitosan and their derivatives.

X-ray photoelectron spectroscopy (XPS) is usually used to determine the bonding energies of C, O and N atoms on the surface of chitosan and its metal chelates, although this is not its only use [44–49]. For example Matienzo and Winnacker [46] presented high-resolution C1s, N1s and O1s XPS spectra for a chitosan film coated on an Al-silicon surface. The films were treated in either an oxygen plasma environment or under UV/ozone irradiation. XPS data showed that hydroxyl and amine entities participated only minimally in the modification. In addition, deposition of chitosan films onto Al-coated silicon wafers produced films with a more ordered chitosan structure. Surface analysis of modified films by XPS also indicated that neither the hydroxyl groups nor the amine segments appeared to participate in surface degradation reactions by either UV/ozone or oxygen plasma during the exposure times chosen for those studies. XPS also provided information regarding the forms of species absorbed on the polymer [48]. For example, a study of the interactions of Cu^{2+}, Mo^{4+} and Cr^{3+} with chitosan beads, cross-linked chitosan beads and chitosan flakes revealed that the adsorption of Mo^{4+} and Cr^{3+} on chitosan flakes and beads was followed by reduction of the Mo^{4+} and Cr^{3+}. Another, example of the use of XPS was presented by Veleshko *et al.* [47]. They examined the complexation between the uranyl group U(VI) and chitosan by means of X-ray photoelectron spectra. The results showed that the interaction of chitosan with the uranyl group yielded complexes containing the nitrogen atom of the amino group and, very probably, oxygen atoms from the chitosan ring and free hydroxyl groups in the equatorial plane.

X-ray emission spectroscopy (XES), also known as X-ray fluorescence spectroscopy (XFS), is a very sensitive probe for examining the local electronic structure and chemical bonding of the emitting atoms. This X-ray technique was used by Kurmaev *et al.* in 2002 [98] to study the chemical bonding in chitosan and chitosan cross-linked with ethylene glycol diglycidyl ether (EGDE). These authors concluded that the changes in the width of resonantly excited O Kα XES were due to site-selective excitation of oxygen atoms belonging to different functional groups (OH and –O–). Comparison of the nitrogen Kα spectra of unmodified and cross-linked chitosan proved that the preferred structural model was the one according to which EGDE was linked via the hydroxyl group.

2.2. Infrared spectroscopy

Infrared (IR) spectroscopy is one of the most important and widely used analytical techniques available to scientists working on chitin and chitosan. It is based on the vibrations of the atoms of a molecule. The infrared spectrum is commonly obtained by passing infrared electromagnetic radiation through a sample that possesses a permanent or induced dipole moment and determining what fraction of the incident radiation is absorbed at a particular energy [99]. The energy of each peak in an absorption spectrum corresponds to the frequency of the vibration of a molecule part, thus allowing qualitative identification of certain bond types in the sample. An IR spectrometer usually records the energy of the electromagnetic radiation that is transmitted through a sample as a function of the wavenumber or frequency. Nowadays, the total spectrum is analyzed by an interference process and converted into the frequency or wavenumber range by means of a mathematical process known as the Fourier transform. Fourier-transform infrared (FTIR) spectroscopy has dramatically improved the quality of infrared spectra and minimized the time required to obtain data [99,100]. Progress in modern infrared spectroscopy is reviewed in literature [101,102].

2.2.1. Typical conditions for the FTIR spectroscopic analysis of chitin, chitosan and their derivatives

FTIR spectra are usually recorded in the middle infrared (4000 cm^{-1} to 400 cm^{-1}) with a resolution of 4 cm^{-1} in the absorbance mode for 8 to 128 scans at room temperature. The samples for FTIR analysis are prepared by grinding the dry blended powders with powdered KBr, often in the ratio of 1:5 (Sample: KBr) and then compressed to form discs. Spectra are sometimes measured using a deuterated triglycerinesulphate detector (DTGS) with a specific detectivity of 1×10^9 $cmHz^{1/2}w^{-1}$ [103] or on films using an attenuated total refraction (ATR) method in an IR spectrometer [104–106]. Diffuse Reflectance Infrared Fourier-Transform (DRIFT) spectroscopic analysis is also applied [107].

2.2.2. Physicochemical characterization of chitin and chitosan using infrared spectroscopy

As already mentioned, natural chitin occurs mainly as α- and β-chitin. The description and interpretation of the infrared spectra of these two forms of chitin have been published by many scientists [108–110]. By way of example, the spectra of α- and β-chitin, and the *Ianthella basta* scaffold after NaOH treatment and H_2O_2 purification are shown in Figure 6 [110].

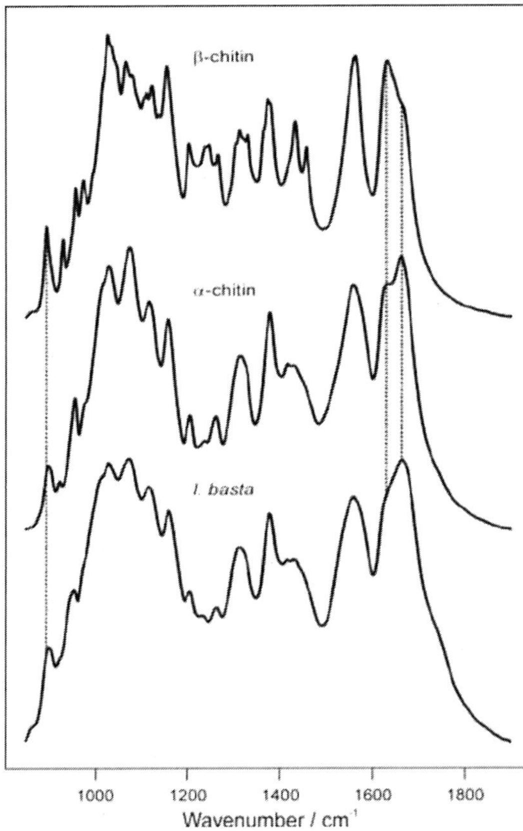

Figure 6. FTIR spectra of β-chitin from *T. rotula*, α-chitin from crabs and *I. basta* chitin after NaOH and H_2O_2 treatment. Dashed vertical lines are drawn to mark characteristic differences between α- and β-chitin. Reprinted from Journal of Structural Biology 168, 2009, Brunner, E., Ehrlich, H., Schupp, P., Hedrich, R., Hunoldt, S., Kammer, M., Machill, S., Paasch, S., Bazhenov, V.V., Kurek, D.V., Arnold, T., Brockmann, S., Ruhnov, M., Born, R., Chitin-based scaffolds are an integral part of the skeleton of the marine demosponge *Ianthella basta*, 539–547, Copyright (2010), with permission from Elsevier [110].

The spectra of α- and β-chitin display a series of narrow absorption bands, typical of crystalline polysaccharide samples. The C=O stretching region of the amide moiety, between 1700 and 1500 cm^{-1}, yields different signatures for α- and β-chitin. For α-chitin, the amide I band is split into two components at 1660 and 1630 cm^{-1} (due to the influence of hydrogen bonding or the presence of an enol form of the amide moiety [109–111]), whereas for β-chitin it is at 1630 cm^{-1} (Figure 6). The amide II band is observed in both chitin allomorphs: at 1558 cm^{-1} for α-chitin and 1562 cm^{-1} for β-chitin [110]. Another characteristic marker is the CH deformation of the β-glycosidic bond. This band shifts from 890 cm^{-1} in β-chitin to 895 cm^{-1}in α-chitin. Infrared spectra of β-chitin reveal two additional bands for CH_x deformations at about 1455 and 1374 cm^{-1} and a greater number of narrower bands in the C–O–C and C–O stretching vibration region (1200–950 cm^{-1}) not observed in α-chitin. As shown in Figure 6, the FTIR spectrum of the chitin isolated from *I. basta* confirmed the finding that this chitin resembles α-chitin more closely than β-chitin [110], demonstrating that FTIR can be used to determine chitin allomorphs.

FTIR spectroscopy has been used to characterize not only isolated chitin [110] but also the source of chitin, e.g., in two species of black coral, *Antipathes caribbeana* and *A. pennacea* [112]. Although FTIR absorption spectra of the natural samples (without deproteinization) showed similar distribution patterns for both species of coral, and confirmed the presence of chitin in both species, small differences were observed (e.g., the intensity of the IR absorption bands in *A. caribbeana* was stronger). The absence of a free hydroxyl in the hydroxymethyl groups CH_2OH in *A. caribbeana* (determined by FTIR analysis) indicated that the chitin chains were organized in sheets, where they were hydrogen-bonded to adjacent chains, a situation that favors a denser fiber packing of chitin. This means that the FTIR measurements permitted an explanation of why natural *A. caribbeana* coral was harder to pulverize and required a longer deproteinization time than *A. pennacea*. The presence of chitin in polyplacophoran sclerites was also confirmed by IR [113].

FTIR spectroscopy has also been used to compare the yield and purity of chitin isolated from pupae of the silkworm (*Bombyx mori*) using two methods of extraction: an open reactor and 1 h of an acidic reaction, and extraction in a closed reactor within 24 h of a basic reaction [114]. The efficiency of chitosan production by the *N*-deacetylation of chitin was also investigated by IR in this work. During the *N*-deacetylation of chitin, the band at 1655 cm^{-1} gradually decreased, while that at 1590 cm^{-1} increased, indicating the prevalence of NH_2 groups. The spectra of chitin and chitosan obtained from the *N*-deacetylation of chitin using a solution of NaOH (40 wt%) in the presence of $NaBH_4$ for 5 h are presented in Figure 7.

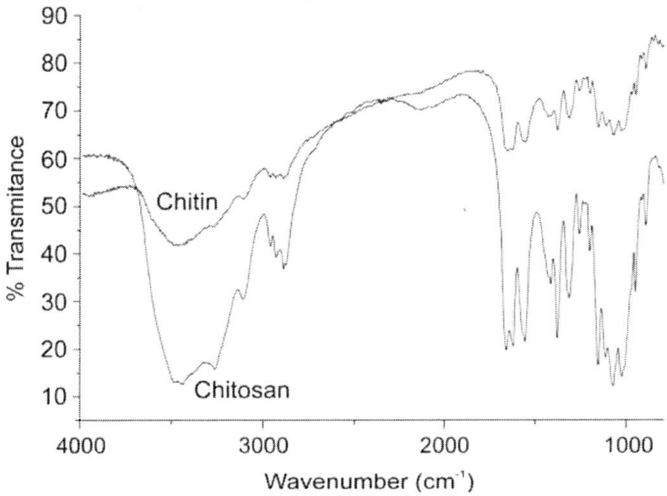

Figure 7. FTIR of chitin (A) and chitosan (B) produced from silkworm pupae; range: 1400–1700 cm^{-1}. Reprinted from Carbohydrate Polymers 64, 2006, Paulino, A.T., Simionato, J.I., Garcia, J.C., Nozaki J., Characterization of chitosan and chitin produced from silkworm crysalides, 98–103, Copyright (2010), with permission from Elsevier [114].

The band at 1590 cm^{-1} displayed a greater intensity than the one at 1655 cm^{-1} and demonstrated the effective deacetylation of chitin. FTIR analyses were also used to find the optimal conditions for the *N*-deacetylation of chitin whiskers (the alkali concentration and the treatment time) using a microwave technique [115].

Prashanhi *et al.* [116] applied IR spectroscopy to observe the changes occurring in the crystallinity and polymorphic nature of chitosan as a function of the *N*-deacetylation of chitin under different conditions: uncontrolled conditions (chitosan A), under an N_2 atmosphere (chitosan B), and with thiophenol (chitosan C). The FTIR spectra of chitosans A, B, and C were similar to each other, but there were subtle differences in the absorption intensities. Apart from the expected decrease in the band at 1665 cm^{-1} (amide I), the vibrational mode of amide II at 1550 cm^{-1} for chitin appeared at 1604 cm^{-1}, 1598 cm^{-1} and 1592 cm^{-1} for chitosan A, B and C respectively [116]. In none of these spectra were there any sharp absorptions at *circa* 3500 cm^{-1}, which confirms that the hydroxyl groups in positions C2 and C6 of the chitosans are involved in intra- and intermolecular hydrogen bonds. The region above 3000 cm^{-1} was centred at 3395 cm^{-1} in chitosan A, at 3407 cm^{-1} in chitosan B and at 3419 cm^{-1} in chitosan C; the shift to the higher frequency demonstrated a higher-order structure for these three chitosans. The CH_2 stretching bands of chitosan B around 1425 cm^{-1} were more intense than those of chitosans A and C. Furthermore, the FTIR spectra exhibited a progressive weakening of the bands at 3265 cm^{-1} and 3100 cm^{-1} during *N*-deacetylation, and the A_{1382}/A_{2920} cm^{-1} ratios of 0.65, 0.56 and 0.46 indicated a higher order structure of the chitosans prepared with thiophenol than those prepared under an N_2 atmosphere. The ratio of the band intensities at 1379 and 2900 cm^{-1} was also used to estimate the crystallinity of chitin and chitosan by Focher *et al.* [75] and Wu *et al.* [117], whereas Prashanth and Tharanathan used the sharp absorption peak around 618 cm^{-1} [118].

The impurities in chitosan preparation were determined by FTIR analysis [119]. The FTIR spectra of control chitosan and low-molecular-weight water-soluble chitosan (LMWSC) prepared in this study were compared to establish the contaminants and prove the synthesis of LMWSC. In the LMWSC spectrum the carboxyl group absorption band derived from lactic acid and impurities formed during enzyme degradation disappeared or significantly decreased.

FTIR spectroscopy has been employed to measure the critical concentration of two chitooligosaccharides form a lyotropic liquid crystalline phase in formic acid (C_1) [120]. Strong interactions between sugar chains and solvent were revealed by the widening of bands attributed to the –OH, –NH, –NHCO– of the chitooligosaccharide, including the C=O of formic acid. FTIR measurements of the shift of seven bands–1. –NH_2, –OH (3390–3418 cm^{-1}), 2. C=O of formic acid (1716–1724 cm^{-1}), 3. amide I (1626–1633 cm^{-1}), 4. amide II (1520–1531 cm^{-1}), 5. 6. C_3–OH (double peaks, 1178–1189 cm^{-1} and 1148–1153 cm^{-1}), 7. C_6–OH (1073–1074 cm^{-1})–enabled the C_1 values of these chitooligosaccharides to be established.

2.2.3. Determination of the degree of *N*-acetylation of chitin and chitosan using infrared spectroscopy

The degree of *N*-acetylation is one of the most important chemical parameters capable of influencing the performance of chitosan and chitin in many of their applications [2,7,121]. Of the various analytical techniques developed for DA determination [122], infrared spectroscopy is at the centre of attention. A convenient and relatively quick technique, it allows the DA values of chitin/chitosan to be determined on the basis of absorption ratios, also in the solid state [123]. Several procedures using different absorption ratios have already been proposed for determining DA for chitin and chitosan samples [124–130]. A review article summarizing the latest literature information on DA determination by IR spectroscopy for chitin and chitosan was published by Kasaai [131]. In that paper, various IR procedures were compared for their performances and limitations, advantages and disadvantages, and different factors affecting the experimental results were discussed, as were the validity data of DA measurements by FTIR spectroscopy. In view of this, the present review will discuss only general information on DA determination by IR.

DA can be determined by IR techniques in the following ways:

- Determination of the A_M/A_R ratio, where A_M is the intensity of the characteristic band of *N*-acetylation, which is a measure of the *N*-acetyl or amine content, and A_R is the intensity of a reference band that does not change with different DA values. The DA parameter of unknown samples can be established by comparing the determined A_M/A_R values with similar ratios of a few reference samples of known DA.

- Drawing a calibration curve by plotting the absorption ratio of chitin/chitosan samples of known DA *versus* their DA as established by IR or a reference method such as ^{1}H NMR spectroscopy. The DA values of unknown samples can then be estimated from the calibration curve.

- Statistical evaluation of several absorption band ratios [131,132].

IR techniques require choosing an appropriate band measure, an appropriate reference band, and drawing a good base line, necessary for measuring the intensity of absorption. The amide I bands at 1655 cm^{-1} (sometimes together with the amide I band at 1630 cm^{-1}) or the amide II band at 1560 cm^{-1} are used as the characteristic band(s) of *N*-acetylation. Among the postulated internal reference bands are the OH stretching band at 3450 cm^{-1} [125,133], the C-H stretching band at 2870–2880 cm^{-1} [134], the –CH_2 bending centred at 1420 cm^{-1} [129], the amide III band at 1315–1320 cm^{-1} [135], the anti-symmetric stretching of the C-O-C bridge at around 1160 cm^{-1} [136], the skeletal vibrations involving the C-O-C stretching bands at 1070 or 1030 cm^{-1} [127,128] and the band at 897 cm^{-1} (C-O-C bridge as well as glycosidic linkage) [137]. The different baselines suggested in the literature [129] are presented in Figure 8.

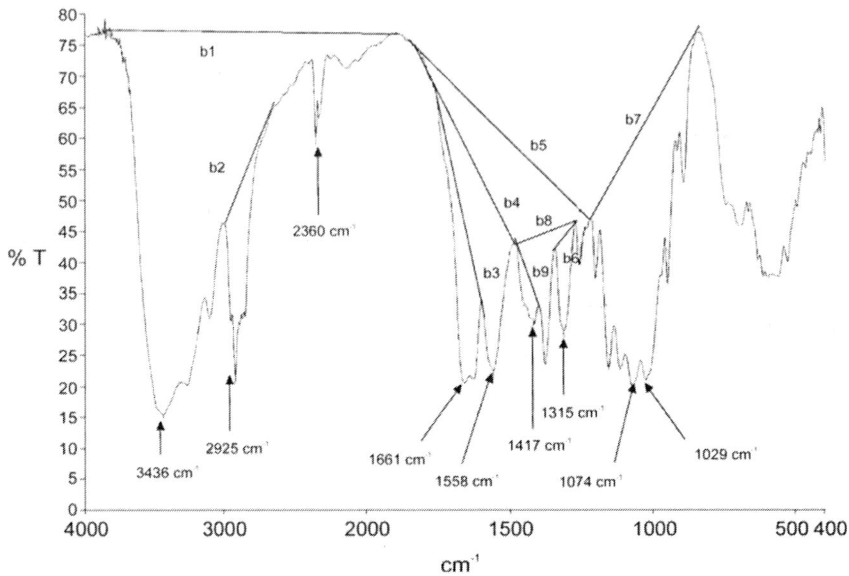

Figure 8. IR spectrum of chitin. Representation of the different baselines mentioned in the literature. Reprinted from Polymer 42, 2001, Brugnerotto, J., Lizardi, J., Goycoolea, F.M., Argüelles-Monal, W., Desbrières, J., Rinaudo, M., An infrared investigation in relation with chitin and chitosan characterization, 3569–3580, Copyright (2010), with permission from Elsevier [129].

Many different absorption band ratios, such as A_{1560}/A_{2875}, A_{1655}/A_{2875}, A_{1655}/A_{3450}, A_{1320}/A_{3450}, A_{1655}/A_{1070}, A_{1655}/A_{1030}, A_{1560}/A_{1160}, A_{1560}/A_{897} and A_{1320}/A_{1420}, have been used to determine DA by FTIR spectroscopy [131]. The different calibration curves proposed in the literature have different baselines and different characteristic bands for measuring the *N*-acetyl content. Moreover, the validity data of these calibrations depend on the absolute technique used to measure DA, and the conditions under which IR spectra are measured may also have a strong influence on the accuracy of DA values. The IR spectra (shown as absorbance) of α-chitin and β-chitin recorded using different sampling techniques are compared in Figure 9 [129].

The ATR spectrum recorded on the film is shown in Figure 9a. This revealed a very low resolution following the collection of 64 scans recorded in typical fashion. The amide I bands (doublet at 1655 and 1625 cm^{-1}) could not be resolved as they were focused into a single band when compared to the standard transmission spectrum of α-chitin obtained in a KBr pellet (Figure 9d). Also, the characteristic band at 1320 cm^{-1} was of very low intensity. By contrast, DRIFT analysis of α-chitin (mixed with HBr) on powder produced a spectrum (Figure 9b) of much better resolution; the amide I band was split into two components. In the transmission spectrum of α-chitin film (Figure 9c) the amide band showed a well-defined peak at 1650 cm^{-1} with a minor shoulder at 1625 cm^{-1}. Knowledge of the source of chitin/chitosan, its nature, water content and level of impurities is also important for the accurate calculation of DA values.

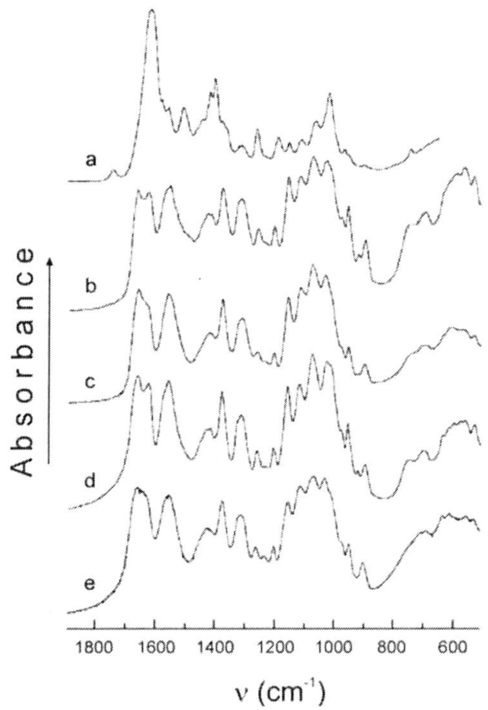

Figure 9. Comparison of IR spectra (shown as absorbance) of α- and β-chitin recorded using different sampling techniques. For α-chitin: (a) ATR on film, (b) DRIFT on powder, (c) Standard transmission on film, (d) Standard transmission on KBr pellet. For β-chitin: (e) Standard transmission on KBr pellets. Reprinted from Polymer 42, 2001, Brugnerotto, J., Lizardi, J., Goycoolea, F.M., Argüelles-Monal, W., Desbrières, J., Rinaudo, M., An infrared investigation in relation with chitin and chitosan characterization, 3569–3580, Copyright (2010), with permission from Elsevier [129].

Near infra-red (NIR) spectroscopy has also been applied to determine DA [138,139]. The spectra were recorded from 9090–4000 cm^{-1} and second-derivative spectra were used to determine DA; D-glucosamine and D-glucosamine hydrochloride were chosen as model compounds. A reference curve was constructed by plotting the predicted DA value (using NIR data)*versus* DA determined by ^{1}H NMR spectroscopy.

As shown here, choosing an appropriate method among the various IR procedures is a difficult task for researchers.

IR techniques are also employed for the quantitative evaluation of DA, for example, during the *N*-deacetylation of chitin by hot alkali [140], the formation of *N*-acetyl chitosan gel [141], the γ-irradiation of chitosan powder [142] or the synthesis of organic-soluble acetylated chitosan [143].

2.2.4. FTIR analysis of chitin and chitosan derivatives

The growing interest in the chemical modification of chitin and chitosan to improve their solubility and applications [23–25] has meant that the most important application of IR spectroscopy in this respect is the structural analysis of chemically modified forms of chitin or chitosan. Typical structural analysis of chitin/chitosan derivatives by FTIR spectroscopy involves: (1) FTIR analysis of chitin/chitosan, (2) FTIR analysis of the chemical reagent(s) used in the reaction, (3) FTIR analysis of the chitin/chitosan derivative obtained, (4) identification of differences between the spectra, (5) interpretation of results.

An example of the application of FTIR spectroscopy in the structural analysis of chitin and chitosan derivatives is the FTIR analysis of chitosan-L-glutamic acid aerogel derivative, which is soluble over a wide pH range [144]. The FTIR spectra obtained during these investigations are presented in Figure 10.

The following different characteristic bands were assigned from the FTIR spectra of chitosan, L-glutamic acid (L-GA) and chitosan-L-glutamic acid derivative (CL-GA) (Figure 10):

- in the IR spectra of chitosan (Figure 10A): 3429 cm^{-1} (O-H stretching overlapping the N-H stretching), 2921 and 2867 cm^{-1} (C-H stretching),

1640 cm^{-1} (amide II band, C-O stretching of the acetyl group), 1592 cm^{-1} (amide II band, N-H stretching) 1485–1380 cm^{-1} (asymmetrical C-H bending of the CH_2 group) and 1035 cm^{-1} (O bridge stretching) of the glucosamine residue.

- in the IR spectra of L-GA (Figure 10B): 2966 cm^{-1} (O-H stretching), 2855 cm^{-1} for (C-H stretching), 1690 cm^{-1} (C=O group) and 1523 cm^{-1} (N-H stretching of the amino group),
- in the IR spectra of CL-GA derivative (Figure 10C): 3110 and 2966 cm^{-1} (axial OH group of chitosan and glutamic acid), 1685 cm^{-1} (amide linkage), 1556 cm^{-1} (N-H bending and stretching) and 1067 cm^{-1} (C-O-C bridge stretching) of the chitosan residue, 1466 cm^{-1} (the asymmetrical deformation of CH_2).

The following changes were observed in the IR spectra of the chitosan derivative: the C-O adsorption peak of the secondary hydroxyl group enlarged and moved to 1067 cm^{-1}, the intensity of primary alcohol 1035 cm^{-1} (C-O stretching vibration) became much smaller than in chitosan, and a new peak appeared at 2966 cm^{-1}, indicating the incorporation of theL-glutamic acid moieties. These FTIR results provided evidence for the formation of an amide linkage between the COOH group of L-GA and the NH_2 group of the main chitosan chain.

FTIR spectroscopy has been applied to confirm the chemical structure of (1) water-soluble chitosan derivatives such as ethylamine hydroxyethyl chitosan (EHCs) [145], *N*-propyl-*N*-methylene phosphonic chitosan [93], *O*-succinyl-chitosan [146],*N*-alkylated chitosan [147] or methoxy poly(ethylene glycol)-grafted chitosan [148]; (2) organic-soluble derivatives, e.g., acetylated chitosans [149,143]; (3) nanoparticle-chitosan composites [150–152]; 4) hydrogels [153,154] or chitosan-graft copolymers [155–157].

Furthermore, FTIR analyses enabled the interactions occurring between chitin/chitosan and the analysed compounds to be studied and explained [107,158–165]. This technique was applied, for example, in an attempt to obtain an explanation for the sorption mechanism of acid dyes [166–168], phenol and *o*-chlorophenol [169] and fluoride [170,171]. FTIR spectroscopy has also been used to establish differences in structure and the degree of substitution of chitin, chitosan and dibutyrylchitin [172,173].

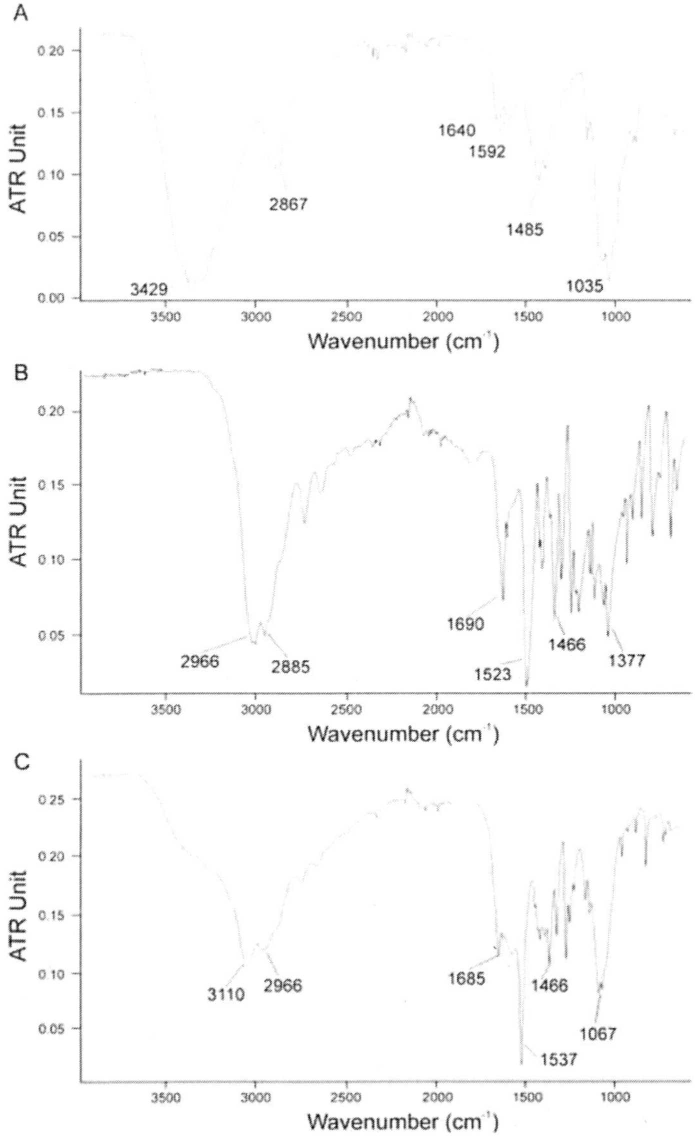

Figure 10. FTIR spectra of (A) chitosan (C), (B) L-glutamic acid (L-GA), and(C) chitosan-L-glutamic acid derivative (CL-GA). Reprinted from Carbohydrate Polymers 76, 2009, Singh, J., Dutta, P.K., Dutta, J., Hunt, A.J., Macquarrie, D.J., Clark, J.H., Preparation and properties of highly soluble chitosan-L-glutamic acid aerogel derivative, 188–195, Copyright (2010), with permission from Elsevier [144].

2.3. UV-Vis spectroscopy

Ultraviolet/visible (UV-Vis) spectroscopy is useful as an analytical technique for two reasons. Firstly, it can be used to identify certain functional groups in molecules, and secondly, it can be used for assaying. Unlike IR spectroscopy, UV-Vis spectroscopy involves the absorption of electromagnetic radiation from the 200–800 nm range and the subsequent excitation of electrons to higher energy states. The absorption of ultraviolet/visible light by organic molecules is restricted to certain functional groups (chromophores) that contain valence electrons of low excitation energy. The UV-Vis spectrum is complex and appears as a continuous absorption band because the superimposition of rotational and vibrational transitions on the electronic transitions gives a combination of overlapping lines. Nowadays, the individual detection of electron transfers without superimposition by neighboring vibrational bands can also be recorded [174]. With UV-Vis spectroscopy it is possible to investigate electron transfers between orbitals or bands of atoms, ions and molecules existing in the gaseous, liquid and solid phase. Analysis of solutions and crystals usually takes place in transmission, whereas powdered samples are often measured in diffuse reflection mode (Diffuse Reflectance Spectroscopy–DRS). Unlike IR spectroscopy, where Fourier transform techniques predominate, dispersive spectrometers are almost exclusively used in UV-Vis spectroscopy because of the large band widths. More details about UV-Vis spectroscopy and its application are presented in many papers and books [174–177].

2.3.1. Typical conditions of UV-Vis measurement

UV-Vis spectra of chitin/chitosan derivatives are usually recorded in aqueous acid (acetic acid, phosphoric acid, perchloric acid, hydrochloric acid) solutions in a 1.0 cm quartz cell at ambient temperature [160,178–180]. Sometimes water [181], an aqueous base [182] or DMSO [144] solutions are applied. The Diffuse Reflectance UV-Visible (DRUV) spectra of powdered or film samples are measured [107,112,183–185]. Analysis in the vacuum ultraviolet through the near-infrared range has also been applied [185,186].

2.3.2. UV-Vis spectra of chitin and chitosan

Chitin and chitosan include various ratios of two far-UV chromophoric groups, *N*-acetyl-glucosamine (GlcNAc) and glucosamine (GlcN); as a result, their extinction coefficients for wavelengths shorter than approximately 225 nm is non-zero. Because GlcNAc and GlcN residues show no evidence of interacting within the chitin/chitosan chain, the monomer units contribute in a simple, additive way to the total absorbance of these polymers at a particular wavelength [180]. The UV spectra of mixtures of *N*-acetyl-glucosamine and glucosamine hydrochloride are quite similar to the spectra of chitosan, and the λ_{max} is 201 nm in 0.1 M HCl solution (Figure 11) [180].

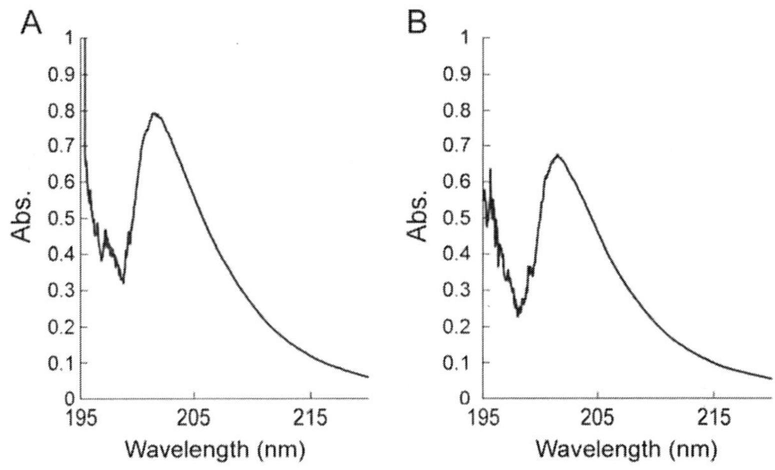

Figure 11. UV spectra: (a) a mixture of *N*-acetyl-glucosamine and glucosamine hydrochloride in 0.1 M hydrochloric acid solution; (b) chitosan in 0.1 M hydrochloric acid solution. Reprinted from Carbohydrate Research 341, 2006, Liu, D., Wei, Y., Yao, P., Jiang, L., Determination of the degree of acetylation of chitosan by UV spectrometry using dual standards, 782–785, Copyright (2010), with permission from Elsevier [180].

UV-Vis spectroscopy can be used for the optical characterization of the source of chitin [112]. The optical spectra for two black coral species, *Antipathes caribbeana* and *A. pennacea*, showed a wide band from 300 nm to 500 nm, with small differences depending on the species.

Additionally, the spectra confirmed the presence of Fe^{+3} in *A. caribbeana* and Mn^{+3} in both species.

2.3.3. Determination of the degree of *N*-acetylation of chitin and chitosan using UV-Vis spectroscopy

The first derivative UV method for DA determination was proposed by Muzzarelli and Rocchetti in 1985 [187]. The principle of this method was based on the absorbance of the intensity of the *N*-acetyl group in chitin or chitosan. The absorbance of *N*-acetyl glucosamine (at maximum wavelength) was linearly dependent on its concentration in the 0.50–5.0 $mg \cdot L^{-1}$ range. The limit of detection of GlcNAc in 0.01 M acetic acid was found to be 0.5 $mg \cdot L^{-1}$. The DA values were determined in acetic acid solutions at 199 nm [187]. The simplicity and convenience of this method were the best among the available methods, because it required very small amounts of sample and did not require expensive equipment. In addition, the first derivative of the spectra was less affected by background noise and impurities. Several modified first-derivative UV methods have been proposed to improve the convenience and accuracy of measurement [178,180,188–191]. Tan *et al.* [178] compared the results of four methods (^{1}H NMR, UV, ninhydrin assay and potentiometric titration) and suggested the first-derivative UV method as a standard method for the routine determination of DA of chitosan. Pedroni *et al.* [188] analysed chitosan solutions in different solvents; 0.1 M HCl solutions turned out to be the best. Hsiao *et al.* [189] suggested concentrated phosphoric acid as the UV-transparent solvent system. In this way, the UV determination could be validated across the whole DA range. Liu *et al.* [180] employed two standards (*N*-acetyl-glucosamine and glucosamine hydrochloride) to determine the degree of *N*-acetylation of chitosan by UV spectrophotometry. Wu and Zivanovic [190] plotted the first derivative UV values at those wavelengths against the concentrations of GlcNAc and found that the best linear regression was obtained at 203 nm (the range of 0–50 μg/mL, $R^2 = 0.996$). Da Silva *et al.* [191] put forward a new mathematical expression. This was derived in such a way that DA could be determined directly from the mass concentration of a chitosan solution and the first derivative of its UV spectrum at 202 nm, thus eliminating the need for the empirical correction of curves for highly deacetylated samples. A procedure was proposed for the accurate mass determination of the hygroscopic chitosan.

In summary, UV techniques for DA determination are more sensitive than IR, ^{13}C NMR and ^{15}N NMR spectroscopy, and they give a better accuracy than NIR, IR and NMR methods [122,192]. The first-derivative UV method is the most sensitive of the proposed techniques. The first derivatives of the spectra are less affected by impurities such as protein, and more accurate data can be obtained. In contrast to other techniques, it is possible to analyse the DA values of a chitin/chitosan sample with a high water content. The first-derivative UV methods are easy to carry out, and the equipment is available in most monitoring and research laboratories. These techniques were used to verify the validity of several methods of DA determination in different laboratories.

2.3.4. Application of UV-Vis spectroscopy to the analysis of chitin/chitosan based compounds

The steps in the structural analysis of chitin/chitosan derivatives by UV-Vis spectroscopy are very similar to those in the FTIR methodology (Section 2.2.4). The main difference is connected with the aim of these analyses. IR spectroscopy is used mostly for determining the molecular structure of chitin and chitosan samples, whereas UV-Vis spectroscopy is more often applied to the study of covalent and non-covalent interactions. Since certain functional groups present in organic molecules absorb light at characteristic wavelengths in the UV-Vis region, this technique is applied qualitatively to identify the presence of these groups in samples, supporting structural information obtained from other spectroscopic methods, especially IR.

As mentioned, one of the most important applications of UV-Vis techniques is the characterization of interactions between chitin or chitosan and the target compounds. For example, the analysis of the interaction between chitosan and cellulose, described by Urreaga and de la Orden [107], will be presented in greater detail. The UV-Vis spectra obtained during that study are shown in Figure 12.

The spectrum of chitosan-treated cellulose (Figure 12) showed the characteristic chitosan absorption below 220 nm revealing the presence of chitosan. Additionally, this spectrum showed very weak absorptions at $\lambda >$ 230 nm, including weak but characteristic bands centred at *circa* 245 nm and 345 nm, which did not appear in the spectra of the starting materials (cellulose and chitin). This observation suggested the existence of chemical interactions between cellulose and chitosan. In order to verify the existence of the above chitosan–cellulose interactions, samples of pure chitosan, pure cellulose and chitosan-coated cellulose were heated at different temperatures and characterized by UV-Vis spectroscopy. The spectrum of the chitosan-treated cellulose was more complex than those of chitin and cellulose and exhibited two new absorptions centred at *circa* 272 and 340 nm. The presence of the band at 272 nm was explained by the independent oxidations of chitosan and cellulose, but the appearance of the 340 nm band again indicated the existence of a chemical reaction in chitosan-treated celluloses, demonstrating that chemical interaction between cellulose and chitosan took place even at moderate temperatures. The new bands, which were observed in materials processed at room temperature, were assigned (also using FTIR and fluorescence spectroscopy) to conjugated imines, produced in the condensation reaction between chitosan amino and carbonyl groups. UV-Vis spectroscopy has also been employed, e.g., to characterize the interactions between chitosan and uranyl ions [160], chitosan and synthetic phospholipid membranes [193], β-cyclodextrin-linked chitosan and p-nitrophenolate [194] and between chitosan and gelatin [195].

UV-Vis techniques were useful for determining the structure of many chitin/chitosan derivatives, for example, chitosan hydrogel membranes obtained by UV- and γ-radiation [196], substituted polyaniline/chitosan composites [163], chitosan-L-glutamic acid aerogel derivatives [144], fluorescent chitosan derivatives containing substituted naphthalimides [197] and dendritic polyaniline nanoparticles synthesized by carboxymethyl chitin templating [103]. They were also employed to determine the actual degree of substitution of fluorescent chitosan (*DS* parameter) and the concentration of the 9-anthrylmethylated unit in modified chitosan [198].

It should also be mentioned that UV-Vis spectroscopy is used to monitor chemical reactions [185,199,200] and the manufacture of complex multilayers, nanoparticles, nanocomposite or multilayer films [183,184, 201–203].

UV-Vis spectroscopy is employed to determine the physicochemical properties of chitin/chitosan-based compounds such as the turbidity (optical dispersion, OD) of their solutions [179], the capacity adsorption of chitosan membranes [204], the lower critical solution temperature (LCST) of a chitosan derivative copolymer solution in PBS [205] or the optical properties of nanoparticles [206].

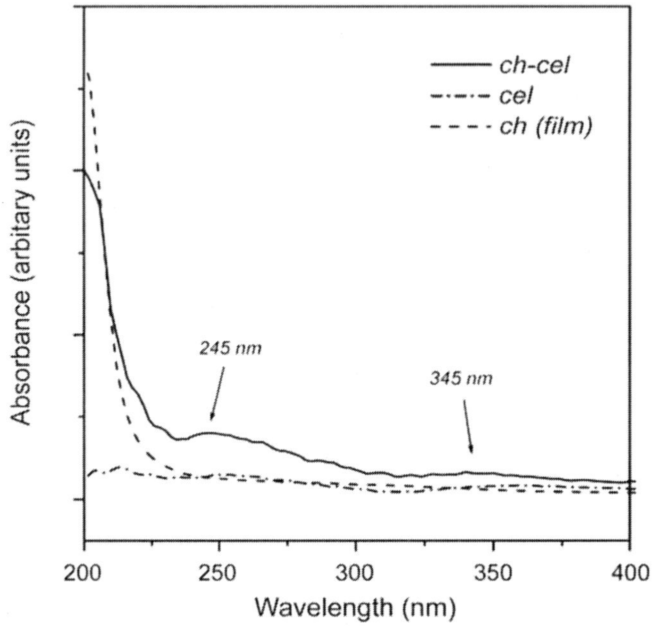

Figure 12. UV–Vis spectrum of a pure chitosan (ch) film measured in transmission mode, and diffuse reflectance UV–Vis spectra of untreated cellulose (cel), and cellulose treated with 1% chitosan (ch–cel). Reprinted from European Polymer Journal 42, 2006, Urreaga, J.M., de la Orden, M.U., Chemical interactions and yellowing in chitosan-treated cellulose, 2606–2616, Copyright (2010), with permission from Elsevier [107].

2.4. Mass spectrometry

Mass spectrometry (MS) is a destructive analytical technique used for measuring the characteristics of individual molecules. The basic information obtained from mass spectrometric analysis is the molecular mass of a compound, which is determined by measuring the mass to charge ratio (m/z) of its ion. With the ionization method, full particulars about a molecule's chemical structure can be found. MS can analyse chemicals with a wide mass range–from small molecules to complicated biomolecules such as carbohydrates, proteins, peptides or nucleic acids. The mass spectrometer is composed of three fundamental parts: an ion source, an analyser and a detector. First, the sample must be introduced to the heart of each mass spectrometer–the ion source. There are two possible ways of doing this: through a direct inlet (septum inlet, syringe, probe, *etc.*), or an indirect inlet (gas chromatograph, liquid chromatograph, or capillary electrophoresis column). In the ion source the sample molecules are ionized by one of a variety of methods (electron impact–EI, chemical ionization–CI, electrospray ionization–ESI, matrix assisted laser desorption ionization–MALDI, *etc.*). Next, the stream of ions is transferred to the analyser, where they are sorted and separated according to m/z. The most commonly used analyzers are quadrupoles, ion trap and time of flight. In tandem mass analysis (MS/MS, MS_n) a combination of two or more analysers is used. The MS can be operated in three modes: full scan (scanning of a selected mass range), more sensitive SIM (selected ion monitoring = SIR - single ion recording), and the most sensitive MRM (multiple reaction monitoring). Finally, the ions reach the detector when their energy is converted to electrical signals that the computer can read. In the end, mass spectra are obtained, and identification of compounds is possible [207–209].

2.4.1. Typical conditions of mass spectrometric analysis of chitin/chitosan and their derivatives

Different types of mass spectrometer can be chosen according to the information required. Spectrometers with various combinations of inlet systems, ion sources and analysers are used for investigations into chitin, chitosan and their derivatives. In most cases the positive ion mode is preferred.

The first method used was pyrolysis-mass spectrometry [210]. That was back in 1982, but combined with gas chromatography it is still a routine technique [211–215]. In GC/MS, the first sample is pyrolyzed (450–615 °C), after which the volatile pyrolysis products are separated on a capillary column (the temperature program from 35–50 °C with different rates to 220–300 °C). In GC/MS analysis, the full scan mode (m/z from 15–35 to 300–650) and the electron impact mode (70 eV) are commonly used. Products are identified on the basis of their GC retention parameters and their mass spectra [210–215].

GLC-MS is used for analyzing chitosan or its impurities. But first, all the target compounds have to be converted to volatile derivatives (hydrolysis, and reduction of chitin [216,217]; this is then followed by the extraction of impurities [218] and trimethylsililation or acetylation). The products are separated using a GLC nonpolar or medium-polar capillary column with a gradient temperature programme (usually from ~60 °C to ~280 °C with different heating rates). For detection, MS with EI ionization is employed [216–218].

Most of the papers dealing with the mass spectrometric analysis of chitin and chitosan report that MALDI was the ionization method selected. This ion source is most often combined with a time of flight analyser (MALDI-TOF) [219–226]. For this type of analysis chitooligosaccharides were dissolved in water [226], MeOH-H_2O (1:1) [224] or 1% acetic acid solution [225]. The matrix most often used is 2,5-dihydroxybenzoic acid (DHB) [219,221–226]; sometimes it is 2-(4-hydroxyphenylazo)benzoic acid (HABA) dissolved in dioxin [224]. To obtain monosodium and monopotassium carbohydrate adducts in the mass spectrum, NaCl and KCl can be added as a cationating agent [224]. The MS ion source is equipped with a nitrogen laser emitting at 337 nm [225,226]. For these analyses an accelerating potential of 20–30 kV is used, and mass spectra are measured in positive ion mode [221,222]. A modified version of MALDI-TOF was introduced for chitin/chitosan investigations–MALDI-TOF PSD MS (postsource decay mass spectrometry) [227]. Another MS technique with a MALDI source is tandem mass spectrometry, for which an ion trap analyser is very useful. One analyzer allows $MS^2$and also MS^3 spectra to be obtained using a collision energy previously established experimentally [228]. MALDI-TOF/TOF equipment can also be applied in chitin and chitosan analysis using tandem mass spectrometry. Precursor ions are accelerated at 8 kV, while their fragment ions are accelerated to 19 kV [229].

Time-of-flight secondary ion mass spectrometry (TOF-SIMS) is another method used during chitosan investigations. This type of spectrometer is equipped with a gallium primary ion beam operated at 15 kV. The secondary ions appearing are extracted (10 kV) and fly through the analyzer (TOF type) to the detector [230,231].

After MALDI, ESI (electrospray ionization) is the second most important ionization technique for chitin/chitosan analysis [219,232–238]. The ESI ion source is usually coupled to an ion trap analyser [235], but a quadrupole type of analyser can also be used [237]. For sample measurements several parameters should be optimized, for example, the source voltage (3.5–4 kV), nitrogen sheath gas rate and capillary temperature. A syringe pump is used (3–4 µL/min) for direct sample injection [235,238]. Tandem mass spectrometry with ESI ionization is preferred for acquiring more structural information. If the triple quad system (QqQ) is used, the first quadrupole selects the precursor ions, the second quadrupole acts as collision chamber, and the third quadrupole analyses the ion products; argon is the neutral collision gas [239,240]. Sometimes, a combination of quadrupole and time-of-flight (Q-TOF) is found useful [241]. Connecting a liquid chromatograph to the MS system is an additional advantage. Separating solutions containing chitosan products on a chromatographic column makes the analysis easier. LC-MS analysis is carried out using the same mass spectrometric modes (positive and negative ion detection, electrospray ionization, scanning and SIM mode, *etc.*). Ion source conditions usually are the same as in classic MS analysis that yields protonated products $[M+H]^+$, sodium adducts $[M+Na]^+$, *etc.* For HPLC separation a different type of column is used, and the mobile phase often consists of acetonitrile or water with added formic acid [242–245]. LC-MS is highly suitable method for investigating chitosan derivatives.

Fast atom bombardment (FAB) is another soft ionization method useful in chitin and chitosan investigations. For this purpose the following compounds can act as the matrix: thioglycerol [246,247], m-nitrobenzyl alcohol [247], glycerol [248–250], dithiothreitol [251], 2,4-dinitrobenzyl alcohol, or dithiothreitol-thioglycerol (1:1, v/v) [223]. The mass spectrometer usually operates in positive-ion mode, most often producing $[M+H]^+$ ions [246,248,249,252] but also $[M+Na]^+$ [250,251] or $[M+K]^+$ [251]. The atom gun is operated at 6–8 kV with xenon as the bombarding gas [246,247,250,251]. This type of ion source cooperates with collision-induced dissociation tandem mass spectrometry (FAB-CID) measurements [246,247,252]. To obtain FAB-CID-MS/MS spectra, the collision is caused by helium, and a 10 kV accelerating voltage is used [246,247].

Sometimes less popular ionization methods are used for chitin/chitosan analysis. One of them is chemical ionization (CI) with the use of ammonia [253] or acetic acid [120] as the reactive gas. In both cases MS works in positive ion mode, generating $[M + NH_4]^+$ and $[M + H]^+$, respectively [120,253]. Another ionization method is plasma desorption by the fission products of californium-252 in a time/flow biochemical mass spectrometer [254].

TC-EA/IRMS (high-temperature conversion-elemental analyser/isotope ratio mass spectrometer) is very rarely used in chitin/chitosan investigations. In this method the first step is the quantitative conversion of oxygen to CO and hydrogen to H_2 at a high temperature. The gaseous products are separated by GC, transferred to IRMS where the ratios of the stable oxygen and hydrogen isotopes in chitin are determined [255,256].

Chitosan TG-MS (thermogravimetry-mass spectrometry) analysis begins with the very rapid heating of the sample from room temperature to 600 °C, after which the gaseous products pass into the mass spectrometer. Ions with m/z in the 40–130 range are analysed by MS with a quadrupole analyzer and Channeltron detector [257].

2.4.2. Mass spectrometric determination of the degree of polymerization of chitin and chitosan

Being a natural co-polymer of glucosamine and *N*-acetyl-glucosamine, chitosan possesses properties like antimicrobial and antitumor activity, and stimulates plant growth [222,238,258]. This activity is closely related to the molecular weight. This fact must be taken into consideration if one is planning to use chitosan in biological or pharmaceutical applications. Chitooligosaccharides can be obtained by depolymerization. This is done physically [226,238] chemically [245] or enzymatically [244,250,259–262]. In each case, MS can be used to determine the degree of depolymerization of chitosan and to track the kinetics.

One of the physical methods for depolymerizing chitosan is sonolysis. It was found that after ultrasound treatment random depolymerization of chitosan occurred, producing many different chitooligosaccharides (COS). ESI-MS [238] and MALDI-TOF MS (Figure 13) [226] were used to detect the products. The primary chemical structure was found to have remained unchanged and that sonolysis was a good enzymatic or alternative chemical method for obtaining oligosaccharides for commercial use.

Depolymerization of chitin/chitosan can be achieved by classical chemical methods. The results of LC-MS analysis proved that hydrolysis of chitin from fungi with 6 M HCl yielded glucosamine monomers [245]. Besides providing qualitative chitin composition, the method also makes it possible to track the kinetics of the reaction by ^{15}N labeling of the glucosamine from the fungal chitin. For this purpose, acid-hydrolyzed chitin from the fungus *P. chrysogenum* growing on a minimal medium and on a medium containing added $(^{15}NH_4)_2SO_4$ (rich medium–Blakeslee's formula) was analyzed by LC-

MS. Then the analysis on the basis of the peak ratios in mass spectrum was performed (Figure 14) [245].

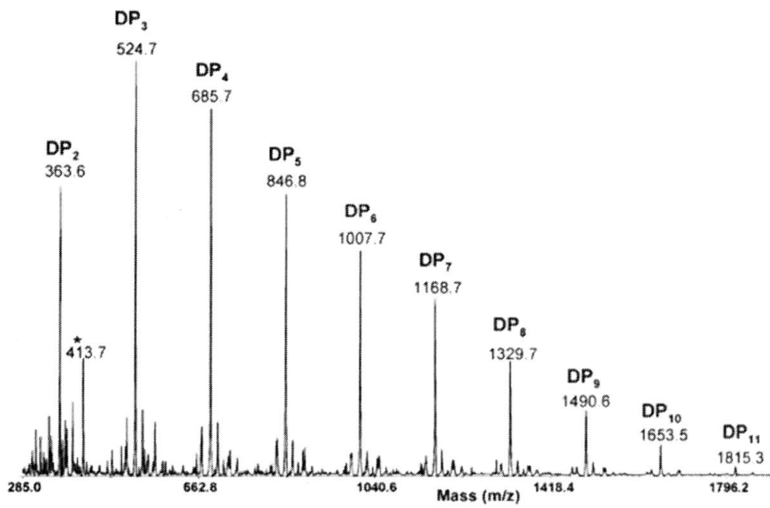

Figure 13. MALDI-TOF MS of deacetylated chitooligosaccharides obtained from ultrasonically treated chitosan. Reprinted with permission from *Biomacromolecules*.**2009**, *10*, 1203–1211. Copyright 2010, American Chemical Society [226].

The enzymatic method of chitin/chitosan depolymerization is the most popular one, and mass spectrometry is often used to detect the products and the efficiency of this process. From MS studies it was found that immobilization of enzymes (papain [259], neutral protease [243]) enhanced depolymerization in the chitosan chain comparing with free enzymes [243,259]. The activities of enzymes isolated from different sources (commercial enzymes [261], isolated from living organisms: *Vibrio harveyi* [244], *Serratia marcensces* [251], *Bacillus circulans* [237], *Amycolatopsis orientalis*, *Streptomyces* sp. [241]) were tested by qualitative MS studies of obtained chitooligosaccharides. It is possible to study enzyme selectivity, and the mechanism of COS bonding to the enzyme. Knowledge of the influence of polymerization and the degree of acetylation is very important in the large-scale production of chitooligosaccharides, and in their use in enzymology or drug design. Table 1 shows a number of examples of the analysis of the degree of polymerization of chitooligosaccharides using mass spectrometry.

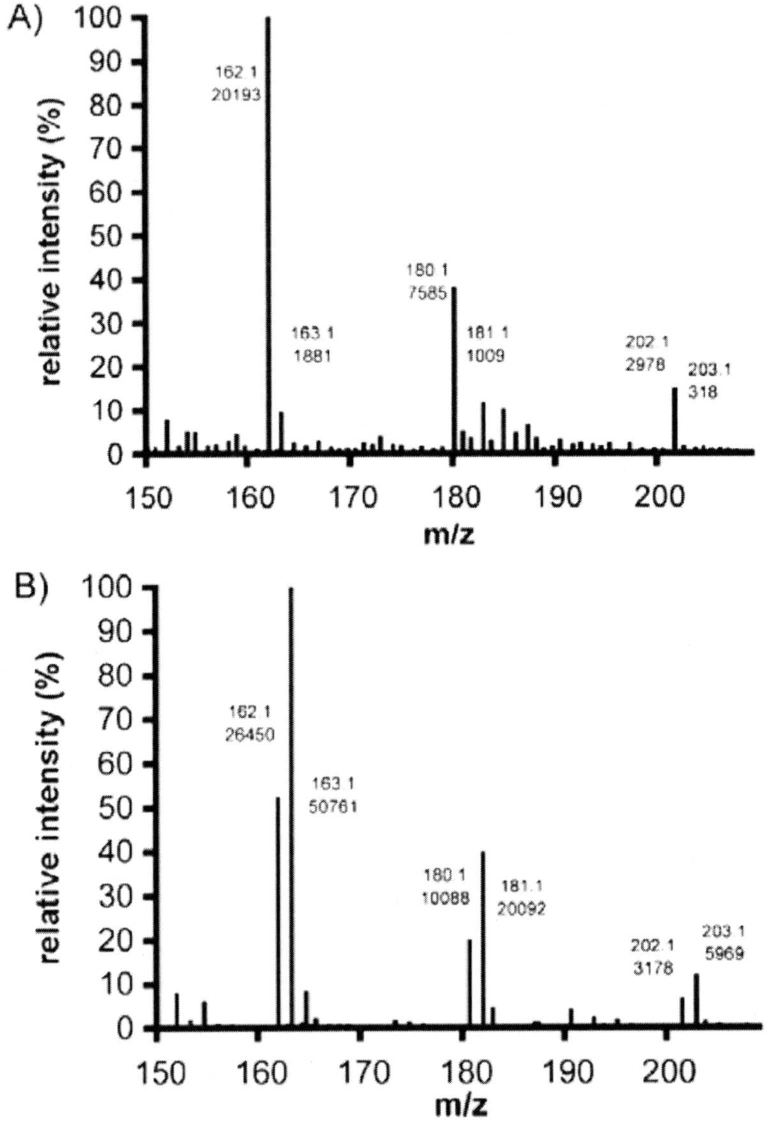

Figure 14. Mass spectra of acid hydrolyzed chitin from *P. chrysogenum* growing on (A) minimal medium, (B) medium with the addition of $(^{15}NH_4)_2SO_4$ (rich medium–Blakeslee's formula). Reprinted with permission from *Biomacromolecules*. **2009**, *10*, 793–797. Copyright 2010, American Chemical Society [245].

Table 1. MS-analyzed chitooligosaccharides obtained after depolymerization using different enzymes.

Enzyme	DP analysis MS method	Chitin/chitosan polymerization degree DP	Chitooligosaccharide applicability	Ref.
Papain	MALDI-TOF	DP 2–DP 8	Antibacterial activity against *Bacillus cereus* and *Escherichia coli*	[222]
Immobilized Papain	MALDI-TOF	DP 3–DP 7	Comparing the depolimerization efficiency between free and immobilized papain	[259]
Pronaze	MALDI-TOF	DP 2–DP 9	Antibacterial activity against *Bacillus cereus* and *Escherichia coli*	[222]
Cellulast (Novozymes)	MALDI-TOF	DP 2–DP 8	Induction of defence response of Oryza sativa L. against Pyricularia grisea (Cooke) Sacc.	[258]
Isozyme of pectinase	MALDI-TOF FAB-MS	DP 2–DP 6	Antimicrobial activity *Bacillus cereus* and *Escherichia coli*	[250]
Chitinase	MALDI-TOF	Q1: DP 3–DP 8 Q2: DP 2–DP 12 Q3: DP 2–DP 10	Affect on fungal (alternaria alternate, *Rhisopus stolnifer, Botrytis cinera, Penicillinum expansum*) growth rate	[262]
Chitosanase	MALDI-TOF	DP 2–DP 8	Investigations of prebiotic effect on the *Bifidobacterium bifidum* and Lactobaccillus sp.	[260]
Chitinase	LC-ESI-MS	DP 2–DP 6	Studies of mechanism of bonding COS to enzyme helpful in the drug-screening program (for drugs in allergic asthma)	[244]

To prepare oligosaccharides, chitosan can be hydrolyzed by a mixture of different enzymes (cellulose, alpha amylase, proteinase). The products are analyzed by MALDI-TOF [220].

Mass spectrometry is a good method for the detection of chitooligosaccharides obtained not only by depolymerization, but also as a result of synthesis. FAB-MS and FAB-MS/MS were used to analyse chitinpentaose, chitintetraose, chitintriose and two modified chitin oligosaccharides produced by Rhizobium NodC protein in *Escherichia coli* [246].

2.4.3. Determination of the degree of *N*-acetylation of chitin and chitosan using mass spectrometry

The degree of *N*-acetylation is a factor that allows chitin and chitosan to be distinguished. Besides UV, IR and NMR spectroscopy, mass spectrometry is one of the spectral techniques used to establish the degree of *N*-acetylation of chitin, chitosan and their derivatives [122].

For this purpose, the pyrolysis-gas chromatography-mass spectrometry (EI ionization) combination is used. From the EI mass spectra it is possible to identify volatile compounds obtained from chitin and chitosan. The peaks at m/z 60 and m/z 42 are derived from *N*-acetyl-glucosamine, while the ions m/z 80 and m/z 67 are from glucosamine. DA is established on the basis of the peak ratios 80:60, 67:60, 80:42, 80:125, 94:125, 80:110. These ratios increase with decrease in the degree of acetylation [122,211,212].

As in the case of investigations into the degree of polymerization of chitin/chitosan, the usual MS method is MALDI-TOF. The mass spectra of the series of oligosaccharides with DA from 25 to 90% obtained with this method are shown in Figure 15[240].

As mentioned above, chitin and chitosan play a significant part in inducing defense responses in plants. Some properties are very selective: for example, chitin oligosaccharides can induce lignifications in wounded wheat leaves, while chitosan COS cannot. Mass spectrometry was used to establish DA in the stereocontrolled synthesis of the chitosan dodecamer (FAB-MS) [263], the synthesis of the chitosan tetramer derivative β-D-GlcNAc-(1→4)- β-D-GlcNAc-(1→4)- β-D-GlcNAc-(1→4)-D-GlcN (FAB-MS) [249], the synthesis of a series of oligosaccharides (MALDI-TOF) [264], and the synthesis of partially *N*-acetylated chitin oligomers using *N*-acylated chitotrioses as substrates in a lysozyme-catalysed transglycosylation reaction (MALDI-TOF, ESI-MS) [219]. All of them are used in biological investigations. The degree of acetylation of enzymatic degradation products of chitin and chitosan is measured. For these investigations chitin deacetylase from *Colletotrichum lindemuthianum* was chosen (FAB-MS, MALDI-TOF) [221,249].

2.4.4. Analysis of chitin and chitosan derivatives using mass spectrometry

Some analytical methods based on mass spectrometry force the conversion of chitin and chitosan into the gaseous state. There are two cases when this is required: during analysis of thermal degradation products, and when GLC-MS is used.

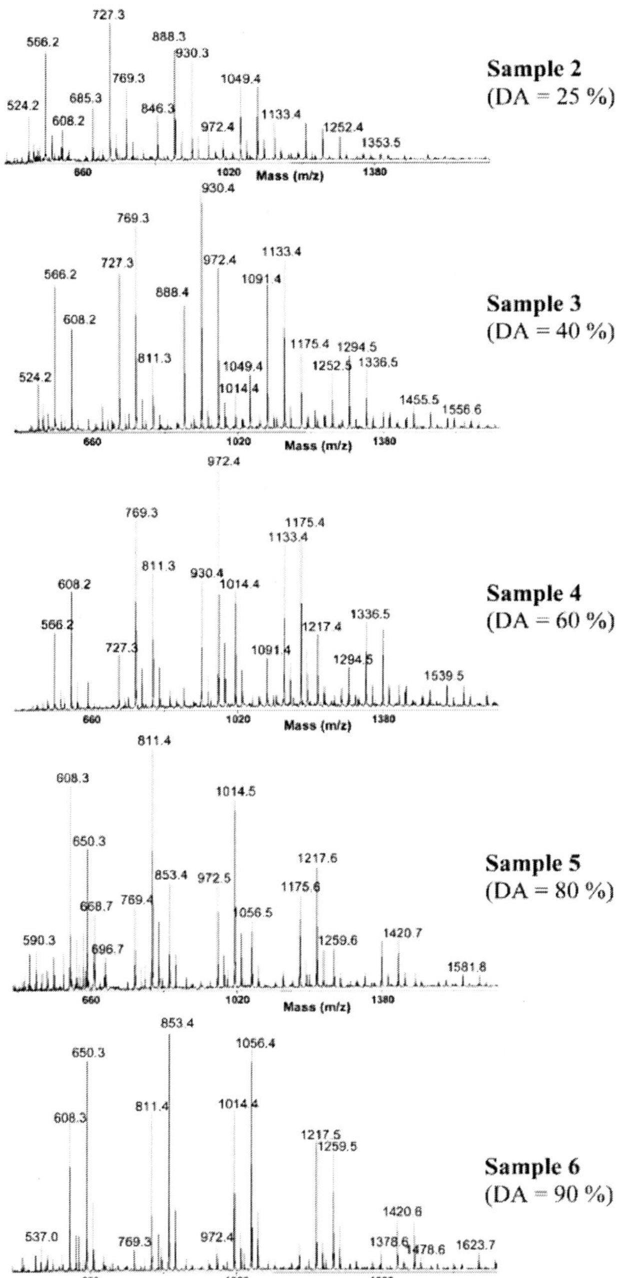

Figure 15. MALDI-TOF MS of chitooligosaccharides with different degrees of acetylation. Reprinted with permission from *Biomacromolecules*. **2008**, *9*, 1731–1738. Copyright 2010, American Chemical Society [240].

The first example is a study of the thermal decomposition products of chitosan, which is done by TG-MS analysis. The chitosan is heated to different temperatures to be degraded, and the gaseous products are detected using a mass spectrometer. From this experiment it is possible to monitor the thermal behavior of chitosan [257].

In comparison with the previous method, using a combination of pyrolysis-gas chromatographymass spectrometry, it is possible to separate a mixture of chitosan or chitosan pyrolysis products. For qualitative analysis the deuterated analogue of*N*-acetyl-glucosamine is used as an internal standard. The main components of the pyrolysates (chitin markers) are acetamide, acetylpyridone and 3-acetamidofuran [213–215,217]. This method is used to confirm the presence of chitin in biological samples [215,255].

For the GLC-MS analysis of chitin and chitosan, these need to be converted to volatile derivatives. For this purpose the chitosan or chitin methanolysis/hydrolysis products are trimethylsilated or acetylated. The results allow the sugar composition of chitin and chitosan to be determined. For qualitative analysis, an internal standard is needed [216,217].

TC-EA/IRMS is a very interesting, but still rarely used method. It is useful in the palaeolimnological study of fossil chironomid head capsules. All organic compounds are converted to CO and H_2. With the isotopic ratio mass spectrometer the ratios of stable oxygen and hydrogen isotopes from chironomid larvae can be determined. Water and diet were also investigated in this experiment to test how the $\delta^{18}O$ and $\delta^{2}H$ in water and the diet influence the $\delta^{18}O$ and $\delta^{2}H$ of chironomid larvae [255,256].

Most of all, however, mass spectrometry, especially LC-MS and LC-MS/MS, is used to monitor the success of chitin and chitosan derivatization reactions. The chitosan is derivatized for different reasons, for example, to sequence mixtures of chitooligosaccharides. For this purpose the reducing end of hetero-chitooligosaccharides was treated with 2-aminoacridone (AMAC) by reductive amination. The products were analyzed using MALDI-TOF-MS (Figure 16), MALDI-TOF/TOF-MS/MS (Figure 17) [229], PSD (postsource decay) MALDI TOF MS [227]. Chitosan sequence determination was a part of an investigation into the inhibition of a family of 18 chitinases. Besides AMAC, 3-(acetylamino)-6-aminoacridine is also used for the reductive amination of chitooligosaccharides. The sequencing information was taken from vMALDI (vacuum MALDI) ion trap MS^n measurements [228].

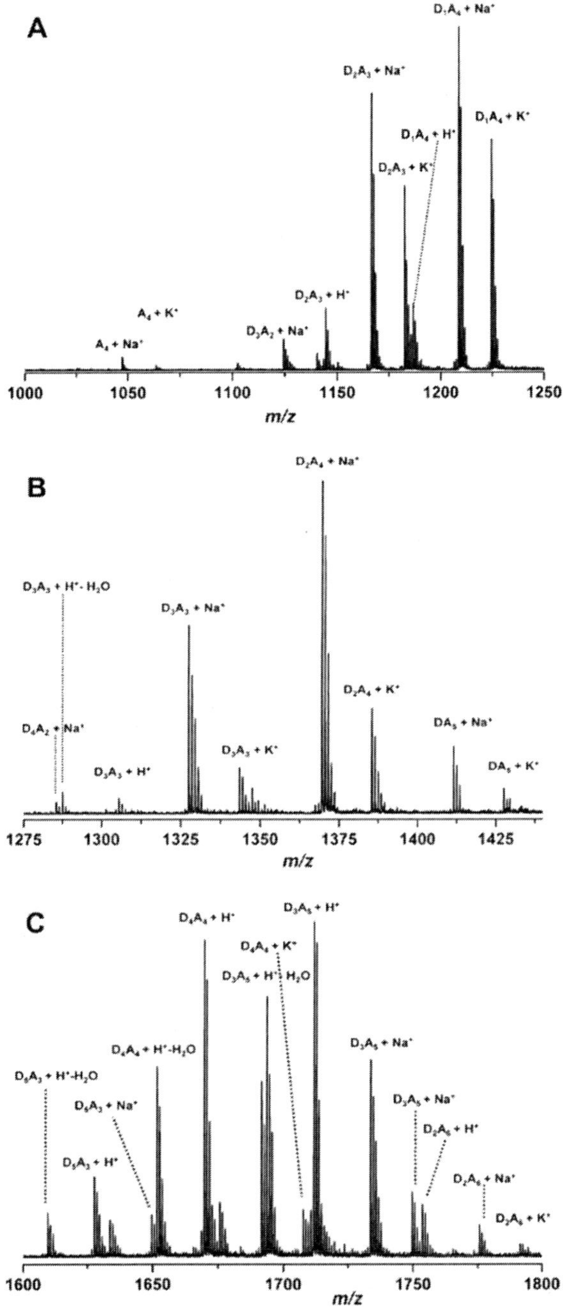

Figure 16. MALDI-TOF-MS of AMAC derivatized chitooligosaccharides: (A) DP 5, (B) DP 6, (C) DP 8. Reprinted from Carbohydrate Polymers 74, 2008, Cederkvist, F.H., Parmer, M.P., Vårum, K.M., Eijsink, V.G.H., Sørlie, M., Inhibition of a family 18 chitinase by chitooligosacharides, 41–49, Copyright (2010), with permission from Elsevier [229].

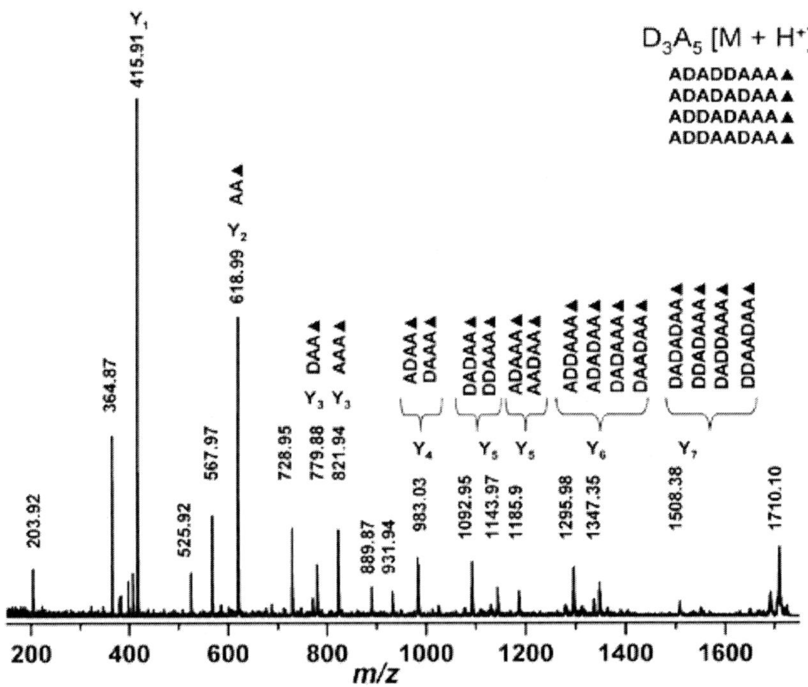

Figure 17. MALDI-TOF/TOF-MS/MS of AMAC derivatized D_3A_5 sugar 1711.80 m/z (▲ - AMAC). Reprinted from Carbohydrate Polymers 74, 2008, Cederkvist, F.H., Parmer, M.P., Vårum, K.M., Eijsink, V.G.H., Sørlie, M., Inhibition of a family 18 chitinase by chitooligosacharides, 41–49, Copyright (2010), with permission from Elsevier [229].

Examples of other chitin/chitosan derivatives analysed by MS are listed in Table 2.

Table 2. MS applications in investigations of chitin/chitosan derivatives.

Chitin and chitosan derivative	MS method	Chitooligosaccharide derivative applicability	Ref.
Copper-chitooligosaccharides complexes	ESI-MS, ESI-MS/MS (triple quadrupole, CAD –colision activated dissociation)	Metal-ligand associations studies	[232]
Lipo-chitin oligosaccharides	ESI-MS (quadrupole), FAB-MS, CID-MS/MS (QTof)	Structural studies of lipo-chitin oligosaccharides isolated from bacteria and their role as signal molecules in symbiosis	[247,252]
Products of electrochemical reaction between caffeic acid and glucosamine	ESI-MS	Studies of chitosan–coated electrodes for bimodal sensing	[236]
Methacrylated chitoligosaccharides	MALDI-TOF	Production of biodegradable biopolymers	[225]
Chitosan/tripolyphosphate nanoparticles	ToF-SIMS	Studies of nanoparticles as drug delivery system	[231]
Catechin-modified chitosan	ESI-MS	Creating polymers for technical applications	[235]
Benzenesulfony chitosan, Dinitrobenzenesulfonyl chitosan	MALDI-TOF	Implications for drug detoxification	[265]
Chitosan-g-PEG=X (where X-Man, cholesterol, coumarin, biotin)	MALDI-TOF	Producing copolymers used in active targeting and antiadhesive therapy	[224]
Multilayers consisting of: chitosan, hyaluronan, and poyethyleneimine	ToF-SIMS	Bioactive coating of endovascular stent	[230]
Dodecyl galate (DDG)-chitosan	FAB-MS, ESI-MS	Peroxidaze catalyzed production of biopolymers	[266]

2.4.5. Application of chitosan in mass spectrometric analysis

Chitin and chitosan may be involved in an analytical process using mass spectrometry not only as an analyte. These polymers can play an indirect role in MS analysis; given the properties of chitin and chitosan, they can be used in pre-MS analysis treatment.

One example is the coating of capillaries for capillary electrophoresis with carboxymethyl chitosan. In contrast to uncoated capillaries, the results are more stable and repeatable. With this type of capillary for CE-MS, a very good separation efficiency, sensitivity and less interference can be achieved [267,268].

Inductively-coupled plasma-mass spectrometry (ICP-MS) is very commonly applied to determine the trace metals in different matrices like water samples. Chitosan-based chelating resins are the basis of the sample pretreatment method. These were tested for the collection and preconcentration of analytes (metals), for matrix elimination, and to find the best adsorption capacity. These novel resins could be:

- *N,N′,N″*-triacetate-type chitosan (EDTriA-type chitosan) [269];
- Cross-linked chitosan with *N*-2-hydroxypropyl iminodiacetic acid groups (CCS-HP/IDA), and cross-linked chitosan with *N,N*- iminodiacetic acid groups (CCS-IDA) [270];
- Cross-linked chitosan modified with catechol and salicylic acid [271];
- Chitosan resin derivatized with 3,4-dihydroxybenzoic acid (CCTS-DHBA) [272];
- Chitosan resin with amino acids [273].

For the same purpose chitosan resin functionalized with histidine was used in a column in combination with an inductively-coupled plasma-atomic emission spectrometer (ICP-AES) [274]. ICP-MS has been used for the meta-analysis of chitosan using samples based on chitosan Chitlacsilver nanoparticles but only to calculate the amount of silver [275].

2.5. NMR spectroscopy

NMR spectroscopy is one of the most powerful techniques for the structural and physicochemical study of organic compounds, both small molecules and polymers. It seems to be highly suitable for studying chitin and chitosan. However, the solubility of these compounds presents a serious problem. As stated earlier, chitin with a higher DA is practically insoluble in all solvents and mixtures of different ones, whereas chitosan is soluble in aqueous acidic solution. Its solubility depends on the degree of *N*-acetylation, the degree of polymerization, the degree of crystallinity, the distribution of GlcNAc and GlcN along the polymer chain, the ionic strength of the solvent and the pH and concentration of chitosan in the solution [276].

Different NMR techniques have been used to study chitin, chitosan and their derivatives, including ^{13}C [277,278] and ^{15}N [279,280] solid-state NMR, and ^{1}H [281–283], ^{31}P [284], and ^{13}C [285] liquid-state NMR. However, only ^{13}C and ^{15}N solid-state NMR can be used to study chitins with a high DA, as this technique does not require the solubilization of the polymer.

2.5.1. Description of NMR techniques

2.5.1.1. ^{1}H NMR spectroscopy

The most important factor is to find a proper solvent, which should have good solubility properties towards the target material. When the sample is only partially soluble, no quantitative analysis is accurate and reproducible [280,286]. Moreover, the residual signal of the solvent should not overlap the signals of the sample. The most common solvents for liquid-state NMR spectroscopy are D_2O/DCl, D_2O/CD_3COOD, D_2O/DCOOD.

Before ^{1}H NMR measurement, a sample should be freeze-dried 2–3 times from 99.99% D_2O in order to remove the residual signal of HOD visible in the ^{1}H spectrum. ^{1}H NMR spectra are usually recorded in aqueous acidic solution (pD ~ 5) at a temperature of 30–80 °C; 8–128 scans are done, depending on the concentration of samples. A typical ^{1}H NMR spectrum of chitosan is shown in Figure 18.

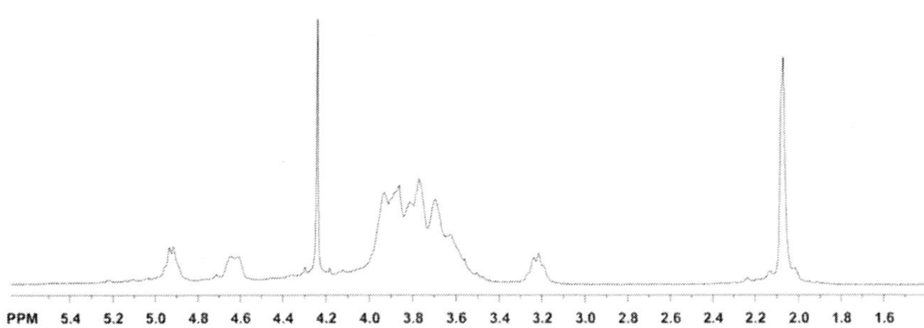

Figure 18. ^{1}H NMR spectrum of chitosan; the DA = 48. Spectrum was recorded at 400 MHz and 80 °C in D_2O (pD ~ 5) relative to internal acetone (δ_H 2.225).

The spectrum shows two signals in the anomeric region, namely H-1 of GlcN (D) at δ ~ 4.9 and H-1 of GlcNAc (A) at δ ~ 4.6. They are shifted to higher values due to the neighboring glycosidic and oxygen atoms of the ring. The resonances of H-3–H-6 ring protons of GlcN and H-2–H-6 of GlcNAc are present in the middle of the spectrum (δ ~ 3.5–4.0), forming a group of broad, overlapping signals. The remaining H-2 ring protons of GlcN residues are shifted to characteristically lower values (δ~ 3.2) because of the adjacent amino group. The characteristic signal of the protons in the *N*-acetyl group (GlcNAc) is at δ ~ 2.1.

2.5.1.2. ^{13}C NMR spectroscopy

There are two kinds of ^{13}C NMR spectroscopy: liquid-state and solid-state. The first one has the same limitations as ^{1}H NMR spectroscopy, namely, the insufficient solubility of most analyzed materials. Moreover, ^{13}C NMR spectroscopy is much less sensitive than ^{1}H NMR spectroscopy owing to the properties of the carbon nucleus and the only 1% abundance of the ^{13}C isotope in nature. Liquid-state ^{13}C NMR spectra are usually recorded from the same kinds of solutions as are used for producing ^{1}H NMR spectra, but the number of scans has to be much higher–this can vary from several dozen to hundreds of thousands. Solid-state ^{13}C NMR spectroscopy is a much more powerful technique. Chitin can be analysed without any special solubilization of the polymer. This means that quite a large amount of sample can be used, which solves the problem of the low sensitivity of ^{13}C NMR spectroscopy. The ^{13}C NMR spectra of solid samples are generally recorded with magic-angle spinning (MAS) and cross-polarization (CP). MAS averages out dipolar interactions and chemical shift anisotropy, producing highly resolved spectra. CP considerably increases the sensitivity of the technique by reducing the relaxation delay due to the magnetization transfer from the ^{1}H to the ^{13}C spins. The intensities of the ^{13}C NMR signals are influenced by the kinetics of the CP process: different contact times affect the intensities of ^{13}C NMR resonances. Hence, it is very important to use the proper contact and relaxation delay times [287–289]. Duarte *et al.* optimized the relaxation delay (5 s) and the contact time (1 ms) during series assignments of the degree of *N*-acetylation of chitin and deacetylated chitin samples. Care should be taken to ensure that the CP build-up is similar for all the carbon atoms used in the quantitative calculation. Moreover, different chitin/chitosan materials often contain paramagnetic centres, which may distort the intensity of some ^{13}C NMR signals in the spectrum [287].

A typical ^{13}C NMR spectrum (Figure 19) of chitin/chitosan contains anomeric (C-1) signals of both monosaccharide residues (GlcN and GlcNAc) at δ ~ 102–105, C-3 and C-5 signals at δ ~ 73–75, C-6 signals at δ ~ 60, C-4 signals shifted to 81–85 due to glycosylation at position 4, and C-2 resonances shifted to δ ~ 55–57 due to the influence of the attached amino group. The remaining signals of the *N*-Ac group of GlcNAc are at δ ~ 23 and δ ~ 174 (methyl group and carboxylic carbon atoms respectively).

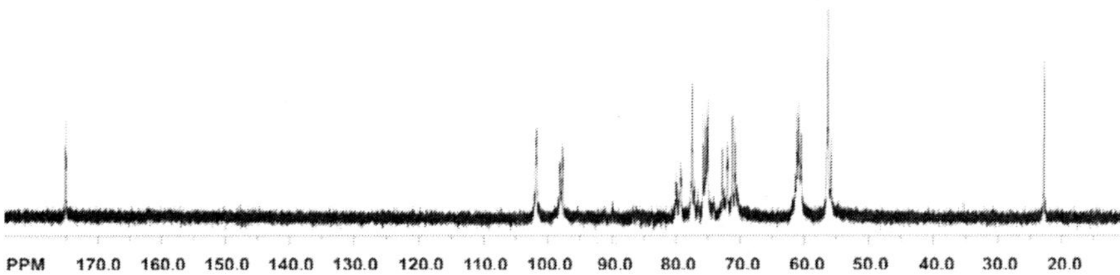

Figure 19. Liquid-state ^{13}C NMR spectrum of chitosan; DA = 48. Spectrum was recorded at 100 MHz and 70 °C in D_2O (pD ~ 5) relative to internal acetone (δ_C 31.45).

The sets of ^{13}C chemical shifts for chitin samples isolated from different sources were found elsewhere [53].

2.5.1.3. ^{15}N NMR spectroscopy

Solid-state ^{15}N NMR spectroscopy, like solid-state ^{13}C NMR spectroscopy, is a very useful technique, especially for studying insoluble chitin, but it is more time-consuming owing to the low natural abundance of the ^{15}N nuclide (<0.4%). Even using high-field NMR spectrometers with large amounts of sample, measurement times are usually of the order of 12–24 h, which makes the technique very expensive. Watson *et al.* developed the ^{15}N labeling of fungal chitin, which makes ^{15}N NMR spectroscopy much more effective [245]. The solid-state ^{15}N NMR spectra for the natural abundance of ^{15}N and for ^{15}N-labeled chitin samples needed 60,000 and only 2,000 repetitions respectively. As with solid-state ^{13}C NMR spectroscopy, care should be taken to ensure that CP build-up is similar for both *N*-acetylated and *N*-deacetylated nitrogen centres.

Two different nitrogen atom signals are present in the chitin/chitosan spectrum: one corresponds to the amine group (NH_2in GlcN; δ ~ 10), the other to the acetamide group ($NH\text{-}CO\text{-}CH_3$ in GlcN; δ ~ 110).

2.5.1.4. ^{31}P NMR spectroscopy

^{31}P NMR spectroscopy is one of the more routine NMR techniques because ^{31}P has an isotopic abundance of 100% and a relatively high magnetogyric ratio. Neither native chitins nor synthesized chitosans contain phosphorus atoms, so they cannot be analysed directly using solid-state ^{31}P NMR spectroscopy. However, it is possible to characterize numerous soluble chitosan derivatives containing phosphorus atoms using both liquid-state and solid-state ^{31}P NMR spectroscopy. Particular types of phosphorus compounds have characteristic chemical shift ranges and are very easily identified in the spectrum.

2.5.1.5. Two-dimensional (2D) NMR spectroscopy

2D NMR spectroscopy provides much more information about the molecule than 1D NMR and is very useful for studying complex molecules. Of the many 2D NMR spectroscopic techniques, COSY (DQF-COSY), NOESY and HSQC are the most common.

COSY (COrrelation SpectroscopY) and DQF-COSY (Double Quantum Filtered COrrelation SpectroscopY) allow one to determine which protons are spin-spin coupled. The signals of protons that are two or three bonds apart are visible. NOESY (Nuclear Overhauser Effect SpectroscopY) shows correlations of all protons which are close enough for dipolar interaction by coupling through space (<5 Å). ^{1}H-^{13}C HSQC (Heteronuclear Single Quantum Coherence) allows one to determine which protons are directly bonded with particular carbon atoms.

2.5.2. Determination of the degree of acetylation (DA)

As already stated, determination of DA of chitin and chitosan is very important from the applicative point of view. Apart from UV/Vis and FTIR, NMR spectroscopy is the most powerful technique for such a study. Review articles summarizing the literature to date on the DA determination of chitin and chitosan using NMR methods have been published by Kasaai [122,131,276]. The DA values of chitin and chitosan can be determined on the basis of integrals from different NMR spectra,*i.e.*, liquid state ^{1}H, and ^{13}C (spectra decoupled from ^{1}H during the acquisition time, without the NOE effect) [290], and solid-state ^{13}C and ^{15}N. For instance, for ^{1}H and ^{13}C NMR spectra, the DA can be calculated by comparing the integral of the methyl carbon/protons of the acetyl group to the integrals of other carbons/protons from the main chains. Many equations for the DA calculation have been put forward for different types of NMR spectroscopy. DA of chitosan can be calculated on the basis of proton integrals from ^{1}H NMR spectra using several equations:

$$\%DA = \frac{1/3 * I_{CH_3(A)}}{1/6 * I_{H2-H6(A+D)}} * 100 \ [178,291] \tag{4}$$

$$\%DA = \frac{7(I_{H1(A)} + I_{CH_3(A)})}{4(I_{H1(D)} + I_{H2-H6(A+D)}) + I_{H1(A)} + I_{CH_3(A)}} * 100 \ [282] \tag{5}$$

$$\%DA = \frac{I_{CH_3(A)}}{3 * I_{H1(A+D)}} * 100 \ [290] \tag{6}$$

$$\%DA = \frac{I_{H1(A)} + I_{CH_3(A)}/3}{I_{H1(D)} + I_{H2(D)} + I_{H1(A)} + I_{CH_3(A)}/3} * 100 \ [212] \tag{7}$$

$$\%DA = \frac{1/3 * I_{CH_3(A)}}{I_{H2(D)}} * 100 \ [292] \tag{8}$$

$$\%DA = 100 - \frac{I_{H1(D)}}{I_{H1(D)} + I_{CH_3(A)}/3} * 100 \ [293] \tag{9}$$

$$\%DA = 100 - \frac{I_{H1(D)}}{I_{H1(D)} + I_{H1(A)}} * 100 \ [293] \tag{10}$$

and the averaged value of three equations [128]:

$$\%DA = 100 - \frac{I_{H2(A)} + I_{H3-H6(A+D)}}{I_{H2(D)}} * 100 \tag{11}$$

$$\%DA = 100 - \frac{I_{H2(A)} + I_{H3-H6(A+D)}}{I_{CH_3(A)}} * 100 \tag{12}$$

$$\%DA = 100 - \frac{I_{H2(D)}}{I_{CH_3(A)}} * 100 \quad (13)$$

On the basis of carbon atom integrals from ^{13}C NMR spectra, DA of chitosan is calculated mainly according to equation [130,278,288,289,294]:

$$\%DA = \frac{I_{CH_3(A)}}{I_{C1-C6(A+D)}/6} * 100 \quad (14)$$

Since solid-state ^{15}N NMR spectra of chitin and chitosan show two very well resolved nitrogen atom signals, DA can be calculated using a simplified equation [279,280]:

$$\%DA = \frac{I_{N(A)}}{I_{N(A)} + I_{N(D)}} * 100 \quad (15)$$

There are many reports on the use of different NMR spectroscopy methods for determining DA of chitin and chitosan. In 1990, Hirai *et al.* suggested using ^1H NMR to study the degree of acetylation of chitosan with different DA (3–40%) [291]. The proposed method was more effective, precise and simple compared to the conventional colloid titration and elemental analysis methods, which were also applied during the study. One of the first assignments of the ^1H and ^{13}C chemical shifts of chitosan using ^1H, ^{13}C, and COSY NMR spectra was also reported [291]. In 1990, Pelletier *et al.* used solid-state ^{13}C NMR to determine DA of chitin and chitosan products [295]. The percentage of deacetylation was calculated by comparing the area of CH_3 resonance to the resonances of the all the remaining carbons. Raymond *et al.* used the same approach [278] to compare DA of chitosan obtained by ^{13}C CP-MAS NMR spectroscopy and conductometric titration. Although both methods yielded similarly good results, the authors concluded that NMR was somewhat limited when DA was low, and likewise conductometric titration when DA values were high [278]. Ottøy *et al.* investigated chitosans fractionated into acid-soluble and acid-insoluble fractions after heterogeneous alkaline deacetylation [294] and obtained the DA values for both groups of fractions using ^{13}C CP-MAS NMR spectroscopy. Comparison of these results with those from ^1H NMR spectroscopy for the acid-soluble fractions showed them to be consistent.

One of the first applications of solid-state ^{15}N NMR spectroscopy to the study of chitin and chitosan was described by Yu *et al.* [279]. The authors used this technique to determine the degree of *N*-acetylation of chitin and chitosan, finding that solid state ^{15}N NMR was the more reliable method, especially for the determination of the extent to which poly(3-hydroxybutyrate) was grafted onto chitosan; the drawback of this method is that it is more time-consuming owing to the low natural abundance of ^{15}N. These authors noted that ^{15}N NMR spectra were easier to analyse than ^{13}C CP-MAS spectra, in which all the oxygen-bearing carbons of both chitin and chitosan overlap; this makes calculating DA more difficult [279].

In 2000 Heux *et al.* compared DA calculated from ^{1}H liquid-state NMR, and ^{13}C and ^{15}N solid-state NMR in the whole range of acetyl content from 0 to 100% [280]. The authors found that all three methods were in good agreement; nevertheless, the limitation of solid-state NMR was reliable detection at DA < 5% due to a small distortion of the spectrum baseline and signal broadening. They also found that ^{15}N CP-MAS NMR spectroscopy was particularly powerful for calculating the acetyl content in complex associations of chitin and other polysaccharides [280]. Sets of ^{13}C and ^{15}N CP-MAS NMR spectra are illustrated in Figures 20 and 21, respectively. The spectra were recorded using the following parameters: for the ^{13}C spectra contact time = 1 ms, recycle delay = 4 s, and 10,000 scans were acquired; for the ^{15}N spectra the corresponding values were 2 ms, 1 s and 100,000 scans [280].

Detailed descriptions of the determination of DA of chitin or chitosan using ^{1}H NMR spectroscopy [128,178,212,282,290,292,293,296], and ^{13}C CP-MAS NMR spectroscopy [130, 287–289,297] can be found elsewhere.

It should be mentioned that ^{1}H NMR spectroscopy is the best choice for determining DA of chitosan. ^{1}H NMR does not require previous calibrations and enables the accurate determination of even low DA values. Moreover, the importance of this method in the study of chitosan is also demonstrated by the fact that ^{1}H NMR data are usually used as standards for calibrating alternative methods [296]. Lavertu *et al.* validated the ^{1}H NMR spectroscopic method for the determination of DA using equations 1, 6a, and 6b. DA values calculated using three different combinations of peak intensities were very close to each other, demonstrating that the technique is also internally consistent. Moreover, the authors discussed the precision, ruggedness, robustness, specificity, stability and accuracy of the technique [293]. It was found to be simple, rapid and more precise than other known techniques like IR or titration.

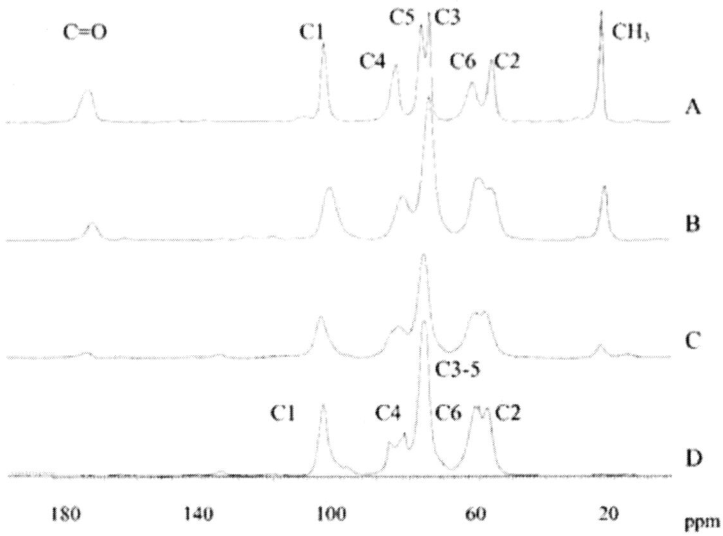

Figure 20. ^{13}C CP-MAS NMR spectra of samples A-D (with decreasing DA; ~100%, ~60%, ~20%, ~0%). Reprinted with permission from *Biomacromolecules*. **2008**, *1*, 746–751. Copyright 2010, American Chemical Society [280].

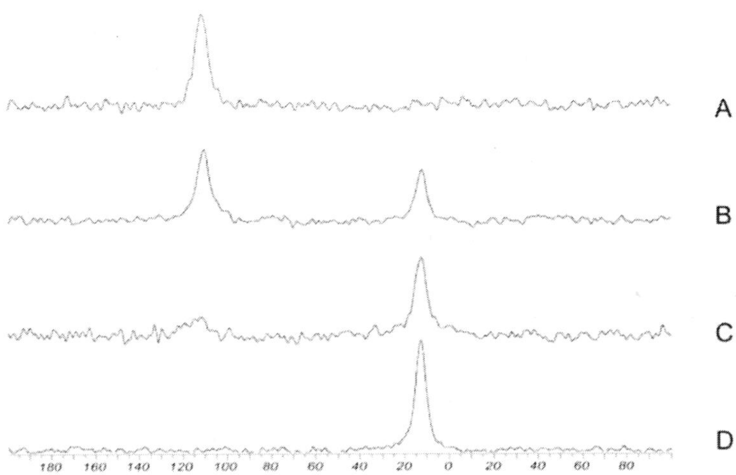

Figure 21. ^{15}N CP-MAS NMR spectra of samples A-D (with decreasing DA; ~100%, ~60%, ~20%, ~0%). Reprinted with permission from Biomacromol. 2000, 4, 746–751. Copyright 2010, ACS Publications [280].

2.5.3. Determination of the pattern of *N*-acetylation

Most properties of chitin and chitosan and further applications depend strongly not only on DA, but probably also on the pattern of acetylation (PA) [2]. PA is a parameter describing the distribution of GlcNAc and GlcN residues along the polysaccharide chain and can be determined by ^{1}H and ^{13}C NMR spectroscopy using appropriate integrals. In the ^{1}H NMR spectrum, the nearest neighbor effect on resonances is visible for the anomeric proton of GlcNAc (sequences AD and AA) and proton H-1 of GlcN (sequences DD and DA) (Figure 22). The rest of the signals in the ^{1}H NMR spectrum are unsuitable for PA assignment as they overlap [282].

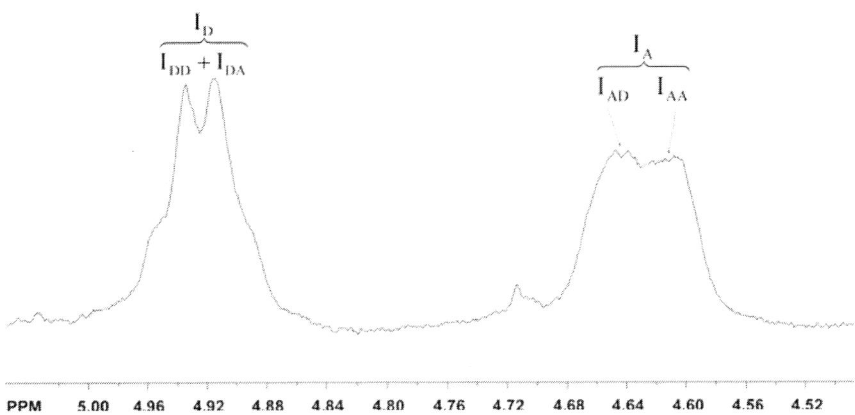

Figure 22. Section of the anomeric region of the ^{1}H NMR spectrum of chitosan; DA = 48. Spectrum was recorded at 400 MHz and 80 °C in D_2O (pD ~ 5) relative to internal acetone (δ_H 2.225). Appropriate intensities of sequences are labeled.

In general, PA in the chitosan samples can be calculated from the C-3, C-5 and C-6 carbon signals in the ^{13}C NMR [297], although detailed inspection of ^{13}C NMR spectra showed that the highest resolution with only low overlapping signals for the four diad frequencies (F_{AA}, F_{AD}, F_{DA} and F_{DD}) were achieved for the carbon C-5 signals [285] (Figure 23). For the other carbon resonances diad, triad and tetrad peaks were superimposed on neighboring peaks, which precludes a quantitative extraction of signal areas.

The relative experimental intensities of ^{1}H or ^{13}C resonances in chitosan samples should be normalized according to Bernoullian statistics [298,299] and can be presented as:

$$F_{AD} = \frac{I_{AD} + I_{DA}}{I_{AD} + I_{DA} + I_{DD} + I_{AA}} \quad (16)$$

$$F_{AA} = \frac{I_{AA}}{I_{AD} + I_{DA} + I_{DD} + I_{AA}} \quad (17)$$

$$F_{DD} = \frac{I_{DD}}{I_{AD} + I_{DA} + I_{DD} + I_{AA}} \quad (18)$$

where I_{AD}, I_{DD}, I_{AA} and I_{DA} are the respective intensities of signals AA, DD, AA and DA in the samples. In the above equation F_{AA} (F_{DD}) is the probability that two A(D) groups are adjacent to each other and F_{AD} the probability that one group A has D as a neighbor, and *vice versa*. The data should be transformed into one parameter PA, which describes the acetylation pattern as described before [298,299]:

$$PA = \frac{F_{AD}}{(2F_{AA}) + F_{AD}} + \frac{F_{AD}}{(2F_{DD}) + F_{AD}} \quad (19)$$

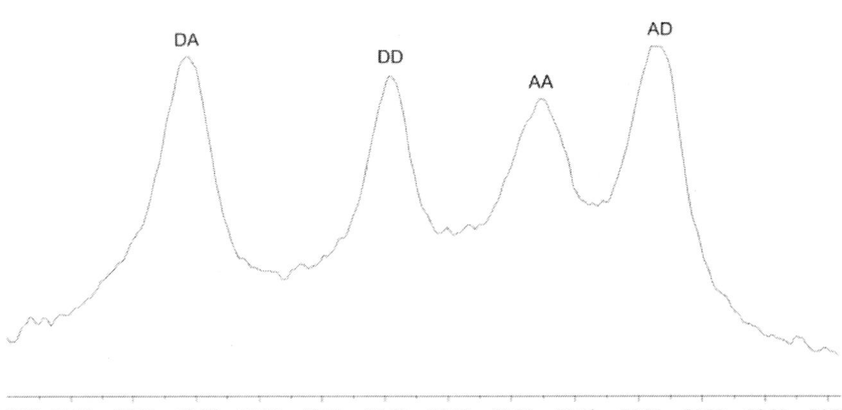

Figure 23. Sections of the ^{13}C NMR spectra of chitosan; DA = 48, showing the C-5 signals region. Spectrum was recorded at 100 MHz and 70 °C in D_2O (pD ~ 5) relative to internal acetone (δ_C 31.45). Appropriate sequences are labeled.

If the statistics are consistent with the Bernoullian model for polymers, the values PA = 0, 1 and 2 respectively indicate a perfect block, random distribution, and the alternating distribution of *N*-acetyl groups along the chitin/chitosan chain (Figure 24).

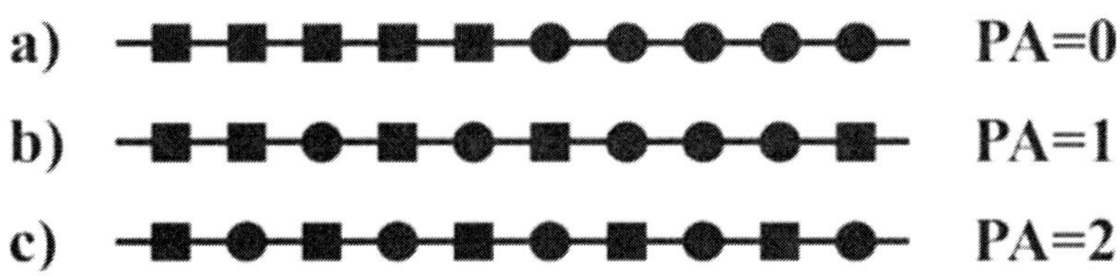

Figure 24. The distribution of GlcNAc (▪) and GlcN (●) residues along the polysaccharide chain at DA = 0.5; (a)–perfect block (PA = 0), (b)–random distribution (PA = 1), and (c)–alternating distribution (PA = 2).

Determination of PA in a chitosan sample using NMR data requires spectra with sufficient resolution of the diad frequencies. In order to get well resolved spectra, the raw chitosan should be depolymerized by nitrous acid. Additionally, pD ~ 4 and higher measurement temperatures (60–80 °C) respectively improve the solubility of chitosan and the reduce viscosity of the solution.

Vårum *et al.* have reported methods for determining the nearest-neighbor (diad) frequencies F_{AA}, F_{AD}, F_{DA}, and F_{DD} of chitosans using ^{1}H and ^{13}C NMR spectroscopy [282,297]. In the first paper, DA and PA of different chitosans obtained by the depolymerization of chitin under homogeneous and heterogeneous conditions were determined by ^{1}H NMR spectroscopy. Chitosan prepared by *N*-deacetylation under homogeneous conditions gave values for the diad frequencies that were roughly consistent with a random distribution of the *N*-acetyl groups. Samples prepared under heterogeneous conditions had a frequency of the AA diad slightly higher than that for a random (Bernoullian) distribution. This means that the GlcNAc units of chitosans prepared by *N*-deacetylation under heterogeneous conditions had a slightly more blockwise distribution than those prepared under homogeneous ones [282]. ^{13}C NMR spectroscopy results revealed that both groups of chitosans (*N*-deacetylated under homogeneous and heterogeneous conditions) gave values for the diad and triad frequencies that were consistent with a random arrangement of GlcN and GlcNAc residues along the chitosan chain [297]. Kumirska *et al.* compared the most frequently used methods for the PA determination of different chitosan samples [300]. These authors proved that ^{13}C NMR spectroscopy based on the analysis of carbon C-5 signals was clearly the most suitable method for determining PA. A number of validation parameters of that method were presented (specificity, sensitivity, repeatability and reproducibility) in that paper [300].

Weinhold *et al.* improved the chitosan pattern determination proposed by Vårum *et al.* [297] through the implementation of a line-fitting procedure for ^{13}C NMR spectra [285]. The authors assigned PA (0.5–1.5) to 32 chitosans and did not find any evidence for the existence of a clear blockwise or clear alternating chitosan preparation, although the various samples were produced under different conditions. Moreover, it was shown for the first time that PA was exponentially correlated with DA:

$$PA = 1.11 - 0.58 * e^{\left(\frac{-DA}{0.13}\right)} \qquad (20)$$

Kumirska *et al.* applied a similar approach for the determination of PA of low-molecular-weight chitosans used in biomedical applications in order to assess the need for the initial depolymerization of the samples [301]. PA values of low-molecular-weight chitosan preparations (<41 kg/mol) were estimated without initial degradation of the native samples, whereas for higher-molecular-weight samples (>41 kg/mol) limited degradation was recommended.

Martinou *et al.* implemented ^{1}H and ^{13}C NMR spectroscopy to monitor the mode of action of chitin deacetylase from *Mucor rouxii* on fully water-soluble, partially *N*-acetylated chitosans [302]. The chitosan was enzymatically deacetylated and the nearest-neighbor frequencies (F_{AA}, F_{AD}, F_{DA} and F_{DD}) were determined by ^{13}C NMR spectroscopy, showing that the transition frequencies F_{AD} and F_{DA} were lower than expected from a random (Bernoullian) distribution in the further enzymatically deacetylated chitosans, while F_{AA} and F_{DD} were higher compared to the random distribution. Such a change in the frequencies associated with the enzymatic deacetylation of chitosans was qualitatively in agreement with an enzyme operating according to a multiple-attack mechanism (chitin deacetylase did not preferentially attack any sequence in the chitosan molecule) [302].

2.5.4. Study of chitin and chitosan derivatives

Chitin and chitosan derivatives can be characterized by all of the above-mentioned NMR techniques in both the liquid and solid states. Very often ^{1}H and ^{13}C NMR spectroscopy are used together to study soluble compounds, and when phosphorus derivatives are involved ^{31}P NMR spectroscopy is additionally invoked. For instance, ^{1}H NMR spectroscopy can be used to assign the degree of substitution (DS) of chitosan by modifying groups. Tommerass *et al.* studied the *N*-alkylated trimer 2-acetamido-2-deoxy-D-glucopyranosyl-β-(1→4)-2-acetamido-2-deoxy-D-glucopyranosyl-β-(1→4)-2,5- anhydro-D-mannofuranose (A–A–M) on a fully *N*-deacetylated chitosan using different NMR techniques. The authors used^{1}H NMR spectroscopy to assign the degree of substitution of chitosan (the DA<0.001) by the A-A-M group [303]. The DS values (0.07, 0.23, 0.40) were calculated from the methyl resonance of the *N*-acetyl group at δ 2.1 compared to H-2 of substituted and unsubstituted chitosan residues at δ 3.0–3.4 (Figure 25).

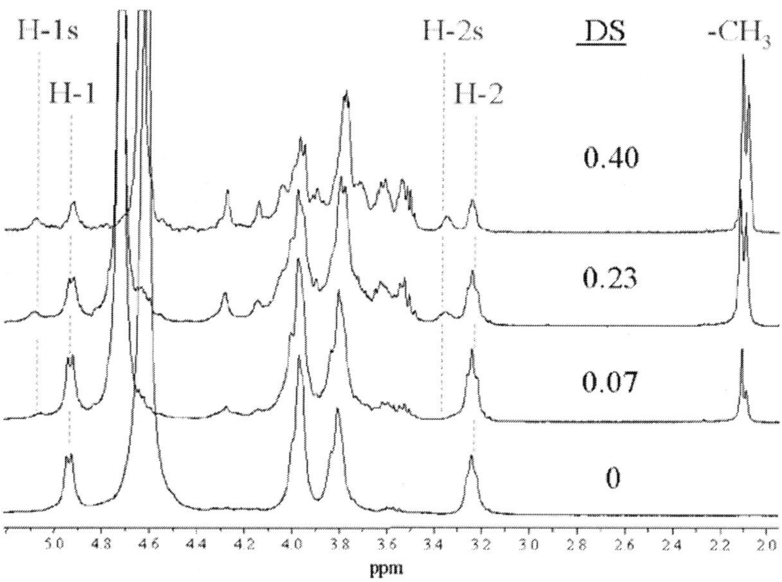

Figure 25. ^{1}H NMR spectra of chitosan (DP = 25, DA < 0.001) *N*-alkylated with the *N*-acetylated trimer A–A–M. Reprinted from Carbohydrate Research 337, 2002, Tømmeraas, K., Köping-Höggård, M., Vårum, K.M., Christensen B.E., Artursson, P., Smidsrød, O., Preparation and characterization of chitosans with oligosacharide branches, 2455–2462, Copyright (2010), with permission from Elsevier [303].

^{1}H NMR spectroscopy was used to characterize a number of chitin and chitosan derivatives, *inter alia*, glycol chitosan [304], hydroxypropyl chitosan [155], *N*-(4-bromonaphthalimide)-chitosan [197], acylated chitosan [143,173], *N*-alkylated chitosan [147], and various *N*-alkyl and *N*-benzyl chitosans [305]. ^{13}C NMR spectroscopy was applied to study many soluble derivatives of chitin and chitosan: 6-carboxychitin [306], 6-*O*-carboxymethyl-chitin and 3,6-*O*-carboxymethyl-chitin [307], and C-6 and C-3 sulphated chitosan [308]. ^{1}H and ^{13}C NMR spectroscopy were utilised to characterize *N,N,N*-trimethyl chitosan chloride [309], 2-*N*-(2-ethoxycarbonylethyl)chitin [310], *O*-succinyl-chitosan [146], chitin-graft-poly(ε-caprolactone) [311], *N*-carboxybutylchitosan and 5-methylpyrrolidinone chitosan [312], and chitosan derivatives with galactose groups [313]. ^{1}H and ^{13}C NMR spectroscopy were supplemented by DQF-COSY NMR spectroscopy for the study of the *N*-alkylated trimer 2-acetamido-2-deoxy-D-glucopyranosyl-β-(1→4)-2-acetamido-2-deoxy-D-glucopyranosyl- β-(1→4)-2,5-anhydro-D-mannofuranose on a fully *N*-deacetylated chitosan [303], and 2D ^{1}H,^{13}C HSQC NMR spectroscopy for chitins sulphated at C-3 and C-6 [314], as well as COSY and ^{1}H,^{13}C HETCOR (hetero-correlation spectroscopy) for the study of *O,N*-carboxymethylchitosans [281].

Solid-state ^{13}C CP-MAS NMR spectroscopy was used to characterize *O*-hydroxypropyl chitin [315], chitin grafted with poly(acrylic acid) [316], derivatives of chitosan and crown ethers [317], dibutyrylchitin [172], and gels obtained by reacting chitosan with diethyl squarate and 1,1,3,3-tetramethoxypropane [318,319].

^{31}P NMR spectroscopy in combination with other NMR techniques was used in the study of numerous chitin and chitosan derivatives containing phosphorus atoms: *N*-(diisopropylphosphono-thiooxomethyl) chitosan, *N*-(2-diethylphosphono-ethyl)-chitosan, and *N*-(2,2-diethylbisphosphono-ethyl)- chitosan [320], phosphorylated chitosan and chitin [321–323], phosphorylcholine chitosan [324], *N*-propyl-*N*-methylene phosphonic chitosan [93], 6-deoxy(diethyl phosphate)-chitin, [325], *O*-ethyl phosphonate chitosan [326], and diethyl phosphate chitosan [327].

2.5.5. Physicochemical characterization of chitin and chitosan

Tanner *et al.* [328] were among the first to attempt to analyze α- and β-chitin polymorphs by NMR spectroscopy. Spectra of different α- and β-chitin were identified, but broad lines and distinct peak asymmetry made it difficult to interpret the spectra, to measure accurate values of chemical shifts, and to make useful comparisons between the spectra of those two polymorphs. In 2004 Jang *et al.* recorded solid-state ^{13}C CP–MAS NMR spectra for α-chitin, β-chitin, and γ-chitin [52]. The NMR spectra of α-chitin and β-chitin were clearly distinguished, but the γ-chitin spectrum was almost the same as that of α-chitin. In the β-chitin spectrum, a signal assigned to C3 and C5 appeared at δ 74, whereas in the α-chitin one, the C3 and C5 appeared as two sharply resolved signals at δ 73 and δ 75, respectively. This was attributed to the different configurations of C3 and C5 resulting from the hydrogen bonds formed. α-Chitin, having an antiparallel structure, has intersheet and intrasheet hydrogen bonding, but β-chitin, having a parallel structure, forms only intrasheet hydrogen bonding. A detailed explanation of the two kinds of hydrogen bonding in α-chitin using ^{13}C CP-MAS NMR spectroscopy was published by Kameda *et al.* [329]. The peaks corresponding to the C3 and C5 carbon atoms of γ-chitin were resolved at δ 73 and δ 75 because γ-chitin formed intersheet hydrogen bonding [52].

In a similar way, Cárdenas *et al.* characterized different polymorphs of chitins isolated from squid, shrimp, prawn, lobsters and king crabs [53].

Using ^{13}C CP-MAS NMR spectroscopy Cortizo *et al.* identified β-chitin isolated from the squid pen *Illex argentinus* [330], and Manni *et al.* isolated α-chitin from shrimp waste [331].

2.5.6. Other applications of NMR techniques

NMR techniques can also be used to examine different chemical and physiochemical processes by the analysis of final or intermediate products. Toffey *et al.* [332] used solid-state ^{13}C-NMR spectroscopy to investigate the thermally-induced conversion into chitin of a water-soluble solid consisting of an ionic complex of chitosan, acetic acid and chitosonium acetate. They recorded the ratio of the signal at δ 179 (the carboxylic atom of the acetate group) and the signal at δ 173 (the carboxylic atom of the *N*-acetyl group) during the conversion.

Holme *et al.* studied the thermal depolymerization of chitosan chloride in the solid state by analyzing the products using^{1}H and ^{13}NMR spectroscopy [333]. The rate of acid hydrolysis of the glycosidic bond in chitosan solutions was found to be in the order A–A ≈ A–D >> D–A ≈ D–D.

Einbu and Vårum reported studies of the depolymerization and de-*N*-acetylation reaction using the chitin dimer (GlcNAc-GlcNAc) and the monomer GlcNAc as respective model substances. The rates of the reactions were determined as a function of acid concentration and temperature using ^{1}H NMR data [334]. In a subsequent paper, the authors described their study of chitin hydrolysis by concentrated and deuterated hydrochloric acid, which was a very suitable solvent for the direct characterization of the chemical composition of chitin by ^{1}H NMR spectroscopy. The rate of hydrolysis of a glycosidic linkage following an *N*-acetylated unit was found to be 54-times higher than the rate of *N*-deacetylation and 115-times higher than the rate of hydrolysis of a glycosidic linkage following a *N*-deacetylated unit [335].

NMR spectroscopy can be utilized to study the enzymatic degradation of chitin. Martinou *et al.* used ^{1}H and ^{13}C NMR spectroscopy to monitor the mode of action of chitin deacetylase from *Mucor rouxii* [302], while Sorbotten *et al.* studied the degradation products of chitosans using chitinase β from *Serratia marcescens* by ^{1}H NMR [336].

NMR spectroscopy, molecular modeling (MM) and molecular dynamics (MD) are excellent tools for studying interactions between chitin/chitosan and proteins or peptides. Colombo *et al.* used molecular dynamics (MD) simulations based on NOE NMR to study the complex between hevein and *N,N'*-diacetylchitobiose and *N,N',N''*-triacetylchitotriose. The results of the simulations showed that a carbohydrate oligomer was able to move on the surface of the relatively flat binding pocket of hevein, therefore occupying different binding subpockets [337]. NMR spectroscopy and MD were also used for studying the interaction of a variety of modified hevein domains with chitooligosaccharides [338]. Aboitiz *et al.* utilized NMR and MM to demonstrate that trisaccharides containing GalNAc and ManNAc residues were also recognized by hevein domains. Their observations indicated that the present nature of the modifications of chitotriose at either the non-reducing end (GalNAc instead of GlcNAc) or at the reducing end (ManNAc instead of GlcNAc) did not modify the mode of binding of the trisaccharide to hevein [339].

2.6. Other spectroscopic methods

Apart from the spectroscopic techniques commonly used in the structural analysis of chitin, chitosan and their derivatives, which have been described above, there are many other spectroscopic methods applied to the analysis of these polymers. They include Scanning Electron Spectroscopy (SEM), Scanning Electron Spectroscopy coupled with Energy Dispersive Spectroscopy (EDAX), Scanning Electron Spectroscopy coupled with Transmission Electron Microscope (SEM-TEM system), Circular Dichroism Spectroscopy (CD) and Inductively Coupled Plasma Spectroscopy (ICP).

SEM is an extremely useful method for the visual confirmation of the morphology and physical state of the surface [44]. SEM has been used, for example, for determining the surface morphology of fungi, and chitin and chitosan in crab shells [340–342], for examining the morphology of deacetylated chitosan powder and chitosan films [140], and for characterizing the surface of chitosan and various types of crosslinked deacetylated chitin before and after metal binding [90]. Nowadays this technique is applied mostly to the determination of new chitin and chitosan derivatives, for example, to study the microstructure of chitosan(chitin)/cellulose biosorbents [49], to characterize the novel nanocomposite scaffold of chitosan and bioactive glass ceramic nanoparticles [343], to characterize the morphology of the surface and the cross-section of chitosan-silica hybrid membranes [344] and to investigate the morphology of vulcanized natural rubber/chitosan blends [91].

Morphological analysis of chitin/chitosan and their derivatives has also been carried out using the SEM-TEM system, for instance, to reveal the structural features of a chitosan/hydroxyapatite nanocomposite [345] or to observe the changes in morphology during the *N*-deacetylation of chitin nanowhiskers to a chitosan nanoscaffold [115].

EDAX is used, among other things, to determine the metal uptake mechanism on chitosan, a complex phenomenon involving nodule formation on the polymer surface, ion adsorption and ion absorption [346]. This technique is also applied to determine the porosity of chitosan beads and membranes, and the diffusion of metal ions through them [44].

Another spectroscopic technique applied to chitin and chitosan investigations is Inductively Coupled Plasma Spectroscopy [ICP] [49]. This has been used to determine the amount of metal ions adsorbed onto chitosan/chitin cellulose biosorbents. The adsorption capacity of these chitosan biosorbents was found to be Cu(II) (0.417 mmol/g) > Zn(II) (0.303 mmol/g) > Cr(VI) (0.251 mmol/g) > Ni(II) (0.225 mmol/g) > Pb(II) (0.127 mmol/g). Inductively-coupled plasma spectrometry was also used for the same reason by Gamage in 2007 [347].

CD spectroscopy measures differences in the absorption of left-handed polarized light *versus* right-handed polarized light that arise due to structural asymmetry [348]. A CD spectrometer records this phenomenon as a function of wavelength. Kittur *et al.* [349] used CD to investigate the degree of *N*-acetylation of low-molecular-weight chitosan (LMWC) prepared from pectolytic hydrolysates of chitosan. Circular dichroism spectra showed a decrease in the segment of *N*-acetylated glucosamine units in LMWC.

In general spectroscopic techniques are very useful and important in the structural analysis of chitin, chitosan and their derivatives, although we need to mention that there are many other non-spectroscopic analytical methods that are routinely used in such studies: viscosity measurements, thermogravimetric analysis (TGA), differential scanning calorimetry (DSC), isothermal titration calorimetry (ITC), cyclic voltamperometry, high-performance liquid chromatography (HPLC) and size exclusion chromatography (SEC).

3. CONCLUSIONS

Chitin and chitosan are natural aminopolysaccharides with unique structures and interesting properties such as biocompatibility and biodegradability; they are non-toxic, have a wide range of applications, and the raw material sources for their production are unlimited. Chemical modification of these polymers results in improved solubility in water or organic solvents, which enhances their biological activities and favors the continuous development of their applications as new functional biomaterials with excellent potential in various fields. In order to acquire a deeper understanding of the mechanism of these properties, it is necessary for chitin/chitosan and their derivatives to be structurally and physicochemically well characterized. Knowledge of the microstructure of these compounds is essential for an understanding of structure–property–activity relationships. Although spectroscopic methods are of great use to scientists working in the chitin/chitosan field, many researchers lack specialist knowledge of these techniques. This paper focuses on the practical aspects of using the five spectroscopic methods most often applied – X-ray spectroscopy, infrared (IR) and UV-Vis-spectroscopy, mass spectrometry (MS) and nuclear magnetic resonance spectroscopy (NMR-spectroscopy) – in the structural investigation and physicochemical characterization of chitin, chitosan and their derivatives. The application of other spectroscopic methods is also discussed. This paper provides hands-on information about these valuable research tools, emphasizing practical aspects such as sample preparation or typical measurement conditions, their limitations and advantages, and interpretation of results. The usefulness of these methods in establishing and confirming molecular structures, determining physicochemical parameters such as the degree of *N*-acetylation, pattern of *N*-acetylation, degree of polymerization, molecular weight, crystallinity, the sequence or degree of substitution, as well as in monitoring reactions, controlling the purity of these compounds and in characterizing their intra- and intermolecular interactions, is discussed. It should also be mentioned that the structural analysis of chitin/chitosan and their derivatives is based on the application of at least a few different types of spectroscopic methods during the same study. The main reason for this is the possibility of obtaining different but compatible and complete information about the structure and physicochemical properties of these compounds, which is impossible using only one spectroscopic technique. Continuous improvements in sampling techniques, analysis software and instrumentation hardware, mean that spectroscopic methods have revolutionized the chemical and physicochemical characterization of chitin/chitosan samples, and are now routinely used by scientists working in the chitin/chitosan field.

ACKNOWLEDGEMENTS

The authors express their gratitude for the financial support provided by the Polish Ministry of Research and Higher Education under grant DS/8200-4-0085-10.

REFERENCES AND NOTES

1. Muzzarelli, RAA; Jeuniaux, C; Gooday, GW. *Chitin in Nature and Technology*; Plenum Publishing Corporation: New York, NY, USA, 1986.

2. Rinaudo, M. Chitin and chitosan: Properties and application. *Prog Polym Sci* **2006**, *31*, 603–632.

3. Roberts, GAF. *Chitin Chemistry*, 1st ed; MacMillan: London, UK, 1992.

4. Austin, PR. Chitin solutions and purification of chitin. *Methods Enzymol* **1988**, *161*, 403–407.

5. Kurita, K. Controlled functionalization of the polysaccharide chitin. *Progr Polym Sci* **2001**, *26*, 1921–1971.

6. Muzzarelli, RAA, Ed.; *Natural Chelating Polymers*; Pergamon Press: New York, NY, USA, 1973; p. 83.

7. Pillai, CKS; Paul, W; Sharma, CP. Chitin and chitosan polymers: Chemistry, solubility and fiber formation. *Progr Polym Sci* **2009**, *34*, 641–678.

8. Ravi Kumar, MNV. A review of chitin and chitosan applications. *React Funct Polym* **2000**, *46*, 1–27.

9. Roberts, GAF. The Road is long... *Adv Chitin Sci* **2007**, *10*, 3–10.

10. Tajik, H; Moradi;, M; Rohani, SMR; Erfani, AM; Jalali, FSS. Preparation of Chitosan from Brine Shrimp (*Artemia urmiana*) Cyst Shells and Effects of Different Chemical Processing Sequences on the Physicochemical and Functional Properties of the Product. *Molecules* **2008**, *13*, 1263–1274.

11. Muzzarelli, RAA; Muzzarelli, C. Chitosan chemistry: Relevance to the biomedical sciences. *Adv Polym Sci* **2005**, *186*, 151–209.

12. Arai, K; Kinumaki, T; Fujita, T. Toxicity of chitosan. *Bull Tokai Reg Fish Lab* **1968**, *43*, 89–94.

13. Di Martino, A; Sittinger, M; Risbud, MV. Chitosan: A versatile biopolymer for orthopaedic tissue-engineering.*Biomaterials* **2005**, *26*, 5983–5990.

14. Crini, G; Badot, P-M. Application of chitosan, a natural aminopolysaccharide, for dye removal from aqueous solutions by adsorption processes using batch studies: A review of recent literature. *Prog Polym Sci* **2008**, *33*, 399–447.

15. Shahidi, F; Arachchi, JKV; Jeon, Y-J. Food applications of chitin and chitosans. *Trends Food Sci Technol* **1999**, *10*, 37–51.

16. Dutta, PK; Tripathi, S; Mehrotra, GK; Dutta, J. Perspectives for chitosan based antimicrobial films in food applications.*Food Chem* **2009**, *114*, 1173–1182

17. Thacharodi, D; Panduranga Rao, K. Development and *in vitro* evaluation of chitosan based transdermal drug delivery systems for controlled delivery of propranolol hydrochloride. *Biomaterials* **1995**, *16*, 145–148.

18. Cho, Y-W; Jang, J; Park, CR; Ko, S-W. Preparation and solubility in acid and water of partially deacetylated chitins.*Biomacromolecules* **2000**, *1*, 609–614.

19. Kim, I-Y; Seo, S-J; Moon, H-S; Yoo, M-K; Park, I-Y; Kim, B-C; Cho, C-S. Chitosan and its derivatives for tissue engineering applications. *Biotechnol Adv* **2008**, *26*, 1–21.

20. Van der Lubben, IM; Verhoef, JC; Borchard, G; Junginger, HE. Chitosan and its derivatives in mucosal drug and vaccine delivery. *Eur J Pharm Sci* **2001**, *14*, 201–207.

21. Alves, NM; Mano, JF. Chitosan derivatives obtained by chemical modifications for biomedical and environmental applications. *Int J Biol Macromol* **2008**, *43*, 401–414.

22. Clasen, C; Wilhelms, T; Kulicke, W-M. Formation and Characterization of Chitosan Membranes. *Biomacromolecules***2006**, *7*, 3210–3222.

23. Sajomsang, W; Tantayanon, S; Tangpasuthadol, V; Thatte, M; Daly, WH. Synthesis and characterization of *N*-aryl chitosan derivatives. *Int J Biol Macromol* **2008**, *43*, 79–87.

24. Silva, SS; Menezes, SMC; Garcia, RB. Synthesis and characterization of polyurethane-g-chitosan. *Eur Polym J* **2003**, *39*, 1515–1519. Zhao, Q; Qian, J; An, Q; Gao, C; Gui, Z; Jin, H. Synthesis and characterization of soluble chitosan/sodium carboxymethyl cellulose polyelectrolyte complexes and the pervaporation dehydration of their homogeneous membranes. *J Membr Sci* **2009**, *333*, 68–78.

25. Al Sagheer, FA; Al-Sughayer, MA; Muslim, S; Elsabee, MZ. Extraction and characterization of chitin and chitosan from marine sources in Arabian Gulf. *Carbohydr Polym* **2009**, *77*, 410–419.

26. Campana-Filho, SP; De Britto, D; Curti, E; Abreu, FR; Cardoso, MB; Battisti, MV; Sim, PC; Lavall, RL. Extraction, structures and properties of α- and β-chitin. *Quim Nova* **2007**, *30*, 644–650

27. Noishiki, Y; Takami, H; Nishiyama, Y; Wada, M; Okada, S; Kuga, S. Alkali-induced conversion of β-chitin to α-chitin.*Biomacromolecules* **2003**, *4*, 896–899.

28. Synowiecki, J; Al-Khateeb, NA. Production, Properties, and Some New Applications of Chitin and Its Derivatives. *Crit Rev Food Sci Nutr* **2003**, *43*, 145–171.

29. Hayes, M; Carney, B; Slater, J; Brück, W. Mining marine shellfish wastes for bioactive molecules: Chitin and chitosan -Part A: Extraction methods. *Biotechnol J* **2008**, *3*, 871–877.

30. Struszczyk, MH. Chitin and Chitosan: Part I. Properties and production. *Polimery* **2002**, *47*, 316–325

31. Aranaz, I; Mengíbar, M; Harris, R; Paños, I; Miralles, B; Acosta, N; Galed, G; Heras, Á. Functional Characterization of Chitin and Chitosan. *Curr Chem Biol* **2009**, *3*, 203–230

32. Shirai, K; Palella, D; Castro, Y; Guerrero-Legarreta, I; Saucedo-Castaneda, G; Huerta-Ochoa, S; Hall, GM. Chen, RH, Chen, HC, Eds.; Characterization of chitins from lactic acid fermentation of prawn wastes. In *Advances in Chitin Science*; Elsevier: Taiwan, 1998; Volume III, pp. 103–110.

33. Kurita, K; Tomita, K; Ishi, S; Nishimura, S-I; Shimoda, K. β-Chitin as a convenient starting material for acetolysis for efficient preparation of *N*-acetylchitooligosaccharides. *J Polym Sci Part A Polym Chem* **1993**, *31*, 2393–2395.

34. Aiba, S. Studies on chitosan: 3. Evidence for the presence of random and block copolymer structures in partially *N*-acetylated chitosans. *Int J Biol Macromol* **1991**, *13*, 40–44.

35. Weinhold, MX; Sauvageau, JCM; Keddig, N; Matzke, M; Tartsch, B; Grunwald, I; Kübel, C; Jastorff, B; Thöming, J. Strategy to improve the characterization of chitosan for sustainable biomedical applications: SAR guided multi-dimensional analysis. *Green Chem* **2009**, *11*, 498–509.

36. Struszczyk, MH; Struszczyk, KJ. *Medical Application of Chitin and Its Derivatives*; Polish Chitin Society: Lotz, Poland, 2007; Monograph XII; pp. 139–147.

37. Guo, T. More power to X-rays: New developments in X-ray spectroscopy. *Laser Photon* **2009**, *6*, 591–622.

38. Broglie, M. Sur une nouveau procédé permettant d'obtenir la photographie des spectres de raies des rayons Röntgen.*Comptes Rendus* **1913**, *157*, 924–926.

39. Siegbahn, K. Electron Spectroscopy for Chemical Analysis (ESCA). *Philos Trans R Soc Lond A Math Phys Sci* **1970**, *268*, 33–57.

40. Cullity, BD. *Elements of X-ray Diffraction*; Addison-Wesley: Reading, MA, USA, 1978.

41. Clark, GL; Smith, AF. X-ray diffraction studies of chitin, chitosan and derivatives. *J Phys Chem* **1937**, *40*, 863–879.

42. Ogawa, K; Yui, T; Miya, M. Dependence on the preparation procedure of the polymorphism and crystallinity of chitosan membranes. *Biosci Biotech Biochem* **1992**, *56*, 858–862.

43. Varma, AJ; Deshpande, SV; Kennedy, JF. Metal complexation by chitosan and its derivatives: a review. *Carbohydr Polym* **2004**, *55*, 77–93.

44. Minke, R; Blackwell, J. The structure of α-chitin. *J Mol Biol* **1978**, *120*, 167.

45. Matienzo, LJ; Winnacker, SK. Dry Processes for Surface Modification of a Biopolymer: Chitosan. *Macromol Mater Eng***2002**, *287*, 871–880

46. Veleshko, AN; Rumyantseva, EV; Veleshko, IE; Teterin, AYu; Maslakov, KI; Teterin, YuA; Kulyukhin, SA; Vikhoreva, GA. X-ray Photoelectron Study of Complexation between Uranyl Group and Chitosan. *Radiochemistry* **2008**, *5*, 523–529.

47. Dambies, L; Guimon, C; Yiacoumi, S; Guibal, E. Characterization of metal ion interactions with chitosan by X-ray photoelectron spectroscopy. *Colloids Surf A* **2001**, *177*, 203–214.

48. Xiaoqi, S; Bo, P; Yang, J; Ji, C; Deqian, L. Chitosan (Chitin)/Cellulose Composite Biosorbents Prepared Using Ionic Liquid for Heavy Metal Ions Adsorption. *Am Inst Chem Eng* **2009**, *55*, 2062–2069.

49. Kurmaev, EZ; Shin, S; Watanabe, M; Eguchi, R; Ishiwata, Y; Takeuchi, T; Moewesd, A; Ederere, DL; Gaof, Y; Iwamig, M; Yanagiharah, M. Probing oxygen and nitrogen bonding sites in chitosan by X-ray emission. *J Electron Spectrosc Relat Phenom* **2002**, *125*, 133–138.

50. Klepka, M; Lawniczak-Jablonska, K; Demchenko, IN; Nedelko, N; Ślawska-Waniewska, A; Rodrigues, CA; Bordini, C. X-ray Absorption Spectroscopy investigation of Fe in metal-chitosan complexes. *J Synchr Rad Nat Sci* **2006**, *5*, 220–223.

51. Jang, MK; Kong, BG; Jeong, YI; Lee, CH; Nah, JW. Physicochemical characterization of α-chitin, β-chitin and γ-chitin separated from natural resources. *J Polym Sci Part A Polym Chem* **2004**, *42*, 3423–3432.

52. Cárdenas, G; Carbrera, G; Taboada, E; Miranda, SP. Chitin characterization by SEM, FTIR, XRD, and 13C cross polarization/mass angle spinning NMR. *J Appl Polym Sci* **2004**, *93*, 1876–1885.

53. Kim, SS; Kim, SH; Lee, YM. Preparation, characterization, and properties of β-chitin and *N*-acetylated β-chitin. *J Polym Sci Part B: Polym Phys* **1996**, *34*, 2367–2374.

54. Yen, MT; Mau, JL. Selected physical properties of chitin prepared from shiitake stipes. *LWT– Food Sci Technol* **2007**, *40*, 558–563.

55. Muzzarelli, C; Francescanheli, O; Tosi, G; Muzzarelli, RAA. Susceptibility of dibutyryl chitin and regenerated chitin fibres to deacylation and depolymerization by lipases. *Carbohydr Polym* **2004**, *56*, 137–146.

56. Wada, M; Saito, Y. Lateral thermal expansion of chitin crystals. *J Polym Sci Part B Polym Phys* **2001**, *39*(1), 168–174.

57. Feng, F; Liu, Y; Hu, K. Influence of alkali-freezing treatment on the solid state structure of chitin. *Carbohydr Res* **2004**,*339*, 2321–2324.

58. Muzzarelli, RAA; Morganti, P; Morganti, G; Palombo, P; Palombo, M; Biagini, G; Mattioli Belmonte, M; Giantomassi, F; Orlandi, F; Muzzarelli, C. Chitin nanofibrils/chitosan composites as wound medicaments. *Carbohydr Polym* **2007**, *70*, 274–284.

59. Abdou, ES; Nagy, KSA; Elsabee, MZ. Extraction and characterization of chitin and chitosan from local sources.*Bioresour Technol* **2008**, *99*, 1359–1367.

60. Samuels, RJ. Solid state characterization of the structure of chitosan films. *J Polym Sci Part B: Polym Phys* **1981**, *19*, 1081–1105.

61. Sakurai, M; Takagi, M; Takahashi, T. Crystal structure of chitosan. I. Unit cell parameters. *Seni Gakkaishi* **1984**, *40*, T-246.

62. Ogawa, K; Hirano, S; Miyanishi, T; Yui, T; Watanabe, T. A new polymorph of chitosan. *Macromolecules* **1984**, *17*, 973–975

63. Ogawa, K; Inukai, S. X-Ray diffraction study of sulfuric, nitric, and halogen acid salts of chitosan. *Carbohydr Res* **1987**,*160*, 425–433.

64. Cairns, P; Miles, MJ; Morris, VJ; Ridout, MJ; Brownsey, GJ; Winter, WT. X-ray fibre diffraction studies of chitosan and chitosan gels. *Carbohydr Res* **1992**, *235*, 23–28.

65. Saito, H; Tabeta, R. High-resolution solid-state ^{13}C NMR study of chitosan and its salts with acids: Conformational characterization of polymorphs and helical structures as viewed from the conformation dependent ^{13}C chemical shifts. *Macromolecules* **1987**, *20*, 2424–2430.

66. Yui, T; Imada, K; Okuyama, K; Obata, Y; Suzuki, K; Ogawa, K. Molecular and Crystal Structure of the Anhydrous Form of Chitosan. *Macromolecules* **1994**, *27*, 7601–7605

67. Mazeau, K; William, WT; Chanzy, H. Molecular and Crystal Structure of a High-Temperature Polymorph of Chitosan from Electron Diffraction Data. *Macromolecules* **1994**, *27*, 7606–7612.

68. Okuyama, K; Noguchi, K; Miyazawa, T; Yui, T; Ogawa, K. Molecular and Crystal Structure of Hydrated Chitosan.*Macromolecules* **1997**, *30*, 5849–5855.

69. Ogawa, K; Oka, K; Miyanishi, T; Hirano, S. *Chitin, Chitosan and Related Enzymes*; Academic Press: Orlando, FL, USA, 1984; pp. 327–345.

70. Ogawa, K. Effect of Heating an Aqueous Suspension of Chitosan on the Crystallinity and Polymorphs. *Agric Biol Chem***1991**, *55*, 2375–2379.

71. Kawada, J; Abe, Y; Yui, T; Okuyama, K; Ogawa, K. Crystalline transformation of chitosan from hydrated to anhydrous polymorph via chitosan monocarboxylic acid salts. *J Carbohydr Chem* **1999**, *18*, 559–571.

72. Lima, IS; Airoldi, C. A thermodynamic investigation on chitosan–divalent cation interactions. *Thermochim Acta* **2004**,*421*, 133–139.

73. Zhang, Y; Xue, C; Xue, Y; Gao, R; Zhang, X. Determination of the degree of deacetylation of chitin and chitosan by X-ray powder diffraction. *Carbohydr Res* **2005**, *340*, 1914–1917.

74. Focher, B; Beltranme, PL; Naggi, A; Torri, G. Alkaline Ndeacetylation of chitin enhanced by flash treatments: reaction kinetics and structure modifications. *Carbohydr Polym* **1990**, *12*, 405–418.

75. Struszczyk, H. Microcrystalline Chitosan. I Preparation and Properties of Microcrystalline Chitosan. *J Appl Polym Sci***1987**, *33*, 177–189.

76. Schiffman, JD; Stulga, LA; Schauer, CL. Chitin and Chitosan: Transformations Due to the Electrospinning Process.*Polym Eng Sci* **2009**, *49*, 1918–1928.

77. Zhang, Y; Xue, C; Li, Z; Zhang, Y; Fu, X. Preparation of half-deacetylated chitosan by forced penetration and its properties. *Carbohydr Polym* **2006**, *65*, 229–234.

78. Fernandez Cerveraa, M; Heinamaki, J; Rasanenc, M; Maunuc, SL; Karjalainenb, M; Nieto Acostaa, OM; Iraizoz Colartea, A; Yliruusib, J. Solid-state characterization of chitosans derived from lobster chitin. *Carbohydr Polym* **2004**,*58*, 401–408.

79. Toffey, A; Glasser, WG. Chitin Derivatives. II. Time–Temperature–Transformation Cure Diagrams of the Chitosan Amidization Process. *J Appl Polym Sci* **1999**, *73*, 1879–1889.

80. Ogawa, K; Yui, T. Crystalllinity of partialy *N*-acetylated chitosan. *Biosci Biotech Biochem* **1993**, *57*, 1466–1469.

81. Seoudi, R; Nada, AMA. Molecular structure and dielectric properties studies of chitin and its treated by acid, base and hypochlorite. *Carbohydr Polym* **2007**, *68*, 728–733.

82. Nishino, T; Matsui, R; Nakamae, K. Elastic Modulus of the Crystalline Regions of Chitin and Chitosan. *J Polym Sci Part B Polym Phys* **1999**, *37*, 1191–1196.

83. Ogawa, K. X-ray study of Chitosan L- and D- Ascorbates. *Chem Mater* **1996**, *8*, 2349–2351.

84. Muzzarelli, RAA. *Chitin*; Pergamon Press: Oxford, UK, 1977.

85. Ogawa, K; Oka, K. X-ray Study of Chitosan-Transition Metal Complexes. *Chem Mater* **1993**, *5*, 726–728.

86. Schlick, S. Binding Sites of Cu2+ in Chitin and Chitosan. An Electron Spin Resonance Study. *Macromolecules* **1986**, *19*, 192–195.

87. El-Sherbiny, IM. Synthesis, characterization and metal uptake capacity of a new carboxymethyl chitosan derivative.*Eur Polym J* **2009**, *45*, 199–210. [

88. Hai-Bing, Li; Yuan-Yin, C; Shi-Lan, L. Synthesis, Characterization, and Metal Ions Adsorption Properties of Chitosan–Calixarenes (I). *J Appl Polym Sci* **2003**, *89*, 1139–1144.

89. Trimukhe, KD; Varma, AJ. A morphological study of heavy metal complexes of chitosan and crosslinked chitosans by SEM and WAXRD. *Carbohydr Polym* **2008**, *71*, 698–702.

90. Jobish, J; Vijayalakshmi, R. Thermal stability, morphology, and X-ray diffraction studies of dynamically vulcanized natural rubber/chitosan blends. *J Mater Sci* **2009**, *44*, 4087–4094.

91. Sahoo, S; Sasmal, A; Nanda, R; Phani, AR; Nayak, PL. Synthesis of chitosan– polycaprolactone blend for control delivery of ofloxacin drug. *Carbohydr Polym* **2010**, *79*, 106–113

92. Zuńiga, A; Debbaudt, A; Albertengo, L; Rodríguez, MS. Synthesis and characterization of *N*-propyl-*N*-methylene phosphonic chitosan derivative. *Carbohydr Polym* **2010**, *79*, 475–480.

93. Mohamed, KR; Mostafa, AA. Preparation and bioactivity evaluation of hydroxyapatite-titania/chitosangelatin polymeric biocomposites. *Mater Sci Eng A* **2008**, *28*, 1087–1099.

94. Nagahama, H; Maeda, H; Kashiki, T; Jayakumar, R; Furuike, T; Tamura, H. Preparation and characterization of novel chitosan/gelatin membranes using chitosan hydrogel. *Carbohydr Polym* **2009**, *76*, 255–260.

95. Tretenichenko, EM; Datsun, VM; Ignatyuk, LN; Nud'ga, LA. Preparation and Properties of Chitin and Chitosan from a Hydroid Polyp. *Russ J Appl Chem* **2006**, *79*, 1341–1346.

96. Aji, PM; Laborie, MPG; Oksman, K. Cross-Linked Chitosan/Chitin Crystal Nanocomposites with Improved Permeation Selectivity and pH Stability. *Biomacromolecules* **2009**, *10*, 1627–1632.

97. Kurmaev, EZ; Shin, S; Watanabe, M; Eguchi, R; Ishiwata, Y; Takeuchi, T; Moewesd, A; Ederere, DL; Gaof, Y; Iwamig, M; Yanagiharah, M. Probing oxygen and nitrogen bonding sites in chitosan by X-ray emission. *J Electron Spectrosc Relat Phenom* **2002**, *125*, 133–138.

98. Stuart, BH. *Infrared Spectroscopy: Fundamentals and Applications (Analytical Techniques in the Sciences (AnTs) *)*; John Wiley & Sons Ltd: Chichester, UK, 2004.

99. Smith, BC. *Fundamentals of Fourier Transform Infrared Spectroscopy*; CRC Press: Boca Raton, FL, USA, 1996.

100. Günzler, H; Gremlich, HU. *IR Spectroscopy: An Introduction*; Wiley-VCH: Weinherm, Germany, 2002.

101. Griffiths, P; De Haseth, JA. *Fourier Transform Infrared Spectrometry*, 2nd ed; Wiley- Interscience: New York, NY, USA, 2007.

102. Thanpitcha, T; Sirivat, A; Jamieson, AM; Rujiravanit, R. Dendritic polyaniline nanoparticles synthesized by carboxymethyl chitin templating. *Eur Polym J* **2008**, *44*, 3423–3429.

103. Bourtoom, T; Chinnan, MS. Preparation and properties of rice starch-chitosan blend biodegradable film. *LWT-Food Sci Technol* **2008**, *41*, 1633–1641.

104. Yang, J; Yao, Z; Tang, C; Darvell, BW; Zhang, H; Pan, L; Liu, J; Chen, Z. Growth of apatite on chitosan-multiwall carbon nanotube composite membranes. *Appl Surf Sci* **2009**, *255*, 8551–8555.

105. Sajomsang, W; Ruktanonchai, UR; Gonil, P; Nuchuchua, O. Mucoadhesive property and biocompatibility of methylated N-aryl chitosan derivatives. *Carbohydr Polym* **2009**, *78*, 945–952.

106. Urreaga, JM; de la Orden, MU. Chemical interactions and yellowing in chitosan-treated cellulose. *Eur Polym J* **2006**, *42*, 2606–2616.

107. Darmon, SE; Rudall, KM. Infra-red and X-ray studies of chitin. *Disc Faraday Soc* **1950**, *9*, 251–260.

108. Pearson, FG; Marchessault, RH; Liang, CY. Infrared spectra of crystalline polysaccharides. V. Chitin. *J Polym Sci* **1960**,*13*, 101–116.

109. Brunner, E; Ehrlich, H; Schupp, P; Hedrich, R; Hunoldt, S; Kammer, M; Machill, S; Paasch, S; Bazhenov, VV; Kurek, DV; Arnold, T; Brockmann, S; Ruhnow, M; Born, R. Chitin-based scaffolds are an integral part of the skeleton of the marine demosponge *Ianthella basta*. *J Struct Biol* **2009**, *168*, 539–547.

110. Focher, B; Naggi, A; Torri, G; Cosani, A; Terbojevich, M. Structural differences between chitin polymorphs and their precipitates from solutions-evidence from CP-MAS ^{13}CNMR, FT-IR and FT-Raman spectroscopy. *Carbohydr Polym***1992**, *17*, 97–102.

111. Juárez-de la Rosa, BA; Ardisson, P-L; Azamar-Barrios, JA; Quintana, P; Alvarado-Gil, JJ. Optical, thermal, and structural characterization of the sclerotized skeleton of two antipatharian coral species. *Mater Sci Eng C* **2007**, *27*, 880–885.

112. Furuhashi, T; Schwarzinger, C; Miksik, I; Smrz, M; Beran, A. Molluscan shell evolution with review of shell calcification hypothesis. *Comp Biochem Physiol B Biochem Mol Biol* **2009**, *154*, 351–371.

113. Paulino, AT; Simionato, JI; Garcia, JC; Nozaki, J. Characterization of chitosan and chitin produced from silkworm chrysalides. *Carbohydr Polym* **2006**, *64*, 98–103.

114. Lertwattanaseri, T; Ichikawa, N; Mizoguchi, T; Tanaka, Y; Chirachanchai, S. Microwave technique for efficient deacetylation of chitin nanowhiskers to a chitosan nanoscaffold. *Carbohydr Res* **2009**, *344*, 331–335.

115. Pranshanth, KVH; Kittur, FS; Tharanathan, RN. Solid state structure of chitosan prepared under different *N*-deacetylating conditions. *Carbohydr Polym* **2002**, *50*, 27–33.

116. Wu, T; Zivanovic, S; Draughon, FA; Conoway, WS; Sams, CE. Physicochemical Properties and Bioactivity of Fungal Chitin and Chitosan. *J Agric Food Chem* **2005**, *53*, 3888–3894.

117. Prashanth, KVH; Tharanathan, RN. Crosslinked chitosan-preparation and characterization. *Carbohydr Res* **2006**, *341*, 169–173.

118. Nah, J-W; Jang, M-K. Spectroscopic Characterization and Preparation of Low Molecular, Water-Soluble Chitosan with Free-Amine Group by Novel Method. *J Polym Sci Part A Polym Chem* **2002**, *40*, 3796–3803.

119. Dong, Y; Wang, H; Zheng, W; Zhao, Y; Bi, D; Zhao, L; Li, X. Liquid crystalline behaviour of chitooligosaccharides.*Carbohydr Polym* **2004**, *57*, 235–240.

120. Li, QD; Dunn, ET; Grandmaison, EW; Goosen, MFA. Applications and Properties of Chitosan. *J Bioact Compat Polym***1992**, *7*, 370–397.

121. Kasaai, MR. Various Methods for Determination of the Degree of *N*-Acetylation of Chitin and Chitosan: A Review. *J Agric Food Chem* **2009**, *57*, 1667–1676.

122. Sabnis, S; Block, LH. Improved infrared spectroscopic method for the analysis of degree of *N*-deacetylation. *Polym Bull***1997**, *39*, 67–71.

123. Muzzarelli, RAA; Tanfani, F; Scarpini, G; Laterza, G. The degree of acetylation of chitins by gas chromatography and infrared spectroscopy. *J Biochem Bioph Methods* **1980**, *2*, 299–306.

124. Domszy, JG; Roberts, GAF. Evaluation of infrared spectroscopic techniques for analysing chitosan. *Die Makromolekulare Chemie* **1985**, *186*, 1671–1677.

125. Baxter, A; Dillon, M; Taylor, KD; Roberts, GAF. Improved method for IR determination of the degree of *N*-acetylation of chitosan. *Inter J Biolog Macromol* **1992**, *14*, 166–169.

126. Shigemasa, Y; Matsuura, H; Sashiwa, H; Saimato, H. Domard, A, Jeuniaux, C, Muzzarelli, RAA, Roberts, GAF, Eds.; An improved IR

spectroscopic determination of degree of deacetylation of chitin. In *Advances in Chitin Science*; André Publisher: Lyon, France, 1996; Volume 1, pp. 204–209.

127. Shigemasa, Y; Matsuura, H; Sashiwa, H; Saimato, H. Evaluation of different absorbance ratios from infrared spectroscopy for analyzing the degree of deacetylation in chitin. *Int J Biol Macromol* **1996**, *18*, 237–242.

128. Brugnerotto, J; Lizardi, J; Goyoolea, FM; Argülles-Monal, W; Desbrières, J; Rinaudo, M. An infrared investigation in relation with chitin and chitosan characterization. *Polymer* **2001**, *42*, 3569–3580.

129. Duarte, ML; Ferreira, MC; Marvão, MR; Rocha, J. An optimized method to determine the degree of acetylation of chitin and chitosan by FTIR spectroscopy. *Int J Biol Macromol* **2002**, *31*, 1–8.

130. Kasaai, MR. A review of several reported procedures to determine the degree of *N*-acetylation for chitin and chitosan using infrared spectroscopy. *Carbohydr Polym* **2008**, *71*, 497–508.

131. Duarte, ML; Ferreira, MC; Marvão, MR; Rocha, J. Uragami, T, Kurita, K, Fukamizo, T, Eds.; Chitin and chitosan: An optimized methodology by FTIR spectroscopy. In *Chitin and Chitosan in Life Science*; Kodansha Scientific Ltd: Tokyo, Japan, 2001; pp. 86–90.

132. Moore, GK; Roberts, GAF. Determination of the degree of *N*-acetylation of chitosan. *Int J Biol Macromol* **1980**, *2*, 115–116.

133. Dong, Y; Xu, C; Wang, J; Wu, Y; Wang, M; Ruan, Y. Influence of degree of deacetylation on critical concentration of chitosan/dichlorocatic acid liquid crystalline solution. *J Appl Polym Sci* **2002**, *83*, 1204–1208.

134. Qin, C; Li, H; Xiao, Q; Liu, Y; Zhu, J; Du, Y. Water-solubility of chitosan and its antimicrobial activity. *Carbohydr Polym***2006**, *63*, 367–374.

135. Miya, M; Iwamoto, R; Yoshikawa, S; Mima, S. I.R spectroscopic determination of CONH content in highly deacetylated chitosan. *Int J Biol Macromol* **1980**, *2*, 323–324.

136. Qu, X; Wirsén, A; Albertsson, A-C. Effect of lactic/glycolic acid side chains on the thermal degradation kinetics of chitosan derivatives. *Polymer* **2000**, *41*, 4841–4847.

137. Rathke, TD; Hudson, SM. Determination of the degree of *N*-acetylation in chitin and chitosan as well as their monomer sugar ratios by near infrared spectroscopy. *J Polym Sci A Polym Chem* **1993**, *31*, 749–753.

138. Vårum, KM; Egelandsdal, B; Ellekjaer, MR. Characterization of partially *N*-acetylated chitosans by near infrared spectroscopy. *Carbohydr Polym* **1995**, *28*, 187–193.

139. Baskar, D; Sampath Kumar, TS. Effect of deacetylation time on the preparation, properties and swelling behavior of chitosan films. *Carbohydr Polym* **2009**, *78*, 767–772.

140. Vachoud, L; Zydowicz, N; Domarad, A. Formation and characterisation of a physical chitin gel. *Carbohydr Res* **1997**,*302*, 169–177.

141. Zainol, I; Akil, HM; Mastor, A. Effect of γ-irradiation on the physical and mechanical properties of chitosan powder.*Mater Sci Eng C* **2009**, *29*, 292–297.

142. Ma, G; Yang, D; Kennedy, JF; Nie, J. Synthesize and characterization of organic-soluble acylated chitosan. *Carbohydr Polym* **2009**, *75*, 390–394.

143. Singh, J; Dutta, PK; Dutta, J; Hunt, AJ; Macquarrie, DJ; Clark, JH. Preparation and properties of highly soluble chitosan–l-glutamic acid aerogel derivative. *Carbohydr Polym* **2009**, *76*, 188–195.

144. Xie, Y; Liu, X; Chen, Q. Synthesis and characterization of water-soluble chitosan derivate and its antibacterial activity.*Carbohydr Polym* **2007**, *69*, 142–147.

145. Zhang, C; Ping, Q; Zhang, H; Shen, J. Synthesis and characterization of water-soluble *O*-succinyl- chitosan. *Eur Polym J***2003**, *39*, 1629–1634.

146. Ma, G; Yang, D; Zhou, Y; Xiao, M; Kennedy, JF; Nie, J. Preparation and characterization of water-soluble *N*-alkylated chitosan. *Carbohydr Polym* **2008**, *74*, 121–126

147. Jeong, Y-I; Kim, D-G; Jang, M-K; Nah, J-W. Preparation and spectroscopic characterization of methoxy poly(ethylene glycol)-grafted water-soluble chitosan. *Carbohydr Res* **2008**, *343*, 282–289.

148. Zong, Z; Kimura, Y; Takahashi, M; Yamane, H. Characterization of chemical and solid state structures of acylated chitosans. *Polymer* **2000**, *41*, 899–906.

149. Kaushik, A; Khan, R; Solanki, PR; Pandey, P; Alam, J; Ahmad, S; Malhotra, BD. Iron oxide nanoparticles–chitosan composite based glucose biosensor. *Biosens Bioelectron* **2008**, *24*, 676–683.

150. Borges, O; Borchard, G; Verhoef, JC; de Sousa, A; Junginger, HE. Preparation of coated nanoparticles for a new mucosal vaccine delivery system. *Int J Pharm* **2005**, *299*, 155–166.

151. Yang, D; Li, J; Jiang, Z; Lu, L; Chen, X. Chitosan/TiO_2 nanocomposite pervaporation membranes for ethanol dehydration. *Chem Eng Sci* **2009**, *64*, 3130–3137.

152. Tan, H; Chu, CR; Payne, KA; Marra, KG. Injectable *in situ* forming biodegradable chitosan–hyaluronic acid based hydrogels for cartilage tissue engineering. *Biomaterials* **2009**, *30*, 2499–2506.

153. Madhumathi, K; Shalumon, KT; Divya Rani, VV; Tamura, H; Furuike, T; Selvamurugan, N; Nair, SV; Jayakumar, R. Wet chemical synthesis of chitosan hydrogel–hydroxyapatite composite membranes for tissue engineering applications. *Int J Biol Macromol* **2009**, *45*, 12–15.

154. Xie, W; Xu, P; Wang, W; Liu, Q. Preparation and antibacterial activity of a water-soluble chitosan derivative. *Carbohydr Polym* **2002**, *50*, 35–40.

155. Wu, C-S. A comparison of the structure, thermal properties, and biodegradability of polycaprolactone/chitosan and acrylic acid graftedpolycaprolactone/chitosan. *Polymer* **2005**, *46*, 47–155.

156. Munro, NH; Hanton, LR; Moratti, SC; Robinson, BH. Synthesis and characterisation of chitosan-graft-poly(OEGMA) copolymers prepared by ATRP. *Carbohydr Polym* **2009**, *77*, 496–505.

157. Wu, Z; Feng, W; Feng, Y; Liu, Q; Xu, X; Sekino, T; Fujii, A; Ozaki, M. Preparation and characterization of chitosan-grafted multiwalled

carbon nanotubes and their electrochemical properties. *Carbon* **2007**, *45*, 1212–1218.

158. Winie, T; Arof, AK. FT-IR studies on interactions among components in hexanoyl chitosan-based polymer electrolytes. *Spectrochim Acta Part A* **2006**, *63*, 677–684.

159. Piron, E; Domarad, A. Interaction between chitosan and uranyl ions. Part 2. Mechanism of interaction. *Int J Biol Macromol* **1998**, *22*, 33–40.

160. Sipos, P; Berkesi, O; Tombácz, E; StPierree, TG; Webb, J. Formation of spherical iron(III) oxyhydroxide nanoparticles sterically stabilized by chitosan in aqueous solutions. *J Inorg Biochem* **2003**, *95*, 55–63.

161. Bhattarai, SR; Bahadur, KCR; Aryal, S; Khil, MS; Kim, HY. *N*-Acylated chitosan stabilized iron oxide nanoparticles as a novel nano-matrix and ceramic modification. *Carbohydr Polym* **2007**, *69*, 467–477.

162. Yavuz, AG; Uygun, A; Bhethanabotla, VR. Substituted polyaniline/chitosan composites: Synthesis and characterization. *Carbohydr Polym* **2009**, *75*, 448–453.

163. Baran, EJ. Spectroscopic investigation of the $VO2^{+}$/chitosan interaction. *Carbohydr Polym* **2008**, *74*, 704–706.

164. Phisalaphong, M; Jatupaiboon, N. Biosynthesis and characterization of bacteria cellulose–chitosan film. *Carbohydr Polym* **2008**, *74*, 482–488.

165. Kamari, A; Wan Ngah, WS; Liew, LK. Chitosan and chemically modified chitosan beads for acid dyes sorption. *J Environ Sci* **2009**, *21*, 296–302.

166. Rosa, S; Laranjeira, MCM; Riela, HG; Fávere, VT. Cross-linked quaternary chitosan as an adsorbent for the removal of the reactive dye from aqueous solutions. *J Hazard Mater* **2008**, *155*, 253–260.

167. Sakkayawong, N; Thiravetyan, P; Nakbanpote, W. Adsorption mechanism of synthetic reactive dye wastewater by chitosan. *J Colloid Interface Sci* **2005**, *286*, 36–42.

168. Nadavala, SK; Swayampakula, K; Boddu, VM; Abburi, K. Biosorption of phenol and *o*-chlorophenol from aqueous solutions on to chitosan–calcium alginate blended beads. *J Hazard Mater* **2009**, *162*, 482–489.

169. Sundaram, CS; Viswanathan, N; Meenakshi, S. Fluoride sorption by nano-hydroxyapatite/ chitin composite. *J Hazard Mater* **2009**, *172*, 147–151.

170. Ma, W; Ya, F-Q; Han, M; Wang, R. Characteristics of equilibrium, kinetics studies for adsorption of fluoride on magnetic-chitosan particie. *J Hazard Mater* **2007**, *143*, 296–302.

171. Van de Velde, K; Kiekens, P. Structure analysis and degree of substitution of chitin, chitosan and dibutyrylchitin by FT-IR spectroscopy and solid state ^{13}C NMR. *Carbohydr Polym* **2004**, *58*, 409–416.

172. Mine, S; Izawa, H; Kaneko, Y; Kadokawa, J. Acetylation of α-chitin in ionic liquids. *Carbohydr Res* **2009**, *344*, 2263–2265.

173. Hunger, M; Weitkamp, J. *In situ* IR, NMR, EPR, and UV/Vis spectroscopy: Tools for new insight into the mechanisms of heterogeneous catalysis. *Angew Chem Int Ed* **2001**, *40*, 2954–2971.

174. Hollas, JM. *Modern Spectroscopy*, 3rd ed; J. Wiley & Sons: Chichester, UK, 1996.

175. Ojeda, CB; Rojas, FS. Recent developments in derivative ultraviolet/visible absorption spectrophotometry. *Anal Chim Acta* **2004**, *518*, 1–24.

176. Förster, H. *UV/VIS Spectroscopy*; Springer: Berlin/Heidelberg, Germany, 2004

177. Tan, SC; Khor, E; Tan, TK; Wong, SM. The degree of deacetylation of chitosan: advocating the first derivative UV-spectrophotometry method of determination. *Talanta* **1998**, *45*, 713–719.

178. De Souza, HKS; Bai, G; do Pilar Gonçalves, M; Bartos, M. Whey protein isolate–chitosan interactions: A calorimetric and spectroscopy study. *Thermochim Acta* **2009**, *495*, 108–114.

179. Liu, D; Wei, Y; Yao, P; Jiang, L. Determination of the degree of acetylation of chitosan by UV spectrophotometry using dual standards. *Carbohydr Res* **2006**, *341*, 782–785.

180. Fu, Z-S; Sun, B-B; Chen, J; Yuan, L. Preparation and photochromism of carboxymethyl chitin derivatives containing spirooxazine moiety. *Dyes Pigm* **2008**, *76*, 515–518.

181. Sahoo, S; Sasmal, A; Nanda, R; Phani, AR; Nayak, PL. Synthesis of chitosan–polycaprolactone blend for control delivery of ofloxacin drug. *Carbohydr Polym* **2010**, *79*, 106–113.

182. Yang, M; Yang, Y; Yang, H; Shen, G; Yu, R. Layer-by-layer self-assembled multilayer films of carbon nanotubes and platinum nanoparticles with polyelectrolyte for the fabrication of biosensors. *Biomaterials* **2006**, *27*, 246–255.

183. Lin, Q-K; Ren, K-F; Ji, J. Hyaluronic acid and chitosan-DNA complex multilayered thin film as surface-mediated nonviral gene delivery system. *Colloids Surf B* **2009**, *74*, 298–303.

184. Ramaprasad, AT; Rao, V; Sanjeev, G; Ramananic, SP; Sabharwal, S. Grafting of polyaniline onto the radiation crosslinked chitosan. *Synth Met* **2009**, *159*, 1983–1990.

185. Nosal, WH; Thompson, DW; Yan, L; Sarkar, S; Subramanian, A; Woollam, JA. UV–vis–infrared optical and AFM study of spin-cast chitosan films. *Colloids Surf B* **2005**, *43*, 131–137.

186. Muzzarelli, RAA; Rocchetti, R. Determination of the degree of acetylation of chitosans by first derivative ultraviolet spectrophotometry. *Carbohydr Polym* **1985**, *6*, 461–472.

187. Pedroni, VI; Gschaider, ME; Schulz, PC. UV spectrophotometry: Improvements in the study of the degree of acetylation of chitosan. *Macromol Biosci* **2003**, *3*, 531–534.

188. Hsiao, HY; Tsai, CC; Chen, S; Hsieh, BC; Chen, RLC. Spectrophotometric determination of deacetylation degree of chitinous materials dissolved in phosphoric acid. *Macromol Biosci* **2004**, *4*, 919–921.

189. Wu, T; Zivanovic, S. Determination of the degree of acetylation (DA) of chitin and chitosan by an improved first derivative UV method. *Carbohydr Polym* **2008**, *73*, 248–253.

190. Da Silva, RMP; Mano, JF; Reis, RL. Straightforward determination of the degree of *N*-acetylation of chitosan by means of first-derivative UV spectrophotometry. *Macromol Chem Phys* **2008**, *209*, 1463–1473.

191. Khan, TA; Peh, KK; Ch'ng, HS. Reporting degree of deacetylation values of chitosan: The influence of analytical methods. *J Pharm Pharm Sci* **2002**, *5*, 205–212.

192. Liu, H; Du, Y; Wang, X; Sun, L. Chitosan kills bacteria through cell membrane damage. *Int J Food Microbiol* **2004**, *95*, 147–155.

193. Tanida, F; Tojima, T; Han, S-M; Nishi, N; Tokura, S; Sakairi, N; Seino, H; Hamada, K. Novel synthesis of a water-soluble cyclodextrin-polymer having a chitosan skeleton. *Polymer* **1998**, *39*, 5261–5263

194. Mi, F-L. Synthesis and characterization of a novel chitosan-gelatin bioconjugate with fluorescence emission.*Biomacromolecules* **2005**, *6*, 975–987.

195. Felinto, MCFC; Parra, DF; da Silva, CC; Angerami, J; Oliveira, MJA; Lugão, AB. The swelling behavior of chitosan hydrogels membranes obtainedby UV- and γ-radiation. *Nucl Instrum Methods Phys Res Sect B* **2007**, *265*, 418–424.

196. Munro, NH; Hanton, LR; Robinson, BH; Simpson, J. Synthesis and characterisation of fluorescent chitosan derivatives containing substituted naphthalimides. *React Funct Polym* **2008**, *68*, 671–678.

197. T∅mmeraas, K; Strand, SP; Tian, W; Kenne, L; Vårum, KM. Preparation and characterisation of fluorescent chitosans using 9-anthraldehyde as fluorophore. *Carbohydr Res* **2001**, *336*, 291–296.

198. Kang, B; Dai, Y; Zhang, H; Chen, D. Synergetic degradation of chitosan with gamma radiation and hydrogen peroxide.*Polym Degrad Stab* **2007**, *92*, 359–362.

199. Sun, Ch; Qu, R; Chen, H; Ji, C; Wang, C; Sun, Y; Wang, B. Degradation behavior of chitosan chains in the 'green' synthesis of gold nanoparticles. *Carbohydr Res* **2008**, *343*, 2595–2599.

200. Huang, H; Yang, X. Synthesis of chitosan-stabilized gold nanoparticles in the absence/presence of tripolyphosphate. *Biomacromolecules* **2004**, *5*, 2340–2346.

201. Chen, X; Zhang, X; Yang, W; Evans, DG. Biopolymer-manganese oxide nanoflake nanocomposite films fabricated by electrostatic layer-by-layer assembly. *Mater Sci Eng C* **2009**, *29*, 284–287.

202. Huang, H; Yang, X. Chitosan mediated assembly of gold nanoparticles multilayer. *Colloids Surf A* **2003**, *226*, 77–86.

203. Larena, A; Cáceres, DA. Variability between chitosan membrane surface characteristics as function of its composition and environmental conditions. *Appl Surf Sci* **2004**, *238*, 273–277.

204. Mao, Z; Ma, L; Yan, J; Yan, M; Gao, C; Shen, J. The gene transfection efficiency of thermoresponsive *N,N,N*-trimethyl chitosan chloride-g-poly(*N*-isopropylacrylamide) copolymer. *Biomaterials* **2007**, *28*, 4488–4500.

205. Zhang, Q; Zhang, L; Li, J. Fabrication and electrochemical study of monodisperse and size controlled Prussian blue nanoparticles protected by biocompatible polymer. *Electrochim Acta* **2008**, *53*, 3050–3055.

206. Johnstone, RAW; Rose, ME. *Mass Spectrometry for Chemists and Biochemists*, 2nd ed; Cambridge University Press: Cambridge, UK, 1996.

207. Siuzdak, G. *The Expanding Role of Mass Spectrometry in Biotechnology*; MCC Press: San Diego, CA, USA, 2006.

208. De Hoffmann, E; Stroobant, V. *Mass Spectrometry: Principles and Applications*, 3rd ed; J. Wiley & Sons: Chichester, UK, 2007.

209. Mattai, J; Hayes, ER. Characterization of chitosan by pyrolysis—mass spectrometry. *J Anal Appl Pyrolysis* **1982**, *3*, 327–334.

210. Nieto, JM; Peniche-Covas, C; Padrón, G. Characterization of chitosan by pyrolysis-mass spectrometry, thermal analysis and differential scanning calorimetry. *Thermochim Acta* **1991**, *176*, 63–68.

211. Sato, H; Mitzutani, S; Tsuge, S. Determination of the degree of acetylation of chitin/chitosan by pyrolysis-gas chromatography in the presence of oxalic acid. *Anal Chem* **1998**, *70*, 7–12.

212. Bierstedt, A; Stankiewicz, BA; Briggs, DEG; Evershed, RP. Quanitative and qualitative analysis of chitin in fossil arthropods using a combination of colorimetric assay and pyrolysis-gas chromatography-mass spectrometry. *Analyst***1998**, *123*, 19–145.

213. Steinbrecht, RA; Stankiewicz, BA. Molecular composition of the wall of insects alfactory sensilla-the chitin question. *J Insect Physiol* **1999**, *45*, 785–790.

214. Furuhashi, T; Beran, A; Blazso, M; Czegeny, Z; Schwarzinger, C; Steiner, G. Pyrolysis GC/MS and IR spectroscopy in chitin Analysis of Molluscan shells. *Biosci Biotechnol Biochem* **2009**, *73*, 93–103.

215. Lee, JW; Deng, F; Yeomans, WG; Allen, AL; Gross, RA; Kaplan, DL. Direct incorporation of glucosamine and *N*-acetylglucosamie into exopolymers by *Gluconacetobacter xylinus* (=*Acetobacter xylinum*) ATCC 10245: production of chitosan-cellulose and chitin-cellulose exopolymers. *Appl Environ Microbiol* **2001**, *67*, 3970–3975.

216. Vesentini, D; Steward, D; Singh, AP; Ball, R; Daniel, G; Franish, R. Chitosan-mediated changes in cell wall composition, morphology and ultrastructure in two wood-inhabiting fungi. *Mycol Res* **2007**, *111*, 875–890.

217. Cunha, AG; Fernandes, SCM; Freire, CSR; Silvestre, AJD; Neto, CP; Gandini, A. what is the real value of chitosan's surface energy? *Biomacromolecules* **2008**, *9*, 610–614.

218. Akiyama, K; Kawazu, K; Kobayashi, A. A novel method for chemo-enzymatic synthesis of elicitor-active chitosan oligomers and partially *N*-deacetylated chitin oligomers using *N*-acylated chitotrioses as substrates in a lysozyme-catalyzed transglycosylation reaction system. *Carbohydr Res* **1995**, *279*, 151–160.

219. Zhang, H; Du, Y; Yu, X; Mitsutomi, M; Aiba, S. Preparation of chitooligosaccharides from chitosan by complex enzyme. *Carbohydr Res* **1999**, *320*, 257–260.

220. Tokuyasu, K; Mitsutomi, M; Yamaguchi, I; Hayashi, K; Mori, Y. Recognition of chitooligosaccharides and their *N*-acetyl groups by putative subsites of chitin deacetylase from a Deuteromycete, *Colletotrichum lindemuthianum*.*Biochemistry* **2000**, *39*, 8837–8843.

221. Vishu Kumar, AB; Varadaraj, MC; Gowda, LR; Tharanathan, RN. Characterization of chito-oligosaccharides prepared by chitosanolysis with the aid of papain and Pronase, and their bactericidal action against *Bacillus cereus* and*Escherichia coli*. *Biochem J* **2005**, *391*, 167–175.

222. Makino, A; Kurosaki, K; Ohmae, M; Kobayashi, S. Chitinase-catalyzed synthesis of alternatingly *N*-deacetylated chitin: A chitin-chitosan hybrid polysaccharide. *Biomacromolecules* **2006**, *7*, 950–957.

223. Fernandez-Megia, E; Novoa-Carballal, R; Quiñoá, E; Riguera, R. Conjugation of bioactive ligands to PEC-grafted chitosan at the distal end of PEG. *Biomacromolecules* **2007**, *8*, 833–842.

224. Boesel, LF; Reis, RL; San Roman, J. Innovative Approach for producing injectable, biodegradable materials using chitooligosaccharides and Green Chemistry. *Biomacromolecules* **2009**, *10*, 465–470.

225. Popa-Nita, S; Lucas, J-M; Ladavière, C; David, L; Domard, A. Mechanisms involved during the ultrasonically induced depolymerization of chitosan: characterization and control. *Biomacromolecules* **2009**, *10*, 1203–1211.

226. Bahrke, S; Einarsson, JM; Gislason, J; Haebel, S; Letzel, MC; Peter-Katalinić, J; Peter, MG. Sequence analysis of chitooligosaccharides by matrix-assisted laser desorption ionization postsource decay mass spectrometry.*Biomacromolecules* **2002**, *3*, 696–704.

227. Haebel, S; Bahrke, S; Peter, MG. Quantitative sequencing of complex mixtures of heterochitooligosaccharides by vMALDI-linear ion trap mass spectrometry. *Anal Chem* **2007**, *79*, 5557–5566.

228. Cederkvist, FH; Parmer, MP; Vårum, KM; Eijsink, VGH; S∅rlie, M. Inhibbition of a family 18 chitinase by chitooligosaccharides. *Carbohydr Polym* **2008**, *74*, 41–49.

229. Thierry, B; Winnik, FM; Merhi, Y; Silver, J; Tabrizian, M. Bioactive coatings of endovascular stents based on polyelectrolyte multilayers. *Biomacromolecules* **2003**, *4*, 1564–1571.

230. Grenha, A; Seijo, B; Serra, C; Remuñán-López, C. Chitosan nanoparticle-loaded mannitol macrospheres: structure and surface characterization. *Biomacromolecules* **2007**, *8*, 2072–2079.

231. Shahgholi, M; Callahan, JH; Rappoli, BJ; Rowley, DA. Investigation of copper-saccharide complexation reactions using potentiometry and electrospray mass spectrometry. *J Mass Spectrom* **1997**, *32*, 1080–1093.

232. Wu, L-Q; Gadre, AP; Yi, H; Kastantin, MJ; Rubloff, GW; Bentley, WE; Payne, GF; Ghodssi, R. Voltage-dependent assembly of the polysaccharide chitosan onto an electrode surface. *Langmuir* **2002**, *18*, 8620–8625.

233. Liew, S-T; Wei, A. Preparation of orthogonally protected chitosan oligosaccharides: observation of an anomalous remote substituent effect. *Carbohydr Res* **2002**, *337*, 1319–1324.

234. Wu, L-Q; Embree, HD; Balgley, BM; Smith, PJ; Payne, GF. Utilizing renewable resources to create functional polymers: chitosan-based associative thickener. *Environ Sci Technol* **2002**, *36*, 3446–3454.

235. Liu, Y; Gaskell, KJ; Cheng, Z; Yu, L; Payne, GF. Chitosan-coated electrodes for bimodal sensing: selective post-electrode film reaction for spectroelectrochemmical analysis. *Langmuir* **2008**, *24*, 7223–7231.

236. Yao, Y-Y; Shrestha, KL; Wu, Y-J; Tasi, H-J; Chen, C-Y; Yang, J-M; Ando, A; Cheng, C-Y; Li, Y-K. Structural simulation and protein engineering to convert an endo-chitosanase to an exo-chitosanase. *Protein Eng Des Sel* **2008**, *21*, 561–566.

237. Li, J; Cai, J; Fan, L. Effect of sonolysis on kinetics and physicochemical properties of treated chitosan. *J Appl Polym Sci***2008**, *109*, 2417–2425.

238. Kerwin, JL; Whitney, DL; Sheikh, A. Mass spectrometric profiling of glucosamine, glucosamine polymers and their catecholamine adducts. Model reactions and cuticular hydrolysates of *Toxorhynchites amboinensis* (Culicidae) pupae.*Insect Biochem Mol Biol* **1999**, *29*, 599–607.

239. Trombotto, S; Ladavière, C; Delolme, F; Domard, A. Chemical preparation and structural characterization of a homogenous series of chitin/chitosan oligomers. *Biomacromolecules* **2008**, *9*, 1731–1738.

240. Dennhart, N; Fukamizo, T; Brzezinski, R; Lacombe-Harvey, M-E; Letzel, T. Oligosaccharide hydrolysis by chitosanase enzymes monitored by real-time electrospray ionization-mass spectrometry. *J Biotechnol* **2008**, *134*, 253–260.

241. Ishimizu, T; Mitsukami, Y; Shinkawa, T; Natsuka, S; Hase, S; Miyagi, M; Sakiyama, F; Norioka, S. Presence of asparagine-linked *N*-acetyloglucosamine and chitobiose in *Pyrus pyrifolia* S-RNases associated with gametophytic self-incompatibility. *Eur J Biochem* **1999**, *263*, 624–634.

242. Li, J; Du, YM; Liang, HB; Yao, PJ; Wei, YA. Effect of immobilized neutral protease on the preparation and physicochemical properties of low molecular weight chitosan and chito-oligomers. *J Appl Polym Sci* **2006**, *102*, 4185–4193.

243. Suginata, W; Pantoom, S; Prinz, H. Substrate binding modes and anomer selectivity of chitinase A from *Vibrio harveyi*. *J Chem Biol* **2009**, *2*, 191–202.

244. Watson, HR; Apperley, DC; Dixon, DP; Edwards, R; Hodgson, DR. An efficient method for ^{15}N-labeling of chitin in fungi. *Biomacromolecules* **2009**, *10*, 793–797.

245. Kamst, E; van der Drift, KMGM; Thomas-Oates, JE; Lugtenberg, BJJ; Spaink, HP. Mass spectrometric analysis of chitin oligosaccharides produced by *Rhizobium* NodC Protein in *Escherichia coli*. *J Bacteriol* **1995**, *177*, 6282–6285.

246. Van der Drift, KMGM; Spaink, HP; Bloemberg, GV; van Brussel, AAN; Lugtenberg, BJJ; Haverkamp, J; Thomas-Oates, JE. *Rhizobium leguminarosum* bv. *trifolii* produces lipo-chitin oligosaccharides with *nodE*-dependent highly unsaturated fatty acyl moieties. *J Biol Chem* **1996**, *271*, 22563–22569.

247. Tokuyasu, K; Ono, H; Ohnishi-Kameyama, M; Hayashi, K; Mori, Y. Deacetylation of chitin oligosaccharides of dp 2–4 by chitin deacetylase

from *Colletotrichum lindemuthianum*. *Carbohydr Res* **1997**, *303*, 353–358.

248. Tokuyasu, K; Ono, H; Mitsutomi, M; Hayashi, K; Mori, Y. Synhesis of a chitosan tetramer derivative, β-D-GlcNAc-(1→4)-β-D-GlcNAc-(1→4)-β-D-GlcNAc-(1→4)-D-GlcN through a partial N-acetylation reaction by chitin deacetylase.*Carbohydr Res* **2000**, *325*, 211–215. Kittur, FS; Vishu Kumar, AB; Varadaraj, MC; Tharanathan, RN. Chitoologosaccharides-preparation with the aid of pectinase isozyme from *Aspergillus niger* and their antibacterial activity. *Carbohydr Res* **2005**, *340*, 1239–1245.

249. Aronson, NN, Jr; Halloran, BA; Alexyev, MF; Amable, L; Madura, JD; Paspulati, L; Worth, C; Van Roey, P. Family 18 chitinase-oligosaccharides substrate interaction: subsite preference and anomer selectivity of *S. Marcescens* chitinase A. *Biochem J* **2003**, *376*, 87–95.

250. Van der Drift, KMGM; Olsthoorn, MMA; Brüll, LP; Blok-Tip, L; Thomas-Oates, JE. Mass spectrometric analysis of lipo-chitin oligosaccharides-signal molecules mediating the host-specific legume-rhizobium symbiosis. *Mass Spectrom Rev***1998**, *17*, 75–95.

251. Kollár, R; Petráková, E; Ashwell, G; Robbins, PW; Cabib, E. Architecture of the yeast cell wall. *J Biol Chem* **1995**, *270*, 1170–1178.

252. Lopatin, SA; Ilyn, MM; Pustobaev, VN; Bezchetnikova, ZA; Varlamov, VP; Davankov, VA. Mass-spectrometric analysis of *N*-acetylchitooligosaccharides prepared through enzymatic hydrolysis of chitosan. *Anal Biochem* **1995**, *227*, 285–288.

253. Verbruggen, F; Heiri, O; Reichart, G-J; De Leeuw, JW; nierop, KGJ; Lotter, AF. Effects of chemical pretreatment on δ^{18}O measurements, chemical composition, and morphology of chironomid head capsules. *J Paleolimnol* **2009**.

254. Wang, YV; O'Brien, DM; Jenson, J; Francis, D; Wooler, MJ. The influence of diet and water on the stable oxygen and hydrogen isotope composition of Chironomidae (Diptere) with paleoecological implications. *Oecologia* **2009**, *160*, 225–233. López, FA; Mercê, ALR; Alguacil, FJ; López-Delgado, A. A kinetic study on the thermal behavior of chitosan. *J Therm Anal Calorim* **2008**, *91*, 633–639.

255. Rodríguez, AT; Ramírez, MA; Cárdenas, RM; Hernández, AN; Velázquez, MG; Bautista, S. Induction of defence response of *Oryza sativa* L. against *Pyricularia grisea* (Cooke) Sacc. by treating seeds with chitosan and hydrolyzed chitosan. *Pestic Biochem Physiol* **2007**, *89*, 206–215.

256. Lin, H; Wang, H; Xue, C; Ye, M. Preparation of chitosan oligomers by immobilized papain. *Enzyme Microb Technol***2002**, *31*, 588–592.

257. Lee, H-W; Park, Y-S; Jung, J-S; Shin, W-S. Chitosan oligosaccharides, dp 2–8, have prebiotic effect on the*Bifidobacterium bifidium* and *Lactobacillus* sp. *Anaerobe* **2002**, *8*, 319–324.

258. Li, J; Du, Y; Yang, J; Feng, T; Li, A; Chen, P. Preparation and characterisation of low molecular weight chitosan and chito-oligomers by a commercial enzyme. *Polym Degrad Stab* **2005**, *87*, 441–448.

259. Oliveira, EN, Jr; El Gueddari, NE; Moerschbacher, BM; Peter, MG; Franco, TT. Growth of phytopathogenic fungi in the presence of partially acetylated chitooligosaccharides. *Mycopathologia* **2008**, *166*, 163–174.

260. Kuyama, H; Nakahara, Y; Nucada, T; Ito, J; Nakahara, Yo; Ogawa, T. Stereocontrolled synthesis of chitosan dodecamer. *Carbohydr Res* **1993**, *243*, C1–C7.

261. Cabrera, JC; Messiaen, J; Cambier, P; Van Cutsem, P. Size, acetylation and concentration of chitooligosaccharide elicitors determine the switch from defence involving PAL activation to cell death and water peroxide production in Arabidopsis cell suspensions. *Physiol Plant* **2006**, *127*, 44–56.

262. Lee, D-W; Baney, RH. Oligochitosan derivatives bearing electron-deficient aromatic rings for adsorption of Amitriptyline: Implications for drug detoxification. *Biomacromolecules* **2004**, *5*, 1310–1315.

263. Vachoud, L; Chen, T; Payne, GF; Vazquez-Duhalt, R. Peroxidase catalyzed grafting of gallate esters onto the polysaccharide chitosan. *Enzyme Microb Technol* **2001**, *29*, 380–385.

264. Fu, X; Huang, L; Zhai, M; Li, W; Liu, H. Analysis of natural carbohydrate biopolimer-high molecular chitosan and carboxymethyl

chitosan by capillary zone electrophoresis. *Carbohydr Polym* **2007**, *68*, 511–516.

265. Fu, X; Liu, Y; Li, W; Pang, N; Nie, H; Liu, H; Cai, Z. Analysis of aristolochic acids by CE-MS with carboxymethyl chitosan-coated capillary. *Electrophoresis* **2009**, *30*, 1783–1789.

266. Katarina, RK; Takayanagi, T; Oshita, K; Oshima, M; Motomizu, S. Sample pretreatment Rusing chitosan-based chelating resin for the determination of trace metals in seawater Samales by inductively coupled plasma-mass spektrometry. *Anal Sci* **2008**, *24*, 1537–1544.

267. Gao, Y; Oshita, K; Lee, K-H; Oshima, M; Motomizu, S. Development of kolumn-pretreatment chelating resins for matrix elimination/multi-element determination by inductively coupled plasma-mass spectrometry. *Analyst* **2002**, *127*, 1713–1719.

268. Oshita, K; Takayanagi, T; Oshima, M; Motomizu, S. Adsorption properties of ionic species on cross-linked chitosans modified with catechol and salicylic acid moieties. *Anal Sci* **2008**, *24*, 665–668.

269. Sabarudin, A; Oshima, M; Takayanagi, T; Hakim, L; Oshita, K; Yun Hua Gao; Motomizu, S. Functionalization of chitosan with 3,4-dihydroxybenzoic acid for the adsorption/collection of uranium in water samples and its determination by inductively coupled plasma-mass spectrometry. *Anal Chim Acta* **2007**, *581*, 214–220. [Google Scholar]

270. Oshita, K; Takayanagi, T; Oshima, M; Motomizu, S. Adsorption behaviour of cationic and anionic species on chitosan resins possessing amino acids moieties. *Anal Sci* **2007**, *23*, 1431–1434.

271. Hosoba, M; Oshita, K; Katarina, RK; Takayanagi, T; Oshima, M; Motomizu, S. Synthesis of novel chitosan resin possessing histidine moiety and its application to the determination of trace silver by ICP-AES coupled with triplet automated-pretreatment system. *Anal Chim Acta* **2009**, *639*, 51–56.

272. Travan, A; Pelillo, C; Donati, I; Marsich, E; Benincasa, M; Scarpa, T; Semeraro, S; Turco, G; Gennaro, R; Paoletti, S. Non-cytotoxic silver nanoparticle-polysaccharide nanocomposites with antimicrobial activity. *Biomacromolecules* **2009**,*10*, 1429–1435.

273. Kasaai, MR. Determination of the degree of *N*-acetylation for chitin and chitosan by various NMR spectroscopy techniques: A review. *Carbohydr Polym* **2010**, *79*, 801–810.

274. Pelletier, A; Lemire, I; Sygusch, J; Chornet, E; Overend, RP. Chitin/chitosan transformation by thermo-mechano-chemical treatment including characterization by enzymatic depolymerization. *Biotechnol Bioeng* **1990**, *36*, 310.

275. Raymond, L; Morin, FG; Marchessault, RH. Degree of deacetylation of chitosan using conductometric titration and solid-state NMR. *Carbohydr Res* **1993**, *246*, 331.

276. Yu, G; Morin, FG; Nobes, GAR; Marchessault, RH. Degree of Acetylation of Chitin and Extent of Grafting PHB on Chitosan Determined by Solid State 15N NMR. *Macromolecules* **1999**, *32*, 518–520

277. Heux, L; Brugnerotto, J; Desbrières, J; Versali, M-F; Rinaudo, M. Solid State NMR for Determination of Degree of Acetylation of Chitin and Chitosan. *Biomacromolecules* **2000**, *1*, 746–751.

278. Rinaudo, M; Le Dung, P; Gey, C; Milas, M. Substituent distribution on *O,N*-carboxymethylchitosans by ^{1}H and ^{13}C NMR. *Int J Biol Macromol* **1992**, *14*, 122–128.

279. Vårum, KM; Anthonsen, MW; Grasdalen, H; Smisr∅d, O. Determination of the degree of *N*-acetylation and the distribution of *N*-acetyl groups in partially *N*-deacetylated chitins (chitosans) by high-field n.m.r. spectroscopy.*Carbohydr Res* **1991**, *211*, 17–23.

280. Desbrières, J; Martinez, C; Rinaudo, M. Hydrophobic derivatives of chitosan: Characterization and rheological behavior. *Int J Biol Macromol* **1996**, *19*, 21–28.

281. Lebouc, F; Dez, I; Madec, P-J. NMR study of the phosphonomethylation reaction on chitosan. *Polym* **2005**, *46*, 319–325.

282. Weinhold, MX; Sauvageau, JCM; Kumirska, J; Thöming, J. Studies on acetylation patterns of different chitosan preparations. *Carbohydr Polym* **2009**, *78*, 678–684.

283. Yang, BY; Montgomery, R. Degree of acetylation of heteropolysaccharides. *Carbohydr Res* **2000**, *323*, 156–162.

284. Duarte, ML; Ferreira, MC; Marvao, MR; Rocha, J. Determination of the degree of acetylation of chitin materials by 13C CP/MAS NMR spectroscopy. *Int J Biol Macromol* **2001**, *28*, 359–363.

285. Tolaimate, A; Desbričres, J; Rhazi, M; Alagui, A; Vincendon, M; Vottero, P. On the influence of deacetylation process on the physicochemical characteristics of chitosan from squid chitin. *Polymer* **2000**, *41*, 2463–2469.

286. Guinesi, LS; Cavalheiro, ETG. The use of DSC curves to determine the acetylation degree of chitin/chitosan samples.*Thermochim Acta* **2006**, *444*, 128–133.

287. Rinaudo, M; Milas, M; Dung, PL. Characterization of chitosan. Influence of ionic strength and degree of acetylation on chain expansion. *Int J Biol Macromol* **1993**, *15*, 281–285.

288. Hirai, A; Odani, H; Nakajima, A. Determination of degree of deacetylation of chitosan by ^{1}H NMR spectroscopy.*Polym Bullet* **1991**, *26*, 87–94.

289. Kasaai, MR; Arul, J; Chin, SL; Charlet, G. The use of intense femtosecond laser pulses for the fragmentation of chitosan. *J Photochem Photobiol A* **1999**, *120*, 201–205.

290. Lavertu, M; Xia, Z; Serreqi, AN; Berrada, M; Rodrrigues, A; Wang, D; Buschmann, MD; Gupta, A. A validated H NMR method for the determination of the degree of deacetylation of chitosan. *J Pharm Biomed Anal* **2003**, *32*, 1149–1158.

291. Ott∅y, MH; Vårum, KM; Smidsr∅d, O. Compositional heterogeneity of heterogeneously deacetylated chitosans.*Carbohydr Polym* **1996**, *29*, 17–24.

292. Pelletier, A; Lemire, I; Sygusch, J; Chornet, E; Overend, RP. Chitin/chitosan transformation by thermo-mechano-chemical treatment including characterization by enzymatic depolymerization. *Biotechnol Bioeng* **1990**, *36*, 310–315.

293. Fernandez-Megia, E; Novoa-Carballal, R; Quiñoá, E; Riguera, R. Optimal routine conditions for the determination of the degree of acetylation of chitosan by ^{1}H-NMR. *Carbohydr Polym* **2005**, *61*, 155–161.

294. Vårum, KM; Anthonsen, MW; Grasdalen, H; Smidsr∅d, O. ^{13}C NMR studies of the acetylation sequences in partially*N*-acetylated chitins (chitosans). *Carbohydr Res* **1991**, *217*, 19–27.

295. Bovey, P; Mireau, A. *NMR of Polymers*; Academic Press: San Diego, CA, USA, 1996.

296. Mirau, A. *A Practical Guide to Understanding the NMR of Polymers*; Academic Press: Hoboken, NJ, USA, 2005.

297. Kumirska, J; Weinhold, MX; Steudte, S; Thöming, J; Brzozowski, K; Stepnowski, P. Determination of the pattern of acetylation of chitosan samples: Comparison of evaluation methods and some validation parameters. *Int J Biol Macromol* **2009**, *45*, 56–60.

298. Kumirska, J; Weinhold, MX; Sauvageau, JCM; Thöming, J; Kaczyński, Z; Stepnowski, P. Determination of the pattern of acetylation of low molecular weight chitosan used in biomedical applications. *J Pharm Biochem Anal* **2009**, *50*, 587–590.

299. Martinou, A; Bouriotis, V; Stokke, BT; Vårum, KM. Mode of action of chitin deacetylase from *Mucor rouxii* on partially*N*-acetylated chitosans. *Carbohydr Res* **1998**, *311*, 71–78.

300. Tommeraas, K; Koping-Hoggard, M; Vårum, KM; Christensen, BE; Artursson, P; Smidsrod, O. Preparation and characterisation of chitosans with oligosaccharide branches. *Carbohydr Res* **2002**, *337*, 2455–2462.

301. Knight, DK; Shapka, SN; Amsden, BG. Structure, depolymerization, and cytocompatibility evaluation of glycol chitosan. *J Biomed Mater Res Part A* **2007**, *83*, 787–798.

302. Rabea, EI; Badawy, ME; Rogge, TM; Stevens, CV; Höfte, M; Steurbaut, W; Smagghe, G. Insecticidal and fungicidal activity of new synthesized chitosan derivatives. *Pest Manag Sci* **2005**, *61*, 951–960.

303. Sun, L; Du, Y; Yang, Y; Shi, X; Li, J; Wang, X; Kennedy, JF. Conversion of crystal structure of the chitin to facilitate preparation of a 6-carboxychitin with moisture absorption–retention abilities. *Carbohydr Polym* **2006**, *66*, 168–175.

304. Chen, L; Du, Y; Wu, H; Xiao, L. Relationship between Molecular Structure and Moisture- Retention Ability of Carboxymethyl Chitin and Chitosan. *J Appl Polym Sci* **2002**, *83*, 1233–1241.

305. Vikhoreva, G; Bannikova, G; Stolbushkina, P; Panov, A; Drozd, N; Makarov, V; Varlamov, V; Gal'braikh, L. Preparation and anticoagulant activity of a low-molecular-weight sulfated chitosan. *Carbohydr Polym* **2005**, *62*, 327–332.

306. Sieval, AB; Thanou, M; Kotzé, AF; Verhoef, JC; Brussee, J; Junginger, HE. Preparation and NMR characterization of highly substituted *N*-trimethyl chitosan chloride. *Carbohydr Polym* **1998**, *36*, 157–165.

307. Aoi, K; Seki, T; Okada, M; Sato, H; Mizutani, S; Ohtani, H; Tsuge, S; Shiogai, Y. Synthesis of a novel *N*-selective ester functionalized chitin derivative and water-soluble carboxyethylchitin. *Macromol Chem Phys* **2000**, *201*, 1701–1708.

308. Detchprohm, S; Aoi, K; Okada, M. Synthesis of a Novel Chitin Derivative Having Oligo(ecaprolactone) Side Chains in Aqueous Reaction Media. *Macromol Chem Phys* **2001**, *202*, 3560–3570.

309. Rinaudo, M; Desbrieres, J; Le Dung, P; Thuy Binh, P; Dong, NT. NMR investigation of chitosan derivatives formed by the reaction of chitosan with levulinic acid. *Carbohydr Polym* **2001**, *46*, 339–348.

310. Zhang, C; Ping, Q; Ding, Y; Cheng, Y; Shen, J. Synthesis, Characterization, and Microsphere Formation of Galactosylated Chitosan. *J Appl Polym Sci* **2004**, *91*, 659–665.

311. Zou, Y; Khor, E. Preparation of sulfated-chitins under homogeneous conditions. *Carbohydr Polym* **2009**, *77*, 516–525.

312. Park, IK; Park, YH. Preparation and Structural Characterization of Water-Soluble *O*-Hydroxypropyl Chitin Derivatives. *J Appl Polym Sci* **2001**, *80*, 2624–2632.

313. Tanodekaewa, S; Prasitsilpa, M; Swasdisonb, S; Thavornyutikarna, B; Pothsreea, T; Pateepasen, R. Preparation of acrylic grafted chitin for wound dressing application. *Biomaterials* **2004**, *25*, 1453–1460.

314. Changhong, P; Weijun, Y; Motang, T. Chemical Modification of Chitosan: Synthesis and Characterization of Chitosan-Crown Ethers. *J Appl Polym Sci* **2003**, *87*, 2221–2225.

315. De Angelis, AA; Capitani, D; Crescenzi, V. Synthesis and ^{13}C CP-MAS NMR Characterization of a New Chitosan-Based Polymeric Network. *Macromolecules* **1998**, *31*, 1595–1601.

316. Capitani, D; De Angelis, AA; Crescenzi, V; Masci, G; Segre, AL. NMR study of a novel chitosan-based hydrogel.*Carbohydr Polym* **2001**, *45*, 245–252.

317. Lebouc, F; Dez, I; Gulea, M; Madec, P-J; Jaffres, P-A. Synthesis of Phosphorus-Containing Chitosan Derivatives.*Phosphorus Sulfur Silicon Relat Elem* **2009**, *184*, 872–889.

318. Jayakumar, R; Nagahama, H; Furuike, T; Tamura, H. Synthesis of phosphorylated chitosan by novel method and its characterization. *Int J Biol Macromol* **2008**, *42*, 335–339.

319. Jayakumar, R; Egawa, T; Furuike, T; Nair, SV; Tamura, H. Synthesis, Characterization, and Thermal Properties of Phosphorylated Chitin for Biomedical Applications. *Polym Eng Sci* **2009**, *49*, 844–849.

320. Wang, X; Ma, J; Wang, Y; He, B. Structural characterization of phosphorylated chitosan and their applications as effective additives of calcium phosphate cements. *Biomaterials* **2001**, *22*, 2247–2255.

321. Meng, S; Liu, Z; Zhong, W; Wang, Q; Du, Q. Phosphorylcholine modified chitosan: Appetent and safe material for cells. *Carbohydr Polym* **2007**, *70*, 82–88.

322. Zou, Y; Khor, E. Preparation of C-6 Substituted Chitin Derivatives under Homogeneous Conditions. *Biomacromolecules***2005**, *6*, 80–87.

323. Palma, G; Casals, P; Cárdenas, G. Synthesis and Characterization of New Chitosan-O-Ethyl Phosphonate. *J Chil Chem Soc* **2005**, *50*, 597–602.

324. Cárdenas, G; Cabrera, G; Taboada, E; Rinaudo, M. Synthesis and Characterization of Chitosan Alkyl phosphates. *J Chil Chem Soc* **2006**, *51*, 815–820.

325. Tanner, SF; Chanzy, H; Vincendon, M; Roux, JC; Gaill, F. High resolution solid-state carbon-13 nuclear magnetic resonance study of chin. *Macromolecules* **1990**, *23*, 3576–3583.

326. Kameda, T; Miyazawa, M; Ono, H; Yoshida, M. Hydrogen Bonding Structure and Stability of α-Chitin Studied by ^{13}C Solid-State NMR. *Macromol Biosci* **2005**, *5*, 103–106

327. Cortizo, MS; Berghoff, CF; Alessandrini, JL. Characterization of chitin from *Illex argentinus* squid pen. *Carbohydr Polym***2008**, *74*, 10–15.

328. Manni, L; Ghorbel-Bellaaj, O; Jellouli, K; Younes, I. Nasri, Extraction and Characterization of Chitin, Chitosan, and Protein Hydrolysates Prepared from Shrimp Waste by Treatment with Crude Protease from *Bacillus cereus* SV1. *Appl Biochem Biotechnol* **2010**.

329. Toffey, A; Samaranayake, G; Frazier, CE; Glasser, WG. Chitin derivatives. I. Kinetics of the heat-induced conversion of chitosan to chitin. *J Appl Polym Sci* **1996**, *60*, 75–85.

330. Holme, HK; Foros, H; Pettersen, H; Dornish, M; Smidsrod, O. Thermal depolymerization of chitosan chloride.*Carbohydr Polym* **2001**, *46*, 287–294.

331. Einbu, A; Vårum, KM. Depolymerization and de-*N*-acetylation of Chitin Oligomers in Hydrochloric Acid.*Biomacromolecules* **2007**, *8*, 309–314.

332. Einbu, A; Vårum, KM. Characterization of Chitin and Its Hydrolysis to GlcNAc and GlcN. *Biomacromolecules* **2008**, *9*, 1870–1875.

333. S∅rbotten, A; Horn, SJ; Eijsink, VGH; Vårum, KM. Degradation of chitosans with chitinase B from *Serratia marcescens*. Production of chito-oligosaccharides and insight into enzyme processivity. *FEBS J* **2005**, *272*, 538–549.

334. Colombo, G; Meli, M; Cañada, J; Asensio, JL; Jiménez-Barberob, J. Toward the understanding of the structure and dynamics of protein–

carbohydrate interactions: molecular dynamics studies of the complexes between hevein and oligosaccharidic ligands. *Carbohydr Res* **2004**, *339*, 985–994.

335. Chávez, MI; Andreu, C; Vidal, P; Aboitiz, N; Freire, F; Groves, P; Asensio, JL; Asensio, G; Muraki, M; Cañada, FJ; Jiménez-Barbero, J. On the Importance of Carbohydrate-Aromatic Interactionsfor the Molecular Recognition of Oligosaccharides by Proteins: NMR Studies of the Structure and Binding Affinity of AcAMP2-like Peptideswith Non-Natural Naphthyl and Fluoroaromatic Residues. *Chem Eur J* **2005**, *11*, 7060–7074.

336. Aboitiz, N; Cañada, FJ; Hušáková, L; Kuzma, M; Křen, V; Jiménez-Barbero, J. Enzymatic synthesis of complex glycosaminotrioses and study of their molecular recognition by hevein domains. *Org Biomol Chem* **2004**, *2*, 1987–1994.

337. Chan, H-Y; Chen, M-H; Yuan, G-F. Fungal chitosan. *Fungal Sci* **2007**, *16*, 39–52.

338. Yen, M-T; Mau, J-L. Physico-chemical characterization of fungal chitosan from shiitake stipes. *LWT–Food Sci Technol***2007**, *40*, 472–479.

339. Yen, M-T; Yang, J-H; Mau, J-L. Physicochemical characterization of chitin and chitosan from crab shells. *Carbohydr Polym* **2009**, *75*, 15–21.

340. Peter, M; Binulal, NS; Soumya, S; Nair, SV; Furuike, T; Tamura, H; Jayakumar, R. Nanocomposite scaffolds of bioactive glass ceramic nanoparticles disseminated chitosan matrix for tissue engineering applications. *Carbohydr Polym* **2010**,*79*, 284–289.

341. Xi, F; Wu, J; Lin, X. Novel nylon-supported organic–inorganic hybrid membrane with hierarchical pores as a potential immobilized metal affinity adsorbent. *J Chromatogr A* **2006**, *1125*, 38–51.

342. Rusua, VM; Nga, C-H; Wilkec, M; Tierscha, B; Fratzld, P; Petera, MG. Size-controlled hydroxyapatite nanoparticles as self-organized organic–inorganic composite materials. *Biomaterials* **2005**, *26*, 5414–5426.

343. Maruca, R; Suder, BJ; Wightman, JP. Interaction of heavy metals with chitin and chitosan. III. Chromium. *J Appl Polym Sci* **1982**, *27*, 4827–4837.

344. Gamage, A; Shahidi, F. Use of chitosan for the removal of metal ion contaminants and proteins from water. *Food Chem* **2007**, *104*, 989–996.

345. Atkins, P; Paula, J. *Elements of Physical Chemistry*, 4th ed; Oxford University Press: New York, NY, USA, 2005.

346. Kittur, FS; Vishu Kumar, AB; Tharanathan, RN. Low molecular weight chitosans-preparation by depolymerization with Aspergillus niger pectinase, and characterization. *Carbohydr Res* **2003**, *338*, 1283–1290.

Chapter 8

Chemical Modeling of Acid-Base Properties of Soluble Biopolymers Derivedfrom Municipal Waste Treatment Materials

Silvia Tabasso [1], Silvia Berto [1,*], Roberta Rosato [1], Janeth Alicia Tafur Marinos [1], Marco Ginepro [1], Vincenzo Zelano [1], Pier Giuseppe Daniele [1] and Enzo Montoneri [2]

[1]Department of Chemistry, University of Turin, Via Pietro Giuria 7, I-10125 Torino, Italy

[2]Biowaste Processing, Via XXIV Maggio 25, 37126 Verona, Italy

ABSTRACT

This work reports a study of the proton-binding capacity of biopolymers obtained from different materials supplied by a municipal biowaste treatment plant located in Northern Italy. One material was the anaerobic fermentation digestate of the urban wastes organic humid fraction. The others were the compost of home and public gardening residues and the compost of the mix of the above residues, digestate and sewage sludge. These materials were hydrolyzed under alkaline conditions to yield the biopolymers by saponification. The biopolymers were characterized by ^{13}C NMR spectroscopy, elemental analysis and potentiometric titration. The titration data were elaborated to attain chemical models for interpretation of the proton-binding capacity of the biopolymers obtaining the acidic sites concentrations and their protonation constants. The results obtained with the models and by NMR spectroscopy were elaborated together in order to better characterize the nature of the macromolecules. The chemical nature of the biopolymers was found dependent upon the nature of the sourcing materials.

KEYWORDS

biopolymers; biowastes; chemical model; potentiometry; protonation constants; NMR

1. INTRODUCTION

Municipal biowastes (MBW) are a readily available cost-effective source [1] of soluble biopolymers (henceforth SBOs, Soluble Bio-Organic substances). These products are composed by a mixture of molecules with 67–463 $kg \cdot mol^{-1}$ molecular weight and 6–53 polydispersity index. They contain long aliphatic C chains and aromatic rings substituted by a wide variety of acid and basic functional groups of different strength. These organic moieties are the likely memory of the main constituents of the sourcing bio-organic refuse matter, which are not completely mineralized during aging under fermentation conditions. As a result of the biological origin of their sourcing materials, in addition to the high polydispersity index, the entire molecular pool is expected to be heterogeneous also for chemical composition. By these features SBO bear chemical similarities with soil and water humic substances (HS) formed under longer aging.

The chemical compositions and properties of SBOs may be different owing to the MBW sourcing material. Recently, several papers have reported that SBO from different sources can find application in diversified fields; e.g., in the formulation of detergents textile dyeing baths, flocculants, dispersants and binding agents for ceramics manufacture [1], emulsifiers [2], auxiliaries for soil/water remediation [3,4,5] and enhanced oil recovery [6], nanostructured materials for chemical [7,8] and biochemical catalysis [9], plastic materials [10], soil fertilizers and plant biostimulants for agriculture [11] and animal feed supplements [12,13]. The wide range of applications arises from the variety of C moieties distributed over the macromolecular pool constituting the SBOs. The C moieties are the memories of the protein, fats, and polysaccharide and lignin proximates constituting the biowastes from which the SBO are obtained. These chemical features are associated to the SBO properties as surfactants, agents for sequestering or carrying small molecules and mineral ions in solution, photosensitizers and reactive biopolymers. Further understanding of the chemical nature of SBOs might help product development for the above uses and other new specific applications. In this work we report the study of the acid-base behavior of these substances. This specific molecular feature has not been studied so far. Yet, in some of the above applications, the performance of SBO has been found to connect to their acidic sites. In agriculture [11], for instance, the SBOs have been found capable to bind and transport soluble mineral nutrients through the soil to enhance plant growth. In the photoremediation of industrial wastewaters [4,5], the capacity of SBOs to bind and keep Fe ions in solution under non-acidic conditions has been claimed enhancing the mineralization of organic pollutants upon exposure to solar light. Specific studies of the acid-base behavior of SBOs are likely to disclose potential for new applications. For instance, due to their protogenic sites, the SBOs may perform as bioadsorbents for the removal of metal ions by exchanging protons with metal ions. The exploitation of the capacity of these substances to bind metal ions at neutral and alkaline pH, and of the property to precipitate at acid pH, may prospect an interesting process for the remediation of water streams containing toxic metals. Other biopolymers of natural origin, such as soil and water humic substances (HS), have been studied for their capacity to exchange protons with metal ions [14,15,16,17,18,19,20,21]. These substances bear chemical similarities with the above SBOs. However, for large scale uses in environmental remediation practices, the SBOs are potentially favored for the cost effectiveness and the environmental benefits arising from the exploitation of their sourcing materials [1].

The evaluation of the proton exchange capacity of bio-macromolecules was not easy. The modeling of protonation equilibria is very complicated, particularly in the case of natural polyelectrolytes with heterogeneous chemical compositions and structures. The literature comprises many articles in which different approaches are proposed to describe the acid-base properties of HS [14,15,16,17,18,19,20,21,22]. The main difficulties when dealing with the protonation parameters of this type of polyelectrolytes are the pH drift during titration and the hysteresis between tandem forward and reverse titration curves. The hysteresis may be caused by base-catalyzed hydrolyses of esters and amides. It can also be due to the dependence of acidity constants on molecular conformation and charge [19,21,22]. The conformation of the macromolecules changes during the titration for the electrostatic repulsion of negative charges, and this affects the proton exchange capacity of functional groups. Moreover, the protonation parameters depend on the effective charge of the polyanion, which is affected by the dissociation degree of the macromolecule [21,22]. Thus, different methods have been proposed and tested for the evaluation of protonation parameters connected to carboxylic and phenolic functional groups in HS.

In this work, alkalimetric and acidimetric titrations of SBO solutions were conducted and were elaborated in order to propose a chemical model that can describe the proton-binding capacity of the macromolecules. The application of software that employs an iterative and convergent numerical method, which is based upon the linear combination of the mass balance equations, allows calculating the acidic sites concentrations and their protonation constants. A discrete model and a diprotic-like model were applied. The first one supposes that a macromolecule contains a series of discrete sites with different dissociation constant values [15]. The diprotic-like model, proposed by Crea *et al.* [22] assumes that the binding capacity of the carboxylic acid groups is described by two protonation constants, whereas the phenolic group behaves as a second ligand in the alkaline pH range. Both model types assume that the protonation parameters are independent of the dissociation degree. Hereinafter, the concentration values of carboxylic groups obtained with the modeling approach are discussed in relation to previously reported values [23] obtained by NMR and elemental analyses.

2. RESULTS AND DISCUSSION

2.1. Chemical Composition and Solution Behavior of Soluble Bio-Organic Substances (SBOs)

In this study three different SBOs, called FORSUD, CVDF, and CV, were investigated. These products were obtained from three different materials. As reported in the Section 3.2, the CV and CVDF were sourced from composts, whereas FORSUD was isolated from the anaerobic fermentation digestate. According to the chemical process described in the same Section 3.2, the SBOs are hydrolysates with molecular cut-off above 5000 Da. They contain water-soluble polymeric molecules, which may range up [23] to several hundred kg·mol^{-1}. Due to their complexity, SBO cannot be characterized as well as synthetic molecules. Their chemical nature can at best be identified by the concentration of C types and functional groups reported inTable 1. The behavior of these polymeric substances cannot be understood as well as synthetic single molecules. The C types and functional groups reported in Table 1 are likely to be not homogeneously distributed over the entire molecular pool. The data indicate that the SBOs contain three main groups capable to bind protons or metal ions; *i.e.*, carboxyl, phenol and amino groups. Indeed, the reduction of the mineral content of these products required a first treatment with HCl only, followed by a second treatment with HF.

For the purpose of this work, it was necessary to understand first the solution behavior of the SBO. From the presence of both lipophilic and hydrophilic, the SBOs were expected to perform typically as surfactants. The behavior of simple surfactants' molecules, such as sodium dodecylbenzene sulfonate (SDBS), is well known [23]. At low concentration, SDBS molecules occupy the air-water interphase, with their polar hydrophilic heads pointing toward the water phase and their lipophilic tails directed toward the air phase. At high concentration, they aggregate to form micelles in the bulk water phase. These may have different shapes, but the polar head are exposed to the water phase and the lipophilic tails are oriented toward the micelle core. Conformation details for the complex mixtures of SBO macromolecules cannot be envisioned as straightforwardly. Two virtual fragments for these polymeric molecules, containing the C types and functional groups listed in Table 1, are reported in a previous work [23]. Starting from these fragments, one could imagine that at low concentrations the single macromolecules occupy the air-water interphase and acquire a more or less flat configuration to point their hydrophilic heads toward the water phase. At higher concentration, upon saturating the air-water interphase, they would occupy the bulk water phase. However, due to their polymeric nature, two conformations are possible in the bulk water phase [24]. The single molecules could be arranged in coil form to yield pseudo-micelles with the external hydrophilic surface pointing toward the water phase and the inner lipophilic molecular segments away from it. At higher concentrations, the single pseudo-micelles might interact to yield micellar aggregates. The aggregation of molecules in pseudo-micelles conformation, due to increasing concentration above 2 $g \cdot L^{-1}$, would occur through H-bonding interaction between the external hydrophilic groups of the pseudo-micelles. As a consequence, in the pseudo-micelles' aggregates some of the hydrophilic phenol and carboxylic groups might end up inside the aggregate core and become less accessible by the titrant. This situation has been proposed based on the results of dynamic light scattering measurements reported in previous work [24]. The size of SBO macromolecules has been found to be more sensitive to pH at 0.2 g·L^{-1} than at 2 g·L^{-1}concentration. At 0.2 g·L^{-1} SBO concentration, the molecular hydrodynamic diameter (D_h) has been found to change from 71–73 nm at pH 5 to 128–168 nm at pH 7 and to 137–218 nm at pH 13, whereas for the 2 g·L^{-1} sample D_h changes from 90–97 nm at pH 5 to 101–123 nm at pH 7–13. The increase of D_h is likely due to the increasing ionization and repulsion between carboxylate groups occurring upon increasing pH. By comparison at the same pH, the ionization and the repulsion in the inner core of the micellar aggregates, formed at concentration above 2 g·L^{-1}, is less due to the lower accessibility of the alkaline cation into the micellar core. This explains the fact that at 0.2 g·L^{-1}, compared to 2 g·L^{-1} concentration, the $D_h$sensitivity *versus* temperature was found up to twice as high, whereas the sensitivity *versus* pH was four to eight times higher. Under these circumstances, it was deemed that a more reliable assessment of the number and nature of SBO protogenic sites could be obtained from measurements performed below the SBO critical micellar concentration (Table 1).

2.2. Proton Binding Capacity of SBOs

The proton binding capacity of the SBOs were characterized first by alkalimetric/acidimetric titrations performed on sample solutions containing the products at 0.30–1.0 g·L^{-1} concentration. The potentiometric titrations were performed in the 2.8–11.5 pH range with 0.1 mol·L^{-1} aqueous KOH or HCl. At acid pH the SBOs precipitate. However, the attainment of the equilibrium conditions was not slowed down. Figure 1 shows an example of the typical alkalimetric titration curves recorded for each type of SBO at 0.5 g·L^{-1} concentration. The equilibrium conditions were difficult to reach in the pH range 5.5–8. Figure 2 reports the direct alkalimetric and reverse acidimetric titration curves for each SBO type. It is evident the hysteresis phenomenon. In alkalimetric titration only one inflection point is evident, while the reverse titration presents two inflection points at pH 8.0–8.5 and 5.0–5.5. Moreover, the back titration points are more acidic than the ones of the direct titration. This phenomenon is not yet entirely understood. Similar behavior was reported for HS [17,20,21]. A plausible explanation is the change of the macromolecule conformation during the titration. On the basis of the solution behavior of SBOs described in the previous paragraph, one can suppose that, in acidic conditions, as at the start of the direct titration, the single molecules were in the compact pseudomicellar coil conformation, while, during titration, the molecules acquired the expanded coil form. In this form, the acidic functional groups became more accessible by the titrant. This change in the acidic groups' arrangement during the titration could be responsible of the slow equilibration. In the reverse titration, the rearrangement of the molecular conformation during the titrant addition could be different compared to that during the direct titration. This might explain the observed hysteresis.

2.2.1. Protonation Constants

The alkalimetric titration results were elaborated in order to calculate the protonation constants and the concentration of protogenic groups. A diprotic-like model (model I and model II) [22] and a discrete model (model III), by Masini *et al.* [15] were applied. The models were compared for their capacity to fit the experimental titration curves, and thus to allow calculation of protonation constants, and evaluation of acidic sites concentration. The results are reported in Table 2. In the diprotic-like model application (model I), the protonation constants of a diprotic site were calculated elaborating the titration data in the pH range 2.8–6.5. A good value of the weighted standard deviation of the fit was obtained for all SBOs. The protonation constant of a less acidic site was also calculated widening the pH range at 10.5. In this case the result was satisfactory only for the FORSUD SBO. Possible reasons might be the presence of more than three acid-base active sites or a high effect of the dissociation degree on the protonation constant. To obtain a better fit of titration curves a widening of the model was applied, increasing the number of protogenic sites (model II). By comparison, model III supposes that a macromolecule contains a series of discrete sites with different independent dissociation constants. The results in Table 2show that all models fit the experimental data. However, based on the values of the weighted standard deviation, model III is proven the best. This result was expected, since model III contains no concentration constraints. Figure 1 shows the comparison between the experimental titration points and the curves calculated with model III.

Table 1. Chemical data and critical micellar concentration (CMC) for soluble bio-organic substances (SBOs).

SBO	Ash (*w/w* %)	C (*w/w* %) [a]	N (*w/w* %) [a]	Aliphatic [b]	NR [b,c]	OR [b,c]	Ar [b]	PhOY [b,d]	COY [b,e]	CMC ($g \cdot L^{-1}$)
CVDF	0.56	48.5	5.2	0.31	0.06	0.16	0.23	0.12	0.11	2.76
CVT230	0.81	52.5	5.2	0.28	0.07	0.02	0.22	0.11	0.11	4.2
FORSUD	0.01	57.5	8.8	0.43	0.09	0.15	0.15	0.06	0.13	1.02

[a] Concentrations values referred to dry matter; [b] C types and functional groups concentration as mole fraction of total organic C; [c] R = H, alkyl or phenyl; [d] Y = OH and/or OR; and [e] Y = OH and/or NR.

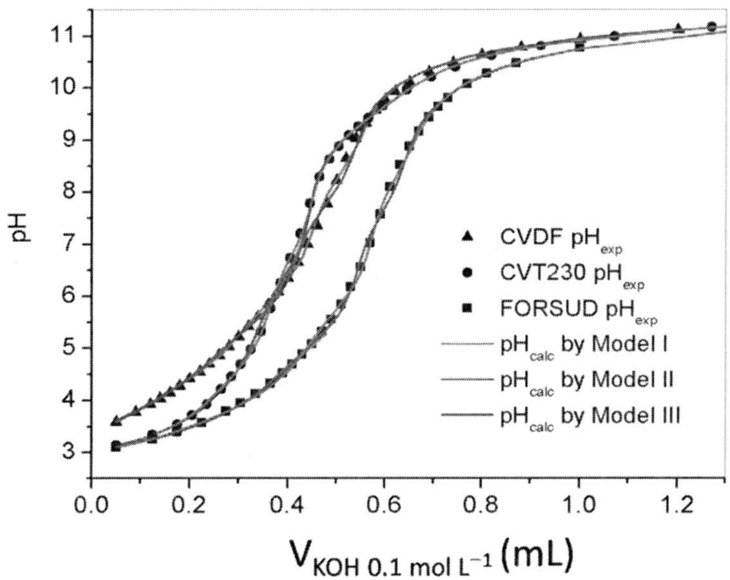

Figure 1. Examples of experimental titration points (scatter plots) and calculated titration curves (colored curves) for CVDF, CVT230 and FORSUD, under the following conditions: 25 mL of solution containing 0.5 g·L^{-1} SBO and 0.1 mol·L^{-1} TEACl, titrated with 0.1 mol·L^{-1} KOH. In the graph, one titration point was reported every other two.

2.2.2. Protogenic Sites Concentrations

Table 2 reports the concentration of the protogenic groups obtained by the alkalimetric titration data elaborated applying the three models. It may be observed that the results obtained by the different models are in good accordance with each other in all cases.

The literature dealing with protonation of HS assumes that the total concentrations of COOH and PhOH functional groups can be obtained as the sum of the concentration values for the species with logK_a < 8 for COOH and with logK_a > 8 for PhOH [17]. In this case, the attribution of log$K$ values to specific functional groups is not allowed because the biopolymers present amines or other nitrogen functions. These functions can contribute to the acid/base reactivity and have log$K$ values dispersed in a wide range as a function of their aromatic or aliphatic nature. However, it is reasonable to suppose that the logK_a < 7 can be attribute to –COOH groups. Table

3 reports the concentration of –COOH calculated as the sum of the concentrations of the species with a $\log K_a < 7$ by the application of Model III (Table 2).

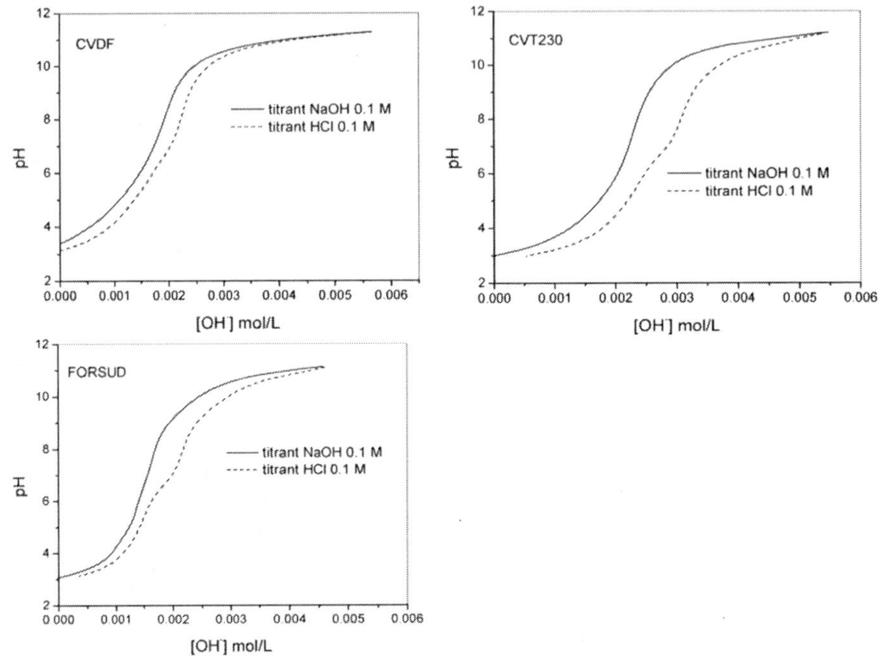

Figure 2. Direct (alkalimetric) and reverse (acidimetric) titration curves of CVDF, CVT230 and FORSUD at 0.5 g·L^{-1} concentration in 0.1 mol·L^{-1} TEACl.

To compare NMR and titration data, it should be known that the ^{13}C NMR spectra of SBOs are characterized by rather broad bands [24]. These do not allow determining selectively concentrations of acidic groups but only total aromatic C bonded to OH or alkoxy or aryloxy moieties (PhOY) from the ^{13}C signal at 140–160 ppm and total carboxyl C (COY) as sum of COOH and CON C from the ^{13}C signal at 160–185 ppm. The contribution of ester C to the carboxyl C NMR signal was excluded, as these functional groups were expected to be hydrolyzed in the alkaline treatment of the SBO sourcing material. For these reasons, we expect that the concentration values of COY and PhOY obtained by ^{13}C NMR are expected higher than those obtained by titration. This expectation was verified for both –COOH and PhOH, where PhOH is the sum of the concentration of all the species with a

$\log K_a > 7$, obtained by the application of Model III (Table 2), as usually done for HS.

The availability of the data in Table 2, coupled to the data in Table 1, allows obtaining the breakdown of the COY data inTable 1 into their components. Table 3 shows the results. These were obtained according to the assumption underlying the following Equations:

$$\mathrm{COY} = \mathrm{COOH} + \mathrm{CONR} \tag{1}$$

$$\mathrm{N} = \mathrm{NR} + \mathrm{CONR} \tag{2}$$

In these equations, COY and N are the total carboxyl and organic N concentration values in Table 1, with R = H, alkyl or aryl, while COOH is the concentration value obtained from titrations and calculated as reported above. The results obtained clearly show the different nature of FORSUD SBO compared to those of CVDF and CVT230 SBO. The former contains lower number and lower concentration of protogenic species than the other two SBOs. These features are likely to be related to the different sourcing materials. The CVDF and CVT230 SBOs are sourced from composted wastes. The higher number and concentration of protogenic species might be the result of the bio-oxidation conditions under which the sourcing CVDF and CVT230 materials were obtained. Moreover, it has been found out that FORSUD SBO has a higher content of amides (Table 3). The higher amount of these functional groups in FORSUD can be traced back to the different nature of its sourcing material relative to CVDF and CVT230. Indeed, coming from the organic humid fraction of urban wastes, fermented under anaerobic conditions, FORSUD is expected to be richer in proteinaceous matter originating from food wastes.

3. EXPERIMENTAL SECTION

3.1. Chemicals

Tetraethyl ammonium chloride (TEACl, purity ≥ 98.5%), the concentrated acids HCl (37%) and HF (≥40%) were from Sigma Aldrich (St. Louis, MO, USA). Standard KOH and HCl solutions were prepared by diluting Merck (Darmstadt, Germania) or Sigma Aldrich concentrated products and standardized against potassium hydrogen phthalate (Sigma Aldrich, purity ≥

99.5%) and sodium carbonate (Sigma Aldrich, purity ≥ 99.5%), respectively. Grade A glassware and deionized water were used for all the solutions.

Table 2. Chemical models and protonation constants of acidic sites obtained from the elaboration of titration data.

SBO	Chemical Model	Acidic Site	logβ [a]	log*K* [a]	Site Concentration ($mmol \cdot g^{-1}$)
CVDF	Model I	Hcvdf1 [b]	5.69 ± 0.02 [c]		3.0 ± 0.4
		H_2cvdf1	9.83 ± 0.03	4.14	
		Weighted standard deviation of the fit [d]	2.14		
	Model II	Hcvdf1	5.42 ± 0.02 [c]		2.6 ± 0.2
		H_2cvdf1	9.41 ± 0.04	3.99	
		Hcvdf2	7.78 ± 0.04		0.8 ± 0.2
		Hcvdf3	10.23 ± 0.02		0.9 ± 0.1
		Weighted standard deviation of the fit	2.55		
	Model III	Hcvdf1	4.34 ± 0.02		1.70 ± 0.08
		Hcvdf2	6.03 ± 0.03		1.0 ± 0.1
		Hcvdf3	8.33 ± 0.04		0.7 ± 0.1
		Hcvdf4	10.40 ± 0.02		0.9 ± 0.2
		Weighted standard deviation of the fit	1.73		

SBO	Chemical Model	Acidic Site	logβ [a]	log*K* [a]	Site Concentration ($mmol \cdot g^{-1}$)
CVT 230	Model I	Hcvt1	5.29 ± 0.01		3.0 ± 0.1
		H_2cvt1	8.86 ± 0.04	3.57	
		Weighted standard deviation of the fit	2.43		
	Model II	Hcvt1	5.10 ± 0.02		2.70 ± 0.05
		H_2cvt1	8.61 ± 0.06	3.51	
		Hcvt2	7.58 ± 0.04		0.8 ± 0.1
		Hcvt3	9.97 ± 0.03		1.3 ± 0.2
		Weighted standard deviation of the fit	3.03		
	Model III	Hcvt1	3.87 ± 0.01		1.76 ± 0.03
		Hcvt2	5.76 ± 0.02		1.07 ± 0.07
		Hcvt3	8.26 ± 0.04		0.71 ± 0.08
		Hcvt4	10.16 ± 0.02		1.2 ± 0.2
		Weighted standard deviation of the fit	1.75		
FORSUD	Model I	Hforsud1	6.12 ± 0.03		1.61 ± 0.02
		H_2forsud1	10.32 ± 0.07	4.20	
		Weighted standard deviation of the fit	1.66		
	Model II	Hforsud1	6.37 ± 0.05		1.67 ± 0.04
		H_2forsud1	10.65 ± 0.08	4.28	
		Hforsud2	9.34 ± 0.02		1.602 ± 0.004
		Weighted standard deviation of the fit	1.80		
	Model III	Hforsud1	4.50 ± 0.04		1.015 ± 0.006
		Hforsud2	6.77 ± 0.06		0.67 ± 0.04
		Hforsud3	9.37 ± 0.02		1.58 ± 0.01
		Weighted standard deviation of the fit	1.43		

[a] The protonation constants are expressed as log of $\beta_r = [AH_r^{-n+r}]/[A^{-n}][H^+]^r$ or $K = [AH_r^{-n+r}]/[H_{r-1}A^{-n+r-1}][H^+]$. The log*K* values are reported only when different from log*β*; [b] Different numbers refer to different acidic sites. The species can be monoprotic or diprotic. In the latter case the species completely protonated are indicated as H_2A and the monoprotonated as HA; [c] ±standard deviation; and [d] Weight for each experimental point is given as $w = 1/s^2$ [25].

Table 3. Breakdown of NMR data into single functional groups concentrations (mmol·g^{-1}).

SBO	PhOY [a]	COY [a]	CONR [b]	NR [b]	COOH [c]
CVDF	4.84	4.55	1.85	1.85	2.7
CVT230	4.76	4.75	1.92	1.78	2.83
FORSUD	2.76	6.16	4.48	1.82	1.68

[a] Data obtained by applying Equation (3) to values reported in Table 1. The acronyms are explained in the text. Section 2.2.2.; [b] Data calculated through Equations (1) and (2); [c] The concentrations of carboxylic groups were obtained from application of chemical modeling processes (model III) and the value reported is the sum of the concentrations of the species with a $\log K_a < 7$ reported in Table 2.

3.2. Preparation and Characterization of SBOs

The SBOs were sourced from materials sampled from three different streams of the ACEA Pinerolese waste treatment plant in Pinerolo (Italy). These were the digestate (FORSUD) recovered from the plant biogas production reactor fed with the organic humid fraction from separate source collection of urban refuse, the compost (CV) obtained from urban private gardens and public park trimming residues aged for 230 days, and the compost (CVDF) obtained from a 35/55/10 *w/w/w*FORSUD/home gardening and parks trimming residues/sewage sludge mix aged for 110 days, composting under aerobic conditions. The sampled materials were processed in a pilot plant made available from Studio Chiono & Associati s.r.l. in Rivarolo Canavese, Italy. This comprised an electrically heated mechanically stirred 500 L reactor, a 102 cm long × 10.1 cm diameter polysulfone ultrafiltration membrane with 5 kDa molecular weight cut-off supplied by Idea Engineering s.r.l. from Lessona (Bi), Italy, and a forced ventilation drying oven. According to the operating experimental conditions, the materials sampled from the Acea plant were reacted 4 h with pH 13 water at 60 °C and 4 *v/w* water/solid ratio. The liquid/solid mix was allowed to settle to separate the supernatant liquid phase containing the soluble hydrolysate from the insoluble residue. The recovered hydrolysate was circulated at 40 L·h^{-1} flow rate through 5 kDa cut off polysulphone ultrafiltration membrane operating with tangential flow at 7 bar inlet and 4.5 bar outlet pressure to yield a retentate with 5%–10% dry matter content. The concentrated retentate was finally dried at 60 °C to yield the solid products in 15%–30% *w/w* yield, relatively to the sourcing material dry matter. These products were found to contain about 15%–28% ashes. The ashes contained, along with the added alkali cation of the hydrolyzing base, also other mineral elements such as Si, Ca, Mg, Al and Fe, which were present in the sourcing materials. To reduce mineral content, the solid SBOs were further washed with HCl, water, HF, again with water, and then dried at 60° to yield the final products which were used in this work. These are hereinafter named after their sourcing materials, *i.e.*, FORSUD, CVDF and CVT230. They were characterized according to a previously reported procedure [23] for the content of ash, elemental C and N, organic C types and functional groups, and critical micellar concentration in water solution. To compare the amount of acid groups obtained from ^{13}C NMR spectra with those obtained by titrations, the maximum concentration ($mmol \cdot g^{-1}$) of carboxyl-like structures and phenolic groups is calculated by the Equation (3):

$$C_i = \frac{C_i \times \%C_{EA}}{1.20011} \quad (3)$$

where C_i is the mole fraction of carboxyl-like structures (δ = 160–190 ppm) or phenolic groups (δ = 140–160 ppm) from NMR analysis in the sample, and $\%C_{EA}$ is the weight percentage of carbon in the sample as determined by elemental analysis.

3.3. Electromotive Force Measurements

Potentiometric measurements were performed using a Metrohm mod. 713 potentiometer (resolution of ±0.1 mV) coupled with a Metrohm 665 Dosimat burette (minimum volume deliverable of ±0.001 cm^3) and equipped with a Metrohm combined glass electrode (mod. 6.0222.100).

For all the potentiometric measurements the electrode couple was standardized, in terms of pH = $-\log[H^+]$, by titrating HCl 10 $mmol \cdot L^{-1}$ solution (at the same ionic strength value as the solution under study) with standard KOH in order to determine the standard potential E^0 before each experiment. All the potentiometric titrations were carried out in a stream of purified nitrogen gently bubbled in the titration cell to avoid O_2 and CO_2 contamination. The measurement cells were thermostated at 25 ± 0.1 °C by means of a water circulation from a thermocryostat (mod. D1-G Haake, Victoria, Australia). The experiments were carried out in TEACl aqueous solutions with ionic strength 0.1 $mol \cdot L^{-1}$. The TEACl was chosen as background salt because it does not interact significantly with polyelectrolytes [22]. Each titration was repeated at least twice.

In order to evaluate the acidic properties of SBOs, a series of alkalimetric titrations in 0.1 $mol \cdot L^{-1}$ TEACl aqueous solutions, was performed. The concentration of SBO in the solution to be titrated were from 0.30 to 1.0 $g \cdot L^{-1}$, lower than their critical micelle concentrations (Table 1) in order to better assess the number and the nature of SBO protogenic sites (seeSection 2.1.). As the SBOs are soluble in high alkaline conditions, the sample was dissolved at 100 $g \cdot L^{-1}$ concentration in 1 $mol \cdot L^{-1}$ KOH. The total alkali content in the solution was in excess relative to the total acidic functions assuming that the maximum content of acid groups could not be greater than that reported in Table 1 for total carboxyl (COY, Y = OH and/or NR) and phenol (PhOY, Y = H and/or OR) groups estimated by ^{13}C NMR spectroscopy. The direct titration of the acidic groups was performed on the 100 $g \cdot L^{-1}$ SBO solution after the dilution and the addition of 0.1 $mol \cdot L^{-1}$ HCl to pH~3.0. The acidified solution was then titrated with the 0.1 $mol \cdot L^{-1}$ KOH solution up to pH 11.5, in order to evaluate the protonation constants and the concentration of both carboxylic and phenolic functions. The reverse titrations were performed by addition of 0.1 $mol \cdot L^{-1}$ HCl. A total of 50 to 120 points was taken for each titration and the potential was read only when its variation was smaller than 0.5 $mV \cdot min^{-1}$. Under this condition, equilibrium was assumed to be reached before each successive alkali/acid addition. The total titration time was about 3 h.

3.4. Data Analysis and Calculations

To determine the electrode parameters from calibration data the calculations were performed by using the non-linear least squares computer program ESAB2M [26]. This program allowed to refine the analytical concentration of the reagents, the electrode formal potential E^0, the coefficient j_a relative to the acidic junction potential ($E_j = j_a\,[H^+]$) and the ionic product of water K_w.

The estimation of the protonation constants and sites concentrations was performed by the BSTAC [25] software elaborating the titration data recorded on SBO solutions. It employs an iterative and convergent numerical method, which is based upon the linear combination of the mass balance equations, minimizes the error squares sum on electromotive force values and takes into account eventual variations of ionic strength among and/or during titrations. The protonation constants are expressed by Equation (4),

$$\beta_r = [AH_r^{-n+r}]/[A^{-n}][H^+]^r \qquad (4)$$

This equation is related to the equilibrium:

$$A^{-n} + rH^+ \leftrightarrow AH_r^{-n+r} \qquad (5)$$

where AH_r^{-n+r} is a weak *r*-protic acid and A^{-n} is the corresponding deprotonated form.

The operator according to a trial and error procedure then selects the most probable species involved in the equilibria. The software refines the constants and the concentration of the acidic sites. Several chemical models were taken into consideration to determine the one fitting best the experimental data.

4. CONCLUSIONS

The elaboration of potentiometric data has allowed attaining consistent chemical models to assess the number, nature and concentration of acidic functions in SBO biopolymers sourced from the digestate and composts of urban biowastes. Three models have been used and all three fitted significantly the experimental data. Several protogenic sites with different acid strength have been found. This is consistent with the fact that SBOs, as products of biological origin, are mixtures of molecules with different chemical composition, molecular weight, structure and solution behavior. The observed hysteresis between tandem forward and reverse titration curves and the values of protonation constants calculated for protogenic functions confirm the similarity of the SBOs with natural humic substances. The alkalimetric titration data have been found consistent with the concentration values for total carboxyl and amine functional groups determined by ^{13}C NMR spectroscopy. The titration data have allowed obtaining the breakdown of total carboxyl and N containing functional groups. Thus, a more detailed characterization of the above biomacromolecules has been possible.

The experimental approach taken in this work has been applied for the first time to the above SBO biopolymers derived from municipal biowaste treatment materials. It has been confirmed a useful tool to understand the nature of complex organic matter from biological source. For the specific case of SBOs, the herewith-reported results may turn useful to foresee and/or explain the behavior of these biomacromolecules in specific applications. The availability of acid-base sites with different strength is highly important in relation to the SBOs capacity to bind and keep metals at neutral and alkaline pH, and precipitate at acid pH. The metal binding capacity and solubility properties can be exploited for uses in different fields, such as agriculture and environmental remediation.

ACKNOWLEDGMENTS

This work was carried partly with Ministero delle Politiche Agricole funds within the Agrienergia project. The authors are grateful to the following private and/or public Italian institutions: (i) Acea Pinerolese Spa in Pinerolo (TO) for supplying the SBO sourcing biomaterials form its urban waste management plant; (ii) Studio Chiono ed Associati in Rivarolo Canavese (TO) for making available pilot equipment and services for the production of the SBO.

AUTHOR CONTRIBUTIONS

Silvia Tabasso and Silvia Berto, organized and performed the experiments, designed and structured the manuscript. They were responsible for writing and submission of the final manuscript. Roberta Rosato, Janeth Alicia Tafur Marinos are responsible for the sample preparation and performed the experiments. Pier Giuseppe Daniele contributed to the potentiometric data interpretation. Marco Ginepro, Vincenzo Zelano and Enzo Montoneri initiated and planned the study. Enzo Montoneri also participated in the preparation of the manuscript.

REFERENCES

1. Montoneri, E.; Mainero, D.; Boffa, V.; Perrone, D.G.; Montoneri, C. Biochemenergy: A project to turn an urban wastes treatment plant into biorefinery for the production of energy, chemicals and consumer's products with friendly environmental impact. *Int. J. Glob. Environ. Issues* **2011**, *11*, 170–196.

2. Vargas, A.K.N.; Savarino, P.; Montoneri, E.; Tabasso, S.; Cavalli, R.; Bianco Prevot, A.; Guardani, R.; Roux, G.A.C. Use of biowaste-derived biosurfactants in the production of emulsions for industrial use. *Ind. Eng. Chem. Res.* **2014**, *53*, 8621–8629.

3. Montoneri, E.; Tomasso, L.; Colajanni, N.; Zelano, I.; Alberi, F.; Cossa, G.; Barberis, R. Urban wastes to remediate industrial sites: A case of polycyclic aromatic hydrocarbons contamination and a new process. *Int. J. Environ. Sci. Technol.* **2014**, *11*, 251–262.

4. Avetta, P.; Bella, F.; Bianco Prevot, A.; Laurenti, E.; Montoneri, E.; Arques, A.; Carlos, L. Waste cleaning waste: photodegradation of monochlorophenols in the presence of waste derived photosensitizer. *ACS Sustain. Chem. Eng.***2013**, *1*, 1545–1550.

5. Gomis, J.; Bianco Prevot, A.; Montoneri, E.; Gonzalez, M.C.; Amat, A.M.; Martire, D.O.; Arques, A.; Carlos, L. Waste sourced bio-based substances for solar-driven wastewater remediation: Photodegradation of emerging pollutants.*Chem. Eng. J.* **2014**, *235*, 236–243.

6. Baxter, M.D.; Acosta, E.; Montoneri, E.; Tabasso, S. Waste biomass-extracted surfactants for heavy oil removal. *Ind. Eng. Chem. Res.* **2014**, *53*, 3612–3621.

7. Boffa, V.; Perrone, D.G.; Magnacca, G.; Montoneri, E. Role of a waste-derived biosurfactant in the sol-gel synthesis of nanocrystalline titanium dioxide. *Ceram. Int.* **2014**, *40*, 12161–12169.

8. Deganello, F.; L.aTummino, M.; Calabrese, C.; Testa, M.L.; Avetta, P.; Fabbri, D.; Bianco Prevot, A.; Montoneri, E.; Magnacca, G. New sustainable LaFeO3 photocatalysts prepared from biowaste sourced soluble substances. *New J. Chem.* **2014**.

9. Magnacca, G.; Laurenti, E.; Vigna, E.; Franzoso, F.; Tomasso, L.; Montoneri, E.; Boffa, V. Refuse derived bio-organics and immobilized soybean peroxidase for green chemical technology. *Process Biochem.* **2012**, *47*, 2025–2031.

10. Franzoso, F.; Tabasso, S.; Antonioli, D.; Montoneri, E.; Persico, P.; Laus, M.; Mendichi, R.; Negre, M. Films made from poly (vinyl alcohol-co-ethylene) and soluble biopolymers isolated from municipal biowaste. *J. Appl. Polym. Sci.* **2014**,*132*, 1301.

11. Sortino, O.; Montoneri, E.; Patanè, C.; Rosato, R.; Tabasso, S.; Ginepro, M. Benefits for agriculture and the environment from urban waste. *Sci. Total Environ.* **2014**, *487C*, 443–451.

12. Montoneri, C.; Montoneri, E.; Tomasso, L.; Piva, A. Compost derived substances decrease feed protein N mineralization in swine cecal fermentation. *J. Agric. Sci.* **2013**, *13*, 31–44.

13. Dinuccio, E.; Biagini, D.; Rosato, R.; Balsari, P.; Lazzaroni, C.; Montoneri, E. Use of acid soluble bio-organic substances extract as rabbits feed additive to reduce manure gases emission during storage. *Adv. Anim. Biosci.* **2013**, *4*, 515.

14. Lubal, P.; Široký, D.; Fetsch, D.; Havel, J. The acidobasic and complexation properties of humic acids: Study of complexation of Czech humic acids with metal ions. *Talanta* **1998**, *47*, 401–412.

15. Masini, J.C.; Abate, G.; Lima, E.C.; Hahn, L.C.; Nakamura, M.S.; Lichtig, J.; Nagatomy, H.R. Comparison of methodologies for determination of carboxylic and phenolic groups in humic acids. *Anal. Chim. Acta* **1998**, *364*, 223–233.

16. Fiol, S.; Lopez, R.; Ramos, A.; Antelo, J.M.; Arce, F. Study of the acid-base properties of three fulvic acids extracted from different horizons of a soil. *Anal. Chim. Acta* **1999**, *385*, 443–449.

17. Santos, E.B.H.; Esteves, V.I.; Rodrigues, J.P.C.; Duarte, A.C. Humic substances' proton-binding equilibria: Assessment of errors and limitations of potentiometric data. *Anal. Chim. Acta* **1999**, *392*, 333–341.

18. Gustafsson, J.P. Modeling the acid-base properties and metal complexation of humic substances with the stockholm humic model. *J. Colloid Interface Sci.* **2001**, *244*, 102–112.

19. Ritchie, J.; Perdue, M. Proton-binding study of standard and reference fulvic acids, humic acids, and natural organic matter. *Geochim. Cosmochim. Acta* **2003**, *67*, 85–96.

20. Cooke, J.D.; Hamilton-Taylor, J.; Tipping, E. On the acid-base properties of humic acid in soil. *Environ. Sci. Technol.***2007**, *41*, 465–470.

21. Ritchie, J.D.; Perdue, E.M. Analytical constrain on acidic functional groups in humic substances. *Organ. Geochem.* **2008**,*39*, 783–799.

22. Crea, F.; de Stefano, C.; Gianguzza, A.; Pettignano, A.; Piazzese, D.; Sammartano, S. Acid-base properties of synthetic and natural polyelectrolytes: Experimental results and models for the dependence on different aqueous media. *J. Chem. Eng. Data* **2009**, *54*, 589–605.

23. Montoneri, E.; Boffa, V.; Savarino, P.; Perrone, D.G.; Montoneri, C.; Mendichi, R.; Acosta, E.J.; Kiran, S. Behavior and properties in aqueous solution of biopolymers isolated from urban refuse. *Biomacromolecules* **2010**, *11*, 3036–3042.

24. Montoneri, E.; Boffa, V.; Savarino, P.; Perrone, D.G.; Musso, G.; Mendichi, R.; Chierotti, M.R.; Gobetto, R. Biosurfactants from urban green waste. *ChemSuSChem* **2009**, *2*, 239–247.

25. De Stefano, C.; Mineo, P.; Rigano, C.; Sammartano, S. Ionic strength dependence of formation constants. XVII. The calculation of equilibrium concentrations and formation constants. *Ann. Chim. (Rome)* **1993**, *83*, 243–277.

26. De Stefano, C.; Princi, P.; Rigano, C.; Sammartano, S. Computer analysis of equilibrium data in solution. ESAB2M: an improved version of the ESAB program. *Ann. Chim. (Rome)* **1987**, *77*, 643–675.

Chapter 9

The Effects of Biopolymer Encapsulation on Total Lipids and Cholesterol in Egg Yolk during in Vitro Human Digestion

Sun-Jin Hur [1,*], Young-Chan Kim [2], Inwook Choi [2] and Si-Kyung Lee [1]

[1]Department of Bioresources and Food Science, Konkuk University, 120 Neungdong-ro, Gwangjin-gu, Seoul 143-701, Korea

[2]Korea Food Research Institute, 1201-62 Anyangpangyo-ro, Bundang-gu, Gyeonggi-do 463-746, Korea

ABSTRACT

The purpose of this study was to examine the effect of biopolymer encapsulation on the digestion of total lipids and cholesterol in egg yolk using an *in vitro* human digestion model. Egg yolks were encapsulated with 1% cellulose, pectin, or chitosan. The samples were then passed through an *in vitro* human digestion model that simulated the composition of mouth saliva, stomach acid, and the intestinal juice of the small intestine by using a dialysis tubing system. The change in digestion of total lipids was monitored by confocal fluorescence microscopy. The digestion rate of total lipids and cholesterol in all egg yolk samples dramatically increased after *in vitro* human digestion. The digestion rate of total lipids and cholesterol in egg yolks encapsulated with chitosan or pectin was reduced compared to the digestion rate of total lipids and cholesterol in other egg yolk samples.

Egg yolks encapsulated with pectin or chitosan had lower free fatty acid content, and lipid oxidation values than samples without biopolymer encapsulation. Moreover, the lipase activity decreased, after *in vitro* digestion, in egg yolks encapsulated with biopolymers. These results improve our understanding of the effects of digestion on total lipids and cholesterol in egg yolk within the gastrointestinal tract.

KEYWORDS

egg yolk; cholesterol; biopolymers; *in vitro* human digestion; lipase activity

1. INTRODUCTION

Eggs are an important part of human diet [1] and contain high amounts of cholesterol. The average egg contains 213 mg of cholesterol [2], which is approximately twice the amount of cholesterol that is found in butter and freeze-dried meat products and about 5–10 times more cholesterol than that found in most dairy products [3]. In general, an increase in the concentration of blood cholesterol is widely recognized as a risk factor for coronary artery disease [4]. Hence, the consumption of table eggs has decreased in many developed countries during the last 4 decades due to these perceptions about cholesterol [1]. However, some studies [5,6] have reported that healthy individuals can consume 1 egg per day without affecting their blood cholesterol levels or increasing their risk of cardiovascular disease.

Numerous studies have demonstrated the ability of biopolymers such as chitosan, pectin, guar gum, xanthan gum, or modified starch to function as dietary fibers by lowering blood cholesterol concentrations and reducing lipid absorption [7]. In a previous study, Lairon *et al.* reported that dietary fibers can alter the breakup and coalescence of lipid droplets in the stomach and small intestine, thereby altering the surface area of emulsified lipids exposed to digestive enzymes [8]. Hur *et al.* reported that dietary fibers from various sources can bind to bile acids as well as mixed micelle components; they also explained particle disruption of the micellization process, which leads to reduced micellar solubilization of lipids [7]. For instance, cationic chitosan can bind to the surface of anionic lipid droplets, which are stabilized by bile salt or phospholipids, and reduce lipase activity by preventing contact between lipase and emulsified lipid substrates [9]. However, how biopolymer encapsulation influences the digestion of total lipids and cholesterol in egg yolk remains to be elucidated. Thus, the purpose of this study was to determine the effects of biopolymer encapsulation on the digestion of total lipids and cholesterol present in egg yolk using an *in vitro* human digestion model.

2. RESULTS

2.1. Digestion Rates of Total Lipids and Cholesterol

Representative confocal images of egg yolks before and after *in vitro* human digestion are presented in Figure 1.

Prior to digestion, lipid droplets appear to be covered by chitosan and pectin, whereas cellulose did not have a covering effect on the egg yolk (Figure 1). The size of the lipid droplets in the control and cellulose-encapsulated samples were much smaller than the size of droplets in the chitosan- or pectin-encapsulated samples. The lipid digestion in the control and cellulose-encapsulated samples was higher than that in the chitosan- or pectin-encapsulated samples (Figure 2).

The cholesterol content decreased in all egg yolk samples after *in vitro* simulated human digestion (Figure 3). After *in vitro*human digestion, the undigested cholesterol content was higher in the egg yolks encapsulated with chitosan or pectin than in the control or cellulose-encapsulated samples. These results suggest that during *in vitro* human digestion, total lipid and cholesterol digestion can be reduced by encapsulating egg yolks with pectin or chitosan. However, the reduction in digestion rate of total lipids and cholesterol in pectin- and chitosan-encapsulated egg yolks was similar.

2.2. Free Fatty Acid

The effect of biopolymer encapsulation on the digestion of total lipids and cholesterol was determined by measuring the level of free fatty acids before and after *in vitro* human digestion (Figure 4). The amount of free fatty acid dramatically increased after *in vitro* human digestion in all egg yolk samples. The amount of free fatty acids after *in vitro* human digestion was lower in egg yolk samples encapsulated with pectin or chitosan than in the other samples, whereas the samples encapsulated with cellulose had the same free fatty acid content as the control samples. The amount of free fatty acids in chitosan- or pectin-encapsulated egg yolks did not differ. The fatty acid composition of palmitic acid and oleic acid changed after *in vitro* human digestion in all egg yolk samples. However, the levels of total saturated and unsaturated fatty acids were not significantly different in any of the egg yolk samples (data are not shown). The release of free fatty acid indicates that lipids are hydrolyzed by digestive enzymes or bile salts during *in vitro* human digestion. In this regard, the decrease in the free fatty acid content, by biopolymer encapsulation, after *in vitro* human digestion may be closely related to the decrease in the digestion rate of total lipids and cholesterol.

2.3. Lipid Oxidation

The effects of biopolymer encapsulation on lipid oxidation are shown in Figure 5. The levels of thiobarbituric acid reactive substances (TBARS; used to measure lipid oxidation) in all egg yolk samples increased after *in vitro* human digestion. Egg yolks encapsulated with biopolymers had lower TBARS levels than the control samples. Among the samples encapsulated with biopolymers, pectin- and chitosan-encapsulated egg yolks had lower levels of TBARS than cellulose-encapsulated egg yolks, although TBARS levels in the chitosan- and pectin-encapsulated samples did not differ. The level of TBARS seems to be closely related to the free fatty acid content.

2.4. Inhibition of Lipase Activity

The effects of biopolymer encapsulation on the inhibition of lipase activity are shown in Figure 6. Lipase activity was lower in all egg yolk samples before digestion. However, the inhibition of lipase activity in egg yolks after *in vitro* digestion was significantly increased due to encapsulation with biopolymers. All biopolymers had an inhibitory effect on lipase activity. Cellulose also inhibited lipase activity, although cellulose encapsulation did not have an inhibitory effect on the digestion of total lipids and cholesterol, and on lipid oxidation or free fatty acid content.

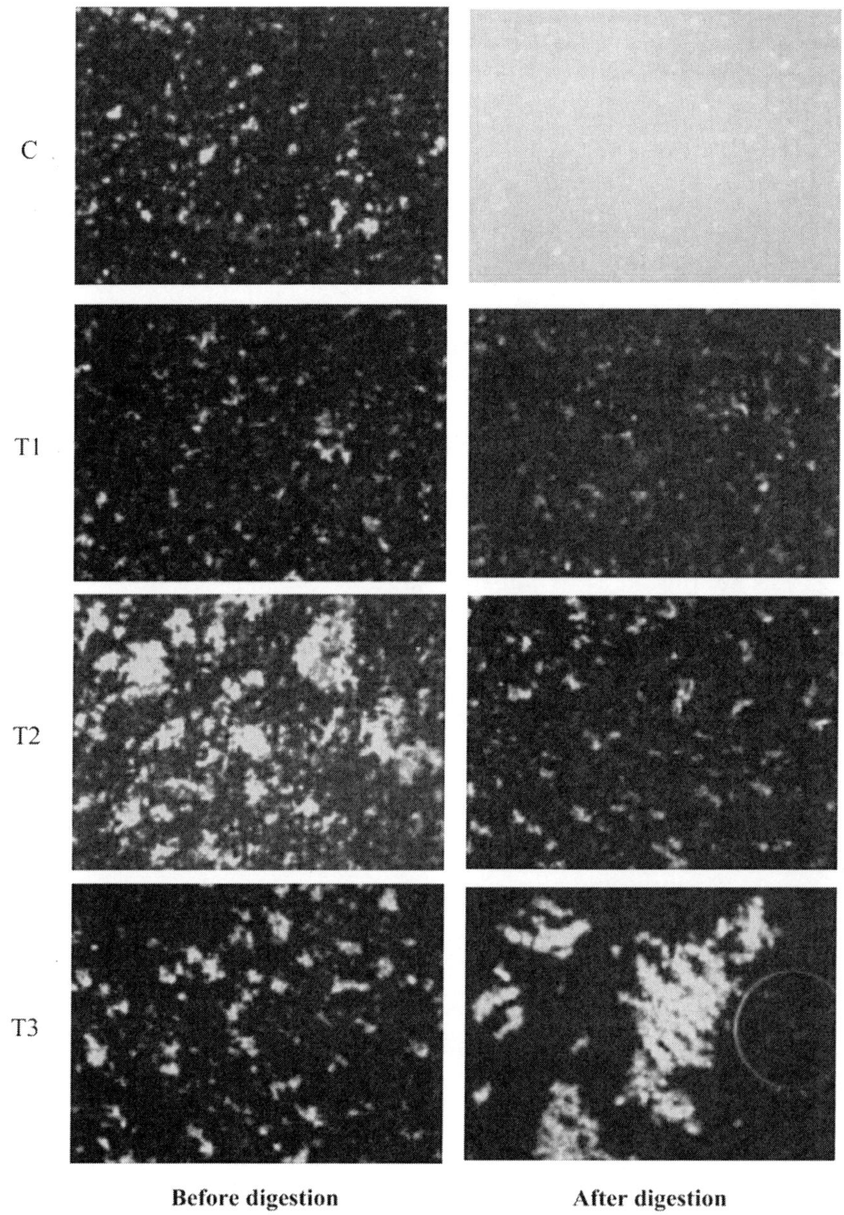

Figure 1. Effect of biopolymers encapsulation on the digestion of total lipids in egg using an *in vitro* human digestion model. C: no encapsulation; T1: encapsulation with cellulose; T2: encapsulation with pectin; T3: encapsulation with chitosan.

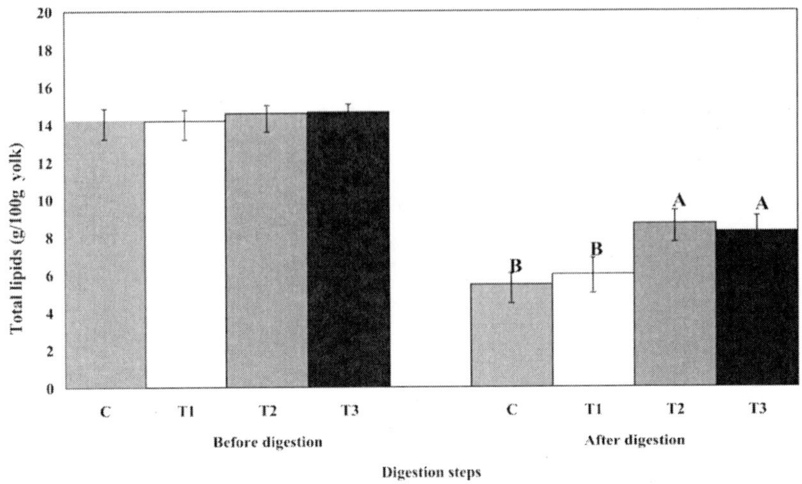

Figure 2. The effect of biopolymer encapsulation on total lipid content in egg yolk during *in vitro* human digestion. C: no encapsulation; T1: encapsulation with cellulose; T2: encapsulation with pectin; T3: encapsulation with chitosan.

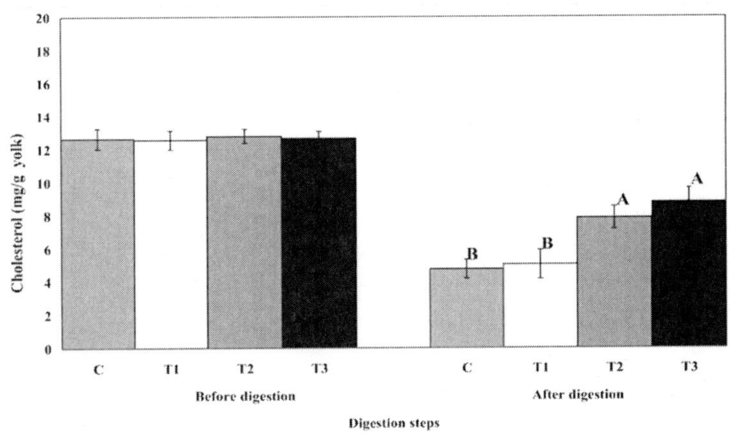

Figure 3. The effect of biopolymer encapsulation on the cholesterol content in egg yolk during *in vitro* human digestion. C: no encapsulation; T1: encapsulation with cellulose; T2: encapsulation with pectin; T3: encapsulation with chitosan. The results of before digestion are total cholesterol content, and results of after digestion are undigested cholesterol content.

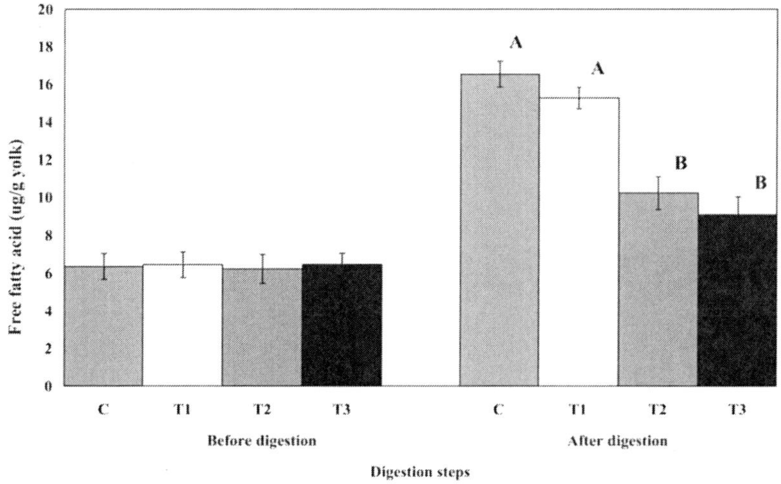

Figure 4. The effect of biopolymer encapsulation on free fatty acid content in egg yolk during *in vitro* human digestion. C: no encapsulation; T1: encapsulation with cellulose; T2: encapsulation with pectin; T3: encapsulation with chitosan.

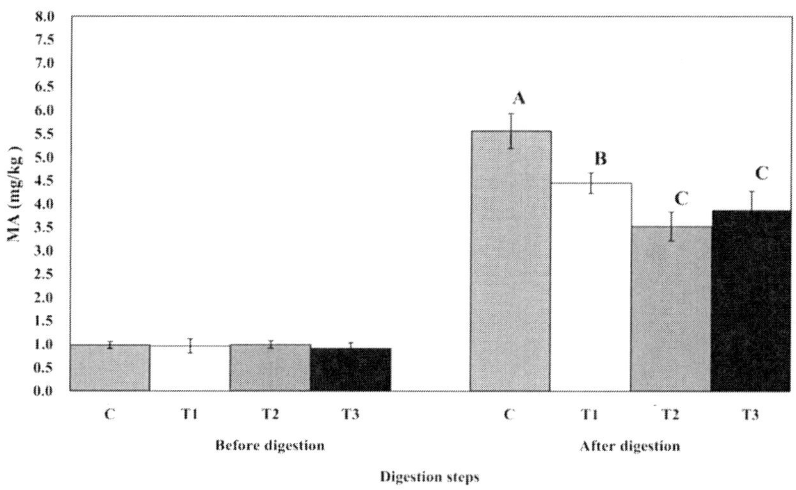

Figure 5. The effect of biopolymer encapsulation of egg yolks on lipid oxidation during*in vitro* human digestion. C: no encapsulation; T1: encapsulation with cellulose; T2: encapsulation with pectin; T3: encapsulation with chitosan. The figure shows the level of total lipid oxidation. MA: Malondialdehyde.

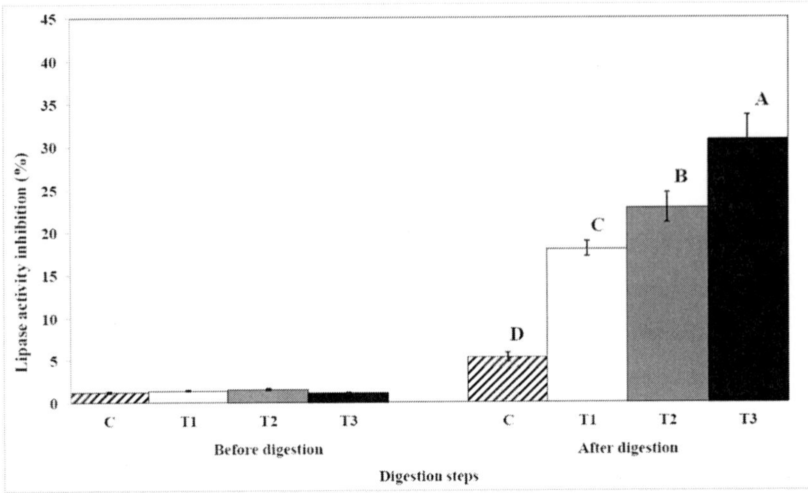

Figure 6. The effect of biopolymer encapsulation of egg yolks on the inhibition of lipase activity during *in vitro* human digestion. C: no encapsulation; T1: encapsulation with cellulose; T2: encapsulation with pectin; T3: encapsulation with chitosan. The inhibition of lipase activity is shown in percentages.

2.5. Particle Size

The effects of biopolymer encapsulation on the particle size are shown in Figure 7. Particle size was higher in all egg yolk samples before digestion. The particle size in the control and cellulose-encapsulated samples were much smaller than the size of droplets in the chitosan- or pectin-encapsulated samples after digestion.

3. DISCUSSION

A number of *in vitro* studies have shown that biopolymers can alter lipid or cholesterol digestion. Dietary fibers can decrease serum cholesterol concentrations by reducing the amount of ingested (exogenous) cholesterol that is adsorbed [10]. Lairon *et al.* reported that some soluble fibers that form viscous solutions drastically reduce the rate of lipid emulsification, with a noticeable decrease in fat lipolysis [11]. Moreover, dietary fiber from various sources can bind to bile acids, mixed micelle components such as monoacylglycerols, free fatty acids, or free cholesterol. This explains the partial disruption of the micellization process, which leads to reduced micellar solubilization of lipid moieties and the reduction in intestinal uptake of lipid moieties and cholesterol [11–13]. In this study, biopolymer-encapsulated egg yolk had lower digestion rates of total lipids and cholesterol in the *in vitro* human digestion model. Moreover, lipase activity was reduced by biopolymer encapsulation. These results indicate that biopolymers, especially pectin and chitosan, easily encapsulate egg yolk lipids and that they can reduce lipase activity. Consequently, lipid and cholesterol digestion was prevented in the *in vitro* human digestion model. In addition, biopolymer encapsulation may interfere with the absorption of total lipids (with cholesterol) into the dialysis tubing membrane (to simulate the villi in the small intestine); or, they may interfere with the ability of lipase to access the lipids contained within them. Beysseriat *et al.* also reported that the ability of dietary fibers to reduce cholesterol digestion and/or absorption through this mechanism depends on their ability to promote droplet aggregation or to adsorb to lipid droplet surfaces [9]. This depends strongly on the electrical charge, molecular weight, and structure of the fibers. For example, cationic chitosan can bind to the surface of anionic lipid droplets that are stabilized by bile salts and/or phospholipids and reduce lipase activity by preventing lipase from coming into contact with the emulsified lipid substrate [9]. The negatively charged lipase may have acted as a bridge between the positively charged chitosan-coated droplets, causing them to clump together, and the interaction of chitosan with lipase may have caused some charge neutralization of the lipid droplets [14].

Figure 8 shows that biopolymers have the ability to aggregate lipid droplets. This aggregation would decrease the transport rate of enzymes to the lipid surfaces, as well as the transport rate of lipid digestion products present within micelles to the gastrointestinal walls. Thereby, the digestion and absorption rate of total lipids and cholesterol would be reduced.

In an animal study, Fernandez reported that pectin was found to be more effective than guar gum in guinea pigs [15], whereas Moundras *et al.* reported that guar gum was found to be more effective than pectin or gum arabic in rats [16]. In this study, chitosan was found to be more effective for the inhibition of total lipid and cholesterol digestion including the inhibition of lipase activity than other biopolymers during *in vitro* human digestion. This result may be due to the higher electrical charge of chitosan and due to the inhibitory effect of lipase activity.

Cellulose had lesser effect on the inhibition of total lipid and cholesterol digestion during *in vitro* human digestion, even though cellulose inhibited lipase activity. This may be because cellulose has a weak electrical charge and, consequently, has no aggregation effect with lipids. In an *in vitro*study, Mun *et al.* reported that chitosan reduced the amount of fatty acids released because lipid droplets were surrounded by cationic chitosan layers or because the lipids droplets were trapped within large chitosan aggregates [17]. They also suggested that chitosan formed a protective layer around the droplets and that it promoted extensive droplet flocculation, both of which inhibited the ability of lipase to interact with the fat inside the droplets. Thus, we assume that the digestion of total lipids and cholesterol would be further reduced in chitosan-encapsulated egg yolks than in those encapsulated with other biopolymers.

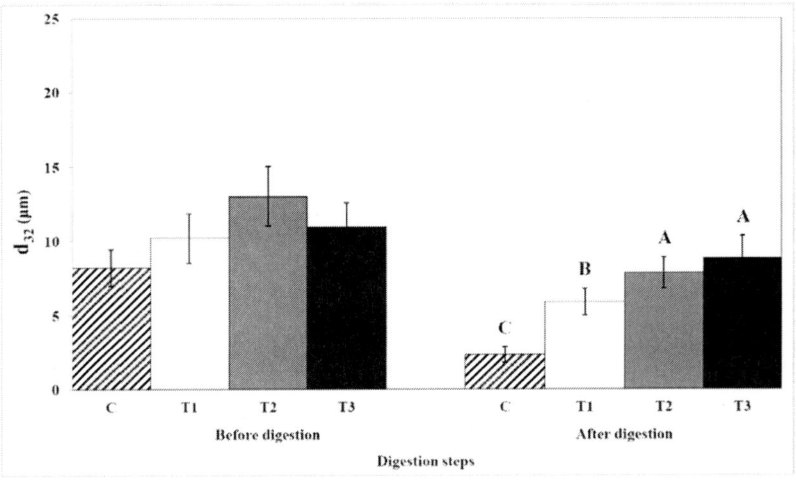

Figure 7. The effect of biopolymer encapsulation of egg yolks on the particle size during *in vitro* human digestion. C: no encapsulation; T1: encapsulation with cellulose; T2: encapsulation with pectin; T3: encapsulation with chitosan. The inhibition of lipase activity is shown in percentages.

Another possible mechanism for the inhibition of lipid digestion is, presumably, the increase in solution viscosity due to biopolymer encapsulation, which causes a reduction in the diffusion of molecules (such as lipase and other enzymes or mixed micelles) and other species in the gastrointestinal tract. As mentioned above, this reduction in diffusion by increasing the solution viscosity would decrease the transport rate of enzymes to the lipid surfaces, thereby reducing the digestion and absorption rates of lipids. Meyer and Doty reported that a high viscosity of the contents of the small intestine may delay lipid digestion, promoting absorption in a more distal part of the small intestine [18]. An increase in solution viscosity is responsible for a decrease in the transit time of ingested food in the gastrointestinal tract. Therefore, an increase in solution viscosity causes a decrease in the transport rate of lipid digestion products present within micelles to the gastrointestinal walls. Khan *et al.* [19] also reported that biopolymers may interfere with the formation of micelles and/or lower the diffusion rate of bile acid and cholesterol-containing micelles through the bolus, consequently diminishing the uptake of cholesterol and bile acids. In our previous study, dietary biopolymers increased the viscosity of the contents of the small intestine [7]. Thus, total lipid and cholesterol digestion in egg yolks can be reduced by biopolymer encapsulation because biopolymers were increased the viscosity of egg yolks in this study (data are not shown). In general, the differences between dietary fibers such as molecular weight and hydrophobicity cause differences in their physicochemical properties such as water solubility, viscosity enhancement, opacity, surface activity, and binding capacity [20]. These differences can cause significant alterations in their effectiveness in reducing cholesterol digestion by interfering with the various physiological processes during digestion and absorption. Yamaguchi *et al.* reported that low molecular weight pectins are more effective in lowering cholesterol than high molecular weight pectins [21]. This has been attributed to the reduction in bile acid binding that occurs when the molecular weight of pectin falls below a critical value [22]. Therefore, the effect of biopolymer encapsulation on total lipid and cholesterol digestion in egg yolk would be largely influenced by the conditions of the biopolymers such as molecular weight, hydrophobicity, viscosity, and pH. Thus, further research is needed to find the most effective biopolymer.

4. EXPERIMENTAL SECTION

4.1. Materials

Potassium chloride, sodium sulfate, sodium hydrogen carbonate, hydrogen chloride, potassium phosphate monobasic, magnesium chloride, hexane, trichloroacetic acid, ether, and ethanol were purchased from Fisher Scientific chemical company (Pittsburgh, PA, USA). Cellulose, pectin, chitosan, bicarbonate, potassium thiocyanate, sodium phosphate dibasic, sodium phosphate monobasic, sodium chloride, calcium chloride, ammonium chloride, urea, glucose sigma, glucuronic acid, glucosamine, α-amylase, uric acid, mucin, bovine serum albumin, pepsin, pancreatin, lipase, bile salt extraction, Nile red, 5α-cholestane, thiobarbituric acid, butylated hydroxyanisole, and phenolphthalein were purchased from Sigma-Aldrich Chemical Co. (St. Louis, MO, USA). Pyridine, bis-[trimethylsilyl] trifluoroacetamide, and trimethylchlorosilane were purchased from Supelco Co. (St. Louis, MO, USA).

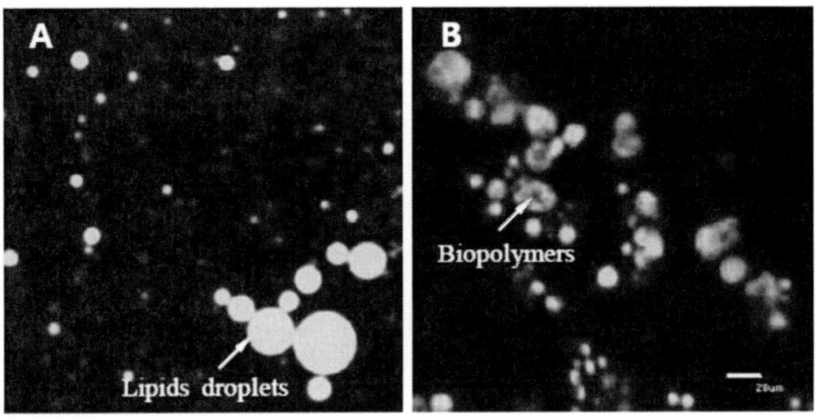

Figure 8. Confocal fluorescence images showing encapsulated lipid with biopolymers in egg yolk. (**A**) Lipid droplets in the egg yolk; (**B**) Lipid droplets with biopolymers in the egg yolk.

4.2. Biopolymer Encapsulation Preparation

The experiments were performed by 5 replications from 10 different egg yolk samples. Eggs were purchased from the local market. Chitosan (10 wt%) was dissolved in acetate buffer solutions (100 mM acetic acid: sodium acetate, pH 3.0, 0–150 mM NaCl). Pectin and cellulose (10 wt% each) were dissolved in phosphate buffer solutions (2 M monobasic sodium phosphate and 2 M dibasic sodium phosphate, pH 7.0). These solutions were stirred for 12 h and then mixed for 3 h using a magnetic stirrer. During mixing, 1 mL of Tween 20 (0.1%, pH 7.6) was added dropwise to reduce surface tension and enhance encapsulation formation. The biopolymer encapsulation was prepared by mixing a final volume of 10 wt% biopolymer solution and egg yolk together for 1 h using a bio-homogenizer. The mixture was continuously stirred for 15 min using power ultrasound at a frequency of 10 MHz (final volume: whole egg yolk mixed with 1% biopolymers). This process was aimed at developing a coating layer around the lipophilic egg yolk. Encapsulation of biopolymers and egg yolk was confirmed using confocal microscopy (Figure 8). The molecular characteristics of biopolymers are listed in Table 1.

Table 1. Molecular characteristics of biopolymer ingredients.

Biopolymer type	Electrical charge	Molecular weight	Hydrophobicity
Cellulose	Non-ionic	100 kDa	Low
Pectin	Anionic (–COO^-) Degree of methylaton: 50	100 kDa	Variable (–CH_3)
Chitosan	Cationic (NH_3^+) Degree of acetylation: 50	100 kDa	Variable (–$COCH_3$)

4.3. In Vitro Human Digestion

An *in vitro* human digestion model that simulated the mouth, stomach, and small intestine was used in this study, which was a modified version of that described by Versantvoort *et al.* [23].

1 Initial system: The initial egg yolk samples encapsulated with 1% biopolymers.

2 Mouth: 10 g of initial egg yolk samples was mixed with 12 mL of simulated saliva solution (pH 6.8) and then stirred for 5 min at 37 °C.

3 Stomach: 24 mL of simulated gastric juice (pH 2) was then added, and the mixture was stirred for 2 h at 37 °C.

4 Small intestine: 24 mL of duodenal juice, 12 mL of bile juice, and 2 mL of HCO_3 solution (pH 6.5 to 7) was then added. The total solution was placed in a 250 mL flask, and then the dialysis tubing (molecular weight cutoff of 50,000, flat width 34 mm, thickness 18 μm, Membrane Filtration Products, Inc. Seguin, TX, USA) containing 10 mL of phosphate buffer (pH 7) was placed in a 250 mL flask, and the mix was stirred for 2 h at 37 °C.

The compositions of the simulated saliva, gastric, duodenal, and bile juices are listed in Table 2. During *in vitro* human digestion, the samples were swirled (60 rpm) on a shaking water bath to simulate the motility of the GI tract (Model 3582, Labline Instruments, Inc., Melrose Park, IL, USA).

Table 2. Constituents and concentrations of the various synthetic juices of the *in vitro*human digestion model representing fed conditions.

	Saliva	Gastric juice	Duodenal juice	Bile juice
Inorganic components	10 mL KCl 89.6 g/L	15.7 mL NaCl 175.3 g/L	40 mL NaCl 175.3 g/L	30 mL NaCl 175.3 g/L
	10 mL KSCN 20 g/L	3.0 mL NaH_2PO_4 88.8 g/L	40 mL $NaHCO_3$ 84.7 g/L	68.3 mL $NaHCO_3$ 84.7 g/L
	10 mL NaH_2PO_4 88.8 g/L	9.2 mL KCl 89.6 g/L	10 mL KH_2PO_4 8 g/L	4.2 mL KCl 89.6 g/L
	10 mL $NaSO_4$ 57 g/L	18 mL $CaCl_2 \cdot 2H_2O$ 22.2 g/L	6.3 mL KCl 89.6 g/L	150 µL HCl 37% g/g
	1.7 mL NaCl 175.3 g/L	10 mL NH_4Cl 30.6 g/L	10 mL $MgCl_2$ 5 g/L	-
	20 mL $NaHCO_3$ 84.7 g/L	6.5 mL HCl 37% g/g	180 µL HCl 37% g/g	20 mL $NaHCO_3$ 84.7 g/L
Organic components	8 mL urea 25 g/L	10 mL glucose 65 g/L	4 mL urea 25 g/L	10 mL urea 25 g/L
		10 mL glucuronic acid 2 g/L		
		3.4 mL urea 25 g/L		
		10 mL glucosamine hydrochloride 33 g/L		
Add to mixture of organic + inorganic components	290 mg α-amylase	1 g BSA	9 mL $CaCl_2 \cdot 2H_2O$ 22.2 g/L	10 mL $CaCl_2 \cdot 2H_2O$ 22.2 g/L
	15 mg uric acid	2.5 g pepsin	1 g BSA	1.8 g BSA
	25 mg mucin	3 g mucin	9 g pancreatin	30 g bile
			1.5 g lipase	
pH	6.8 ± 0.2	1.30 ± 0.02	8.1 ± 0.2	8.2 ± 0.2

The inorganic and organic components are augmented to 500 mL with distilled water. If necessary, the pH of the juices was adjusted to the appropriate interval.

4.4. Total Lipid and Cholesterol Contents

Total lipids were extracted with chloroform and methanol as described by Folch *et al.* [24]. Cholesterol was determined by the modified method of Russo and others [25]. Briefly, extracted lipid (50 mg) was added into a 50 mL tube with 10 mL of saponification reagent (30% KOH and ethanol with the ratio of 6:94) and 0.5 mL internal standard (2 mg 5 α-cholestane/sample), and then capped and incubated for 1 h at 60 °C. After cooling the sample, 8 mL of deionized distilled water and 3 mL hexane were added and mixed thoroughly and allowed to separate. The top layer (hexane layer) was removed and dried in scintillation vials, and 100 µL of bis-[trimethylsilyl]trifluoroacetamide + 1% trimethylchlorosilane (Supelco

Co., Bellefonte, PA, USA) and 200 µL of pyridine were added and mixed. The samples were left to set overnight and then analyzed by gas chromatography (Agilent 6890). A ramped oven temperature condition (180 °C for 2.5 min, increased to 230 °C at 2.5 °C/min, then held at 230 °C for 7.5 min) was used. Temperatures of both the inlet and detector were 280 °C. Helium was the carrier gas at linear flow of 1.1 mL/min. Detector (flame ion detector) air, H_2, and make-up gas (He) flows were 350, 35, and 43 mL/min, respectively.

4.5. FREE FATTY ACID CONTENT

Free fatty acid content was determined by the modified method of AOAC [26]. Free fatty acid contents were weighed by titrimetry. Briefly, 5 g of sample was weighed into a 50-mL test tube and homogenized with 15 mL of deionized distilled water using a Polytron homogenizer (IKA, Model T25, Staufen, Germany) for 10 s at the highest speed. Then, 2 mL of egg yolk homogenate was transferred to a 300 mL flask, and 100 mL of ether/ethanol solution (ether:ethanol, 1:1, *v*/*v*) was added. Several drops of phenolphthalein were added, and the free fatty acids were titrated with 0.1 M KOH. Free fatty acid (KOH/g) = 5.611 $\times A \times F$/Sample weight (g), A: volume (mL) of 0.1 M KOH solution; F: titer of KOH.

4.6. Thiobarbituric Acid-Reactive Substances

TBARS were determined by the modified method of Buege and Aust [27]. Briefly, 5 g of egg yolk sample was weighed into a 50-mL test tube and homogenized with 15 mL of deionized distilled water using a Polytron homogenizer for 10 s at the highest speed (before digestion samples only). Then, 1 mL of egg yolk homogenate was transferred to a disposable test tube (3 × 100 mm), and butylated hydroxyanisole (50 µL, 10%) and thiobarbituric acid/trichloroacetic acid (TBA/TCA) (2 mL) were added. The mixture was vortexed and then incubated in a boiling water bath for 15 min to develop color. The sample was cooled in cold water for 5 min, vortexed again, and centrifuged for 15 min at 2000× *g*. The absorbance of the resulting supernatant solution was determined at 531 nm against a blank containing 1 mL of DDW and 2 mL of TBA/TCA solution. The amounts of TBARS were expressed as milligrams of malondialdehyde (MA) per kilogram of egg yolk samples.

4.7. Lipase Activity

Lipase activity was determined by the modified method of Gooda Sahib *et al.* [28]. Porcine lipase was dissolved in 0.01 M Tris-HCl buffer (25 units/mL). *In vitro* digested solutions were dissolved in 0.01 M Tris-HCl buffer at different concentrations (7.81–250 ppm). Olive oil (10% *v/v*) was mixed with Arabic gum (10 g) mixture (10% *w/v* in 0.1 M), (1.57 g) Tris-HCl buffer, pH 2, 0.5 M (2.92 g) NaCl, and 20 mM (0.02 g) $CaCl_2$ using a homogenizer. Then, 0.2 mL of lipase solution (25 units) was allowed to react with 0.5 mL of *in vitro* digested solution for 30 min at 4 °C. Reconstituted substrate emulsion (2 mL) was then added, and the mixture was incubated for 30 min at 37 °C. The reaction was stopped using acetone and ethanol mixture (1:1/*v:v*) and titrated with 0.02 M NaOH to pH 9.4. Titrations were carried out using an automatic titrator (785 DMP Titrino, Metrohm, Herisau, Switzerland).

4.8. Confocal Laser Scanning Microscopy

The total lipid content and biopolymer encapsulation were analyzed through confocal laser scanning microscopy. A confocal scanning fluorescence microscope (Carl Zeiss, LSM 5 Live, GmbH, Jena, Germany) with a 20× objective lens was used to capture confocal images. Nile red (a lipid fluorescent dye) was excited with a 488-nm argon laser line. The fluorescence emitted from the sample was monitored using a fluorescence detector (543 nm) with a pinhole size of 150 μm. The resulting images consisted of 512 × 512 pixels, with a pixel size of 414 nm and a pixel dwell time of 5 s.

4.9. Particle Size

The particle size of the sample was measured using a laser light scattering instrument (Mastersizer X, Malvern Instruments Ltd., Malvern, UK). This instrument is based on diffraction of a monochromatic beam of laser light (λ = 632.8 nm) when it is scattered by the droplets in a dilute egg yolk. The instrument measures the angular dependence of the intensity of laser light diffraction and finds the particle size that gives the best fit to the experimental measurements and predictions based on light scattering theory. The mean particle size was reported as the surface-weighted mean diameter, d_{32} (= $\sum n_i d_i^3 \sum n_i d_i^2$), where n_i is the number of particles with diameter d_i.

4.10. Statistics

The effect of biopolymer encapsulation on the digestion of total lipids and cholesterol in egg yolks during *in vitro* human digestion were analyzed using SAS software (SAS Inst. Inc., Cary, NC, USA) by the generalized linear model procedure. The Student-Newman-Keuls multiple range test was used to compare differences between means.

5. CONCLUSIONS

The study results show that encapsulation of egg yolk with biopolymers affected their physical stability and digestibility when they were passed through an *in vitro* human digestion model. The decrease in the diameters of lipid droplets as the droplets moved from the mouth to the stomach and then to the small intestine suggests that encapsulation of egg yolks with biopolymers such as pectin and chitosan can reduce the digestion of total lipids and cholesterol during *in vitro* human digestion. However, encapsulation with various biopolymers had a limited effect on the microstructure changes that occurred during *in vitro* human digestion of lipid droplets in egg yolk. Biopolymers vary widely in their electrical and hydrophobic characteristics, and the postprandial metabolism resulting from the digestion and absorption of available nutrients is a highly complex process involving numerous potential interactions. Therefore, the reduction capacities of total lipids and cholesterol digestion can be expected to vary widely. Thus, further research is needed to understand the effect of biopolymer encapsulation that occurs during *in vitro* human digestion and how biopolymers are associated with the changes in the digestion of total lipids and cholesterol.

ACKNOWLEDGMENTS

This paper was supported by the SMART Research Professor Program of Konkuk University.

REFERENCES

1. Majumder, K.; Wu, J. Angiotensin I converting enzyme inhibitory peptides from simulated *in vitro* gastrointestinal digestion of cooked eggs. *J. Agric. Food Chem* **2009**, *57*, 471–477.

2. *Anonymous Composition of Foods, Dairy and Egg Product, Raw-Processed-Prepared*; U.S. Department of Agriculture, U.S. Government Printing Office: Washington, DC, USA, 1989.

3. Nourooz-Zadeh, J.; Appelqvist, L.A. Cholesterol oxides in Swedish foods and food ingredients: Milk powder products.*J. Food Sci* **1988**, *53*, 74–79

4. Cohn, J.; Kamili, A.; Wat, E.; Chung, R.W.; Tandy, S. Dietary phospholipids and intestinal cholesterol absorption.*Nutrients* **2010**, *2*, 116–127.

5. Siri-Tarino, P.; Sun, Q.; Hu, F.; Krauss, R. Saturated fatty acids and risk of coronary heart disease: Modulation by replacement nutrients. *Curr. Atheroscler. Rep* **2010**, *12*, 384–390.

6. Lee, A.; Griffin, B. Dietary cholesterol, eggs and coronary heart disease risk in perspective. *Nutr. Bull* **2006**, *31*, 21–27.

7. Hur, S.J.; Kim, D.H.; Chun, S.C.; Lee, S.K. Effects of dietary conjugated linoleic acid and biopolymer encapsulation on lipid metabolism in mice. *Inter. J. Mol. Sci* **2013**, *14*, 6848–6862.

8. Lairon, D. Soluble fibers and dietary lipids, in dietary fiber in health and disease. *Diet. Fiber Health Dis* **1997**, *427*, 99–108.

9. Beysseriat, M.; Decker, E.A.; McClements, D.J. Preliminary study of the influence of dietary fiber on the properties of oil-in-water emulsions passing through an *in vitro* human digestion model. *Food Hydrocoll* **2006**, *20*, 800–809.

10. Jenkins, D.J.A.; Kendall, C.W.C.; Ransom, T.P.P. Dietary fiber, the evolution of the human diet and coronary heart disease. *Nutr. Res* **1998**, *18*, 633–652.

11. Lairon, D.; Play, B.; Jourdheuil-Rahmani, D. Digestible and indigestible carbohydrates: Interactions with postprandial lipid metabolism. *J. Nutr. Biochem* **2007**, *18*, 217–227.

Index

G

H

I

K

L

M

N

P

R

S

T